D0782144

volume 49

problems and propositions in analysis

Gabriel Klambauer

PROBLEMS AND PROPOSITIONS IN ANALYSIS

PURE AND APPLIED MATHEMATICS

A Program of Monographs, Textbooks and Lecture Notes

Contributions to *Lecture Notes in Pure and Applied Mathematics* are reproduced by direct photography of the author's typewritten manuscript. Potential authors are advised to submit preliminary manuscripts for review purposes. After acceptance, the author is responsible for preparing the final manuscript in camera-ready form, suitable for direct reproduction. Marcel Dekker, Inc. will furnish instructions to authors and special typing paper. Sample pages are reviewed and returned with our suggestions to assure quality control and the most attractive rendering of your manuscript. The publisher will also be happy to supervise and assist in all stages of the preparation of your camera-ready manuscript.

LECTURE NOTES
IN PURE AND APPLIED MATHEMATICS

1. *N. Jacobson,* Exceptional Lie Algebras
2. *L.-Å. Lindahl and F. Poulsen,* Thin Sets in Harmonic Analysis
3. *I. Satake,* Classification Theory of Semi-Simple Algebraic Groups
4. *F. Hirzebruch, W. D. Newmann, and S. S. Koh,* Differentiable Manifolds and Quadratic Forms
5. *I. Chavel,* Riemannian Symmetric Spaces of Rank One
6. *R. B. Burckel,* Characterization of C(X) Among Its Subalgebras
7. *B. R. McDonald, A. R. Magid, and K. C. Smith,* Ring Theory: Proceedings of the Oklahoma Conference
8. *Y.-T. Siu,* Techniques of Extension of Analytic Objects
9. *S. R. Caradus, W. E. Pfaffenberger, and B. Yood,* Calkin Algebras and Algebras of Operators on Banach Spaces
10. *E. O. Roxin, P.-T. Liu, and R. L. Sternberg,* Differential Games and Control Theory
11. *M. Orzech and C. Small,* The Brauer Group of Commutative Rings
12. *S. Thomeier,* Topology and Its Applications
13. *J. M. López and K. A. Ross,* Sidon Sets
14. *W. W. Comfort and S. Negrepontis,* Continuous Pseudometrics
15. *K. McKennon and J. M. Robertson,* Locally Convex Spaces
16. *M. Carmeli and S. Malin,* Representations of the Rotation and Lorentz Groups: An Introduction
17. *G. B. Seligman,* Rational Methods in Lie Algebras
18. *D. G. de Figueiredo,* Functional Analysis: Proceedings of the Brazilian Mathematical Society Symposium
19. *L. Cesari, R. Kannan, and J. D. Schuur,* Nonlinear Functional Analysis and Differential Equations: Proceedings of the Michigan State University Conference
20. *J. J. Schäffer,* Geometry of Spheres in Normed Spaces
21. *K. Yano and M. Kon,* Anti-Invariant Submanifolds
22. *W. V. Vasconcelos,* The Rings of Dimension Two
23. *R. E. Chandler,* Hausdorff Compactifications
24. *S. P. Franklin and B. V. S. Thomas,* Topology: Proceedings of the Memphis State University Conference
25. *S. K. Jain,* Ring Theory: Proceedings of the Ohio University Conference
26. *B. R. McDonald and R. A. Morris,* Ring Theory II: Proceedings of the Second Oklahoma Conference
27. *R. B. Mura and A. Rhemtulla,* Orderable Groups
28. *J. R. Graef,* Stability of Dynamical Systems: Theory and Applications
29. *H.-C. Wang,* Homogeneous Banach Algebras
30. *E. O. Roxin, P.-T. Liu, and R. L. Sternberg,* Differential Games and Control Theory II
31. *R. D. Porter,* Introduction to Fibre Bundles
32. *M. Altman,* Contractors and Contractor Directions Theory and Applications
33. *J. S. Golan,* Decomposition and Dimension in Module Categories
34. *G. Fairweather,* Finite Element Galerkin Methods for Differential Equations
35. *J. D. Sally,* Numbers of Generators of Ideals in Local Rings
36. *S. S. Miller,* Complex Analysis: Proceedings of the S.U.N.Y. Brockport Conference
37. *R. Gordon,* Representation Theory of Algebras: Proceedings of the Philadelphia Conference
38. *M. Goto and F. D. Grosshans,* Semisimple Lie Algebras
39. *A. I. Arruda, N. C. A. da Costa, and R. Chuaqui,* Mathematical Logic: Proceedings of the First Brazilian Conference

40. *F. Van Oystaeyen,* Ring Theory: Proceedings of the 1977 Antwerp Conference
41. *F. Van Oystaeyen and A. Verschoren,* Reflectors and Localization: Application to Sheaf Theory
42. *M. Satyanarayana,* Positively Ordered Semigroups
43. *D. L. Russell,* Mathematics of Finite-Dimensional Control Systems
44. *P.-T. Liu and E. Roxin,* Differential Games and Control Theory III: Proceedings of the Third Kingston Conference, Part A
45. *A. Geramita and J. Seberry,* Orthogonal Designs: Quadratic Forms and Hadamard Matrices
46. *J. Cigler, V. Losert, and P. Michor,* Banach Modules and Functors on Categories of Banach Spaces
47. *P.-T. Liu and J. G. Sutinen,* Control Theory in Mathematical Economics: Proceedings of the Third Kingston Conference, Part B
48. *C. Byrnes,* Partial Differential Equations and Geometry
49. *G. Klambauer,* Problems and Propositions in Analysis

PROBLEMS AND PROPOSITIONS IN ANALYSIS

Gabriel Klambauer

Department of Mathematics
University of Ottawa
Ottawa, Ontario, Canada

MARCEL DEKKER, INC. New York and Basel

Library of Congress Cataloging in Publication Data

Klambauer, Gabriel.
Problems and propositions in analysis.

(Lecture notes in pure and applied mathematics ; 49)
1. Mathematical analysis--Problems, exercises,
etc. I. Title.
QA301.K53 515 .076 79-15854
ISBN 0-8247-6887-6

MARCEL DEKKER, INC.

270 Madison Avenue, New York, New York 10016

Current printing (last digit):

10 9 8 7 6 5 4 3 2 1

PRINTED IN THE UNITED STATES OF AMERICA

To

Helga Funk

PREFACE

Solving problems is an essential activity in the study of mathematics. Instructors pose problems to define scope and content of knowledge expected of their students; mathematical competitions and written qualifying examinations are designed to test the participant's ability and ingenuity in solving unusual problems. Moreover, it is a familiar fact of mathematical instruction that a single good problem can awaken a dormant mind more readily than highly polished lectures do.

This book contains problems with solutions and the reader is invited to produce additional solutions. To ensure a wide appeal I have concentrated on basic matters of real analysis and have consulted problem sections in various mathematical journals and the collected works of some great mathematicians.

Dr. John Abramowich aroused my interest to write this book and Dr. Edward L. Cohen has encouraged me throughout the project; I am grateful to both these personal friends. I am pleased to express my gratitude to Mrs. Wendy M. Coutts, my technical typist, for her fine work and to the administration of the University of Ottawa for the generous support that I have enjoyed in connection with this and two other book writing projects. My warmest thanks are due to my family.

Gabriel Klambauer

CONTENTS

PROBLEMS AND PROPOSITIONS IN ANALYSIS

CHAPTER 1

ARITHMETIC AND COMBINATORICS

PROBLEM 1. Let A and B denote positive integers such that $A > B$. Suppose, moreover, that A and B expressed in the decimal system have more than half of their digits on the left-hand side in common. Show that

$$\sqrt[p]{A} - \sqrt[p]{B} < \frac{1}{p}$$

holds for $p = 2,3,4,\ldots$

Solution. Since

$$\frac{x^p - y^p}{x - y} = x^{p-1} + x^{p-2}y + \cdots + y^{p-1} > py^{p-1}$$

for $y < x$, we obtain, on setting $x^p = A$ and $y^p = B$,

$$\sqrt[p]{A} - \sqrt[p]{B} < \frac{1}{p}\sqrt[p]{\frac{(A - B)^p}{B^{p-1}}}.$$

Let k be the number of digits of $A - B$. Then B has at least $2k + 1$ digits and so $A - B < 10^k$ and $B > 10^{2k}$. Thus

$$\frac{(A - B)^p}{B^{p-1}} < \frac{10^{pk}}{10^{(2p-2)k}} = \frac{1}{10^{(p-2)k}} \leq 1$$

because p is at least equal to 2. Thus $\sqrt[p]{A} - \sqrt[p]{B} < 1/p$.

PROBLEM 2. Show that, for $n = 1,2,3,\ldots,$

$$\left\{1 + \frac{1}{n}\right\}^{n} < \left\{1 + \frac{1}{n+1}\right\}^{n+1} \quad \text{and} \quad \left\{1 + \frac{1}{n}\right\}^{n+1} > \left\{1 + \frac{1}{n+1}\right\}^{n+2}.$$

Solution. Since, for $0 \le a < b$,

$$\frac{b^{n+1} - a^{n+1}}{b - a} < (n + 1)b^{n} \quad \text{or} \quad b^{n}[(n + 1)a - nb] < a^{n+1},$$

setting $a = 1 + 1/(n + 1)$ and $b = 1 + 1/n$ we obtain the first inequality. Note also, taking $a = 1$ and $b = 1 + 1/(2n)$, we get

$$\left\{1 + \frac{1}{2n}\right\}^{n} \frac{1}{2} < 1 \quad \text{or} \quad \left\{1 + \frac{1}{2n}\right\}^{2n} < 4.$$

To verify the second inequality, we observe that, for $0 \le a < b$,

$$\frac{b^{n+1} - a^{n+1}}{b - a} > (n + 1)a^{n};$$

taking $a = 1 + 1/(n + 1)$ and $b = 1 + 1/n$ yields

$$\left\{1 + \frac{1}{n}\right\}^{n+1} > \left\{1 + \frac{1}{n+1}\right\}^{n+2} \left[\frac{n^{3} + 4n^{2} + 4n + 1}{n(n + 2)^{2}}\right].$$

But the term in square brackets is at least 1.

PROBLEM 3. Show that, for $n = 1,2,3,\ldots,$

$$\left\{1 + \frac{1}{n}\right\}^{n} < 3.$$

Solution. Since

$$\left\{1 + \frac{1}{n}\right\}^{n} = 1 + \binom{n}{1}\frac{1}{n} + \cdots + \binom{n}{k}\frac{1}{n^{k}} + \cdots + \binom{n}{n}\frac{1}{n^{n}},$$

where

$$\binom{n}{k} = \frac{n(n - 1)(n - 2)\cdots(n - k + 1)}{k!} \quad \text{with} \quad k! = 1\cdot2\cdot3\cdots(k - 1)k,$$

and, for $2 \le k \le n$,

$$\binom{n}{k}\frac{1}{n^{k}} = \frac{1}{k!}\left(1 - \frac{1}{n}\right)\left(1 - \frac{2}{n}\right)\cdots\left(1 - \frac{k - 1}{n}\right) \le \frac{1}{2\cdot3\cdots k} \le \frac{1}{2^{k-1}},$$

we have, for $n \ge 2$,

$$\left\{1 + \frac{1}{n}\right\}^n < 1 + 1 + \frac{1}{2} + \frac{1}{4} + \cdots + \frac{1}{2^{n-1}} < 1 + \frac{1}{1 - \frac{1}{2}} = 3.$$

PROBLEM 4. For $n = 3,4,5,\ldots$, show that

$$\sqrt{n} < \sqrt[n]{n!} < \frac{n + 1}{2}.$$

Solution. We begin by showing that $(n!)^2 > n^n$ for $n = 3,4,5,\ldots$ Consider

$$(n!)^2 = [1 \cdot n][2(n - 1)][3(n - 2)]\cdots[(n - 1)2][n \cdot 1].$$

But the first and the last factors in square brackets are equal and are less than the other factors in square brackets because, for $n - k > 1$ and $k > 0$ we have

$$(k + 1)(n - k) = k(n - k) + (n - k) > k \cdot 1 + (n - k) = n.$$

Thus $(n!)^2 > n^n$ which is equivalent with $\sqrt{n} < \sqrt[n]{n!}$.

To verify that

$$n! < \left\{\frac{n + 1}{2}\right\}^n \quad \text{for } n = 2,3,4,\ldots,$$

we first note that

$$\left\{\frac{n + 2}{n + 1}\right\}^{n+1} = \left\{1 + \frac{1}{n + 1}\right\}^{n+1} > 2 \quad \text{for } n = 1,2,3,\ldots$$

by the first inequality in Problem 2. Thus

$$\frac{\left\{\frac{n + 2}{2}\right\}^{n+1}}{\left\{\frac{n + 1}{2}\right\}^n} = \left\{\frac{n + 2}{n + 1}\right\}^{n+1} \frac{n + 1}{2} > n + 1$$

or

$$\frac{(n + 1)^{n+1}}{2^n} < \left\{\frac{n + 2}{2}\right\}^{n+1} \quad \text{for } n = 1,2,3,\ldots$$

We now proceed by induction: if

$$n! < \left\{\frac{n + 1}{2}\right\}^n$$

holds, then

$$n!(n + 1) < \frac{(n + 1)^{n+1}}{2^n}$$

follows. But we have shown already that

$$\frac{(n + 1)^{n+1}}{2^n} < \left\{\frac{n + 2}{2}\right\}^{n+1} \qquad \text{for } n = 1,2,3,\ldots$$

Hence

$$(n + 1)! < \left\{\frac{n + 2}{2}\right\}^{n+1}.$$

However

$$n! < \left\{\frac{n + 1}{2}\right\}^{n}$$

is obviously true for n = 2.

Remark. For a generalization of the result in Problem 4 see Problem 5 in Chapter 2.

PROBLEM 5. Show that

$$\sqrt{2} > \sqrt[3]{3} > \sqrt[4]{4} > \sqrt[5]{5} > \cdots > \sqrt[n]{n} > \sqrt[n+1]{n + 1} > \cdots$$

Solution. We have

$$\frac{\sqrt[n+1]{n + 1}}{n^n} = \sqrt[n+1]{\frac{(n + 1)^n}{n^{n+1}}} = \sqrt[n(n+1)]{\left(1 + \frac{1}{n}\right)^n \frac{1}{n}} < \sqrt[n(n+1)]{\frac{3}{n}}$$

(see Problem 3). But, for $n \geq 3$, we have $3/n \leq 1$. Therefore

$$\frac{\sqrt[n+1]{n + 1}}{\sqrt[n]{n}} < 1 \qquad \text{for } n \geq 3.$$

PROBLEM 6. Show that

$$2 > \sqrt{3} > \sqrt[3]{4} > \sqrt[4]{5} > \cdots > \sqrt[n-1]{n} > \sqrt[n]{n + 1} > \cdots$$

Solution. We have to show that

$$\frac{\sqrt[n]{n+1}}{\sqrt[n-1]{n}} < 1 \qquad \text{for } n = 2,3,4,\ldots$$

But we have, by Problem 3,

$$\sqrt[n(n-1)]{\frac{(n+1)^{n-1}}{n^n}} = \sqrt[n(n-1)]{\left[1 + \frac{1}{n}\right]^{n-1} \frac{1}{n}} = \sqrt[n(n-1)]{\left[1 + \frac{1}{n}\right]^{n} \frac{1}{n+1}}$$

$$= \sqrt[n(n-1)]{\frac{3}{n+1}} < 1.$$

PROBLEM 7. Show that the number M which in the decimal system is expressed by means of 91 unities is a composite number, that is, $M = K \cdot L$, where K and L are integers different from 1.

Solution. Since

$$M = 1 + 10 + 10^2 + \cdots + 10^{90} = \frac{10^{91} - 1}{10 - 1}$$

and

$$\frac{10^{91} - 1}{10 - 1} = \frac{(10^7)^{13} - 1}{10^7 - 1}\,\frac{10^7 - 1}{10 - 1} = \frac{(10^{13})^7 - 1}{10^{13} - 1}\,\frac{10^{13} - 1}{10 - 1}$$

we note that the claim is true because

$$\frac{(10^7)^{13} - 1}{10^7 - 1}, \quad \frac{10^7 - 1}{10 - 1} \quad \text{and} \quad \frac{(10^{13})^7 - 1}{10^{13} - 1}, \quad \frac{10^{13} - 1}{10 - 1}$$

are integers as can be seen from the identity

$$\frac{x^n - y^n}{x - y} = x^{n-1} + x^{n-2}y + \cdots + y^{n-1} \qquad \text{for } x \neq y.$$

PROBLEM 8. Let n be a positive integer. Show that n^k ($k \geq 2$ an integer) can be represented as a sum of n successive odd numbers.

Solution. We have to verify that for n and k as given we can find an integer s such that

$$(2s + 1) + (2s + 3) + \cdots + (2s + 2n - 1) = n^k.$$

But the expression on the left-hand side of the last equation equals $(2s + n)n$. It therefore remains to prove that it is possible to find an integer s such that

$$(2s + n)n = n^k, \quad \text{that is,} \quad s = \frac{n(n^{k-2} - 1)}{2}.$$

But n can be either even or odd. In both cases s will be an integer, however.

PROBLEM 9. Show that a sum of positive integers in the decimal system is divisible by 9 if and only if the sum of all digits of those numbers is divisible by 9.

Solution. First we observe that the difference between a positive integer a and the sum s of its digits is divisible by 9; this is clear by noting that

$$a = C_0 + C_1 \cdot 10 + C_2 \cdot 10^2 + C_3 \cdot 10^3 + \cdots$$

$$s = C_0 + C_1 + C_2 + C_3 + \cdots$$

and thus

$$a - s = C_1 \cdot 9 + C_2 \cdot 99 + C_3 \cdot 999 + \cdots.$$

Next, let $a_1, a_2, \ldots, a_n$ denote positive integers and $s_1, s_2, \ldots, s_n$ the respective sums of their digits. In the identity

$$a_1 + a_2 + \cdots + a_n = [(a_1 - s_1) + (a_2 - s_2) + \cdots + (a_n - s_n)] + (s_1 + s_2 + \cdots + s_n)$$

the component on the right-hand side in square brackets is divisible by 9 because $a_k - s_k$ (for $k = 1,2,\ldots,n$) is divisible by 9, as we know already. Consequently, $a_1 + a_2 + \cdots + a_n$ is divisible by 9 if and only if we have that $s_1 + s_2 + \cdots + s_n$ is divisible by 9.

PROBLEM 10. Let a and h be real numbers, and n some positive integer. We introduce the notation:

$$a^{n|h} = a(a - h)(a - 2h)\cdots[a - (n - 1)h];$$

when n = 0 we define $a^{0|h} = 1$. Thus, in particular, $a^{n|0} = a^n$, $a^{1|h} = a$. Moreover, we set

$$\binom{n}{0} = 1, \quad \binom{n}{k} = \frac{n!}{k!(n - k)!} \quad \text{for } k = 1,2,\dots,n.$$

Prove that

$$(a + b)^{n|h} = \sum_{k=0}^{n} \binom{n}{k} a^{(n-k)|h} b^{k|h}.$$

The foregoing result is called the *Factorial Binomial Theorem*; it contains the ordinary *Binomial Theorem* as a special case (when h = 0).

Solution. We prove the claim by induction on n. When n = 0 both sides of the formula reduce to 1, and therefore the claim is true in that case. Now suppose that the claim is true for some integer n ≥ 0, that is,

$$(a + b)^{n|h} = a^{n|h} + \binom{n}{1} a^{(n-1)|h} b^{1|h} + \binom{n}{2} a^{(n-2)|h} b^{2|h} + \cdots + b^{n|h} \tag{E.1}$$

is valid. We must then show that the claim is also true for n + 1. To do this, we multiply both sides of equation (E.1) by a + b - nh. On the left-hand side we obtain $(a + b)^{(n+1)|h}$, as can be seen directly from the definition. On the right-hand side we obtain a sum whose k-th term (where k runs from 0 to n) is

$$\binom{n}{k} a^{(n-k)|h} b^{k|h} (a + b - nh)$$

$$= \binom{n}{k} a^{(n-k)|h} b^{k|h} [a - (n-k)h] + \binom{n}{k} a^{(n-k)|h} b^{k|h} (b - kh)$$

$$= \binom{n}{k} a^{(n-k+1)|h} b^{k|h} + \binom{n}{k} a^{(n-k)|h} b^{(k+1)|h}.$$

Summing over k and using the relation

$$\binom{n}{k - 1} + \binom{n}{k} = \binom{n + 1}{k}$$

we obtain

$$\binom{n}{0} a^{(n+1)|h} + \binom{n}{1} a^{n|h} b^{1|h} + \cdots + \binom{n}{n} a^{1|h} b^{n|h}$$

$$+ \binom{n}{0} a^{n|h} b^{1|h} + \cdots + \binom{n}{n-1} a^{1|h} b^{n|h} + \binom{n}{n} b^{(n+1)|h}$$

$$= a^{(n+1)|h} + \binom{n+1}{1} a^{n|h} b^{1|h} + \binom{n+1}{2} a^{(n-1)|h} b^{2|h}$$

$$+ \cdots + b^{(n+1)|h}. \tag{E.2}$$

We have therefore shown that $(a + b)^{(n+1)|h}$ is equal to the expression in (E.2), which is precisely the statement of the claim for $n + 1$.

PROBLEM 11. Use the result in Problem 10 to evaluate the following sums:

(a) $\binom{n}{0}\binom{m}{j} + \binom{n}{1}\binom{m}{j-1} + \binom{n}{2}\binom{m}{j-2} + \cdots + \binom{n}{j}\binom{m}{0}$,

(b) $\binom{m}{0}\binom{n}{j} - \binom{m+1}{1}\binom{n}{j-1} + \binom{m+2}{2}\binom{n}{j-2} - \cdots + (-1)^j \binom{m+j}{j}\binom{n}{0}$;

here $j \le \min\{m,n\}$ in Part (a), and $j \le n$ in Part (b).

Solution. Since

$$\binom{n}{k} = \frac{n(n-1)\cdots(n-k+1)}{k!} = \frac{n^{k|1}}{k!}$$

we obtain

$$\binom{n}{k}\binom{m}{j-k} = \frac{n^{k|1}}{k!}\frac{m^{(j-k)|1}}{(j-k)!} = \frac{1}{j!}\frac{j!}{k!(j-k)!} n^{k|1} m^{(j-k)|1}$$

$$= \frac{1}{j!}\binom{j}{k} n^{k|1} m^{(j-k)|1}.$$

Hence the sum to be evaluated in Part (a) equals

$$\frac{1}{j!}\left\{\binom{j}{0} m^{j|1} + \binom{j}{1} m^{(j-1)|1} n^{1|1} + \binom{j}{2} m^{(j-2)|1} n^{2|1}\right.$$

$$\left. + \cdots + \binom{j}{j} n^{j|1}\right\}$$

$$= \binom{m+n}{j}.$$

To evaluate the sum in Part (b), we observe that

$$(-1)^k \binom{m+k}{k}\binom{n}{j-k} = (-1)^k \frac{(m+k)^{k|1}}{k!}\frac{n^{(j-k)|1}}{(j-k)!}$$

$$= \frac{1}{j!}\binom{n}{k}(-1)^k (m+k)^{k|1}\, n^{(j-k)|1}.$$

But

$$(-1)^k (m+k)^{k|1} = (-1)^k (m+k)(m+k-1)\cdots(m+1)$$

$$= (-m-1)(-m-2)\cdots(-m-k+1)(-m-k)$$

$$= (-m-1)^{k|1}.$$

Thus

$$(-1)^k \binom{m+k}{k}\binom{n}{j-k} = \frac{1}{j!}\binom{j}{k}(-m-1)^{k|1}\, n^{(j-k)|1}$$

and so our sum equals

$$\frac{1}{j!}\left\{\binom{j}{0} n^{j|1} + \binom{j}{1} n^{(j-1)|1}(-m-1)^{1|1} + \binom{j}{2} n^{(j-2)|1}(-m-1)^{2|1} + \cdots + \binom{j}{j}(-m-1)^{j|1}\right\}$$

$$= \frac{(n-m-1)^{j|1}}{j!} = \frac{(n-m-1)(n-m-2)\cdots(n-m-j)}{j!}.$$

If $n - m - 1 \geq j$, this equals $\binom{n-m-1}{j}$.

Observe that since $\binom{m+k}{k} = \binom{m+k}{m}$ we can rewrite this identity in the form

$$\binom{m}{m}\binom{n}{j} - \binom{m+1}{m}\binom{n}{j-1} + \binom{m+2}{m}\binom{n}{j-2} - \cdots + (-1)^j\binom{m+j}{m}\binom{n}{0}$$

$$= \frac{(n-m-1)(n-m-2)\cdots(n-m-j)}{j!}.$$

PROBLEM 12. Let m and j be positive integers and $j \leq m$. Put

$$(m,j) = \frac{(1-x^m)(1-x^{m-1})\cdots(1-x^{m-j+1})}{(1-x)(1-x^2)\cdots(1-x^j)}.$$

Show the following results (due to Gauss):

(i) $(m,j) = (m,m-j)$;

(ii) $(m,j+1) = (m-1,j+1) + x^{n-j-1}(m-1,j)$;

(iii) $(m,j+1) = (j,j) + x(j+1,j) + x^2(j+2,j) + \cdots + x^{m-j-1}(m-1,j)$;

(iv) (m,j) is a polynomial in x;

(v) $1 - (m,1) + (m,2) - (m,3) + \cdots + (-1)^m(m,m)$

$$= \begin{cases} (1 - x)(1 - x^2)\cdots(1 - x^{m-1}) & \text{if } m \text{ is even,} \\ 0 & \text{if } m \text{ is odd.} \end{cases}$$

Solution. Part (i) is clear from the fact that (m,j) equals

$$\frac{(1 - x)(1 - x^2)\cdots(1 - x^{m-j})(1 - x^{m-j+1})\cdots(1 - x^{m-1})(1 - x^m)}{(1 - x)(1 - x^2)\cdots(1 - x^j)(1 - x)(1 - x^2)\cdots(1 - x^{m-j})}.$$

To obtain Part (ii) we note that

$$(m,j+1) = (m-1,j+1)\,\frac{1 - x^m}{1 - x^{m-j-1}} = (m-1,j+1)\left[1 + x^{m-j-1}\,\frac{1 - x^{j+1}}{1 - x^{m-j-1}}\right]$$

$$= (m-1,j+1) + x^{m-j-1}(m-1,j).$$

To prove Part (iii) we make use of the result in Part (ii) and get

$$(m,j+1) = (m-1,j+1) + x^{m-j-1}(m-1,j),$$

$$(m-1,j+1) = (m-2,j+1) + x^{m-j-2}(m-2,j),$$

$$\cdots$$

$$(j+2,j+1) = (j+1,j+1) + x(j+1,j),$$

$$(j+1,j+1) = (j,j).$$

Adding these equalities termwise, we find

$$(m,j+1) = (j,j) + x(j+1,j) + \cdots + x^{m-j-1}(m-1,j).$$

To verify Part (iv) we observe that

$$(m,1) = \frac{1 - x^m}{1 - x} = 1 + x + x^2 + \cdots + x^{m-1}$$

and so $(m,1)$ is a polynomial in x for any positive integer m. Assuming that

(m,j) is a polynomial in x for $k \le j$, we get by Part (iii) that $(m,j+1)$ is also a polynomial in x and so the claim follows by induction.

We introduce the notation

$$f(x,m) = 1 - (m,1) + (m,2) - (m,3) + \cdots + (-1)^m(m,m)$$

to prove Part (v). Since

$$\begin{aligned}
1 &= 1,\\
(m,1) &= (m-1,1) + x^{m-1},\\
(m,2) &= (m-1,2) + x^{m-2}(m-1,1),\\
(m,3) &= (m-1,3) + x^{m-3}(m-1,2),\\
&\cdots\\
(m,m-1) &= (m-1,m-1) + x(m-1,m-2),\\
(m,m) &= (m-1,m-1),
\end{aligned}$$

we get, upon multiplying these equalities successively by ± 1 and adding,

$$\begin{aligned}
f(x,m) = {} & (1 - x^{m-1}) - (m-1,1)(1 - x^{m-2}) + (m-1,2)(1 - x^{m-3})\\
& + \cdots + (-1)^{m-2}(m-1,m-2)(1 - x).
\end{aligned}$$

But

$$\begin{aligned}
(1 - x^{m-2})(m-1,1) &= (1 - x^{m-1})(m-2,1),\\
(1 - x^{m-3})(m-1,2) &= (1 - x^{m-1})(m-2,2),\\
&\cdots
\end{aligned}$$

Therefore

$$\begin{aligned}
f(x,m) &= (1 - x^{m-1})\left\{1 - (m-2,1) + (m-2,2) - \cdots + (-1)^{m-2}(m-2,m-2)\right\}\\
&= (1 - x^{m-1})f(x,m-2).
\end{aligned}$$

Thus

$$\begin{aligned}
f(x,m) &= (1 - x^{m-1})f(x,m-2),\\
f(x,m-2) &= (1 - x^{m-3})f(x,m-4),\\
&\cdots
\end{aligned}$$

We first assume that m is an even number. We get

$$f(x,m) = (1 - x^{m-1})(1 - x^{m-3})\cdots(1 - x^3)f(x,2).$$

But

$$f(x,2) = 1 - (2,1) + (2,2) = 2 - \frac{1 - x^2}{1 - x} = 1 - x.$$

This shows that

$$f(x,m) = (1 - x^{m-1})(1 - x^{m-3})\cdots(1 - x^3)(1 - x)$$

when m is even.

Finally, when m is odd,

$$f(x,m) = (1 - x^{m-1})(1 - x^{m-3})\cdots(1 - x^2)f(x,1).$$

But $f(x,1) = 0$, consequently $f(x,m) = 0$ for any odd number m.

PROBLEM 13. Show the following result (due to Euler):

$$(1 + xz)(1 + x^2z)\cdots(1 + x^nz) = F(n),$$

where

$$F(n) = 1 + \sum_{k=1}^{n} \frac{(1 - x^n)(1 - x^{n-1})\cdots(1 - x^{n-k+1})}{(1 - x)(1 - x^2)\cdots(1 - x^k)} x^{\frac{k(k+1)}{2}} z^k.$$

Solution. A straightforward calculation shows that

$$F(n+1) - F(n) = zx^{n+1}F(n),$$

that is,

$$F(n+1) = (1 + zx^{n+1})F(n).$$

Therefore

$$F(n) = (1 + zx^n)F(n-1),$$

$$F(n-1) = (1 + zx^{n-1})F(n-2),$$

$$\cdots$$

$$F(3) = (1 + zx^3)F(2),$$

$$F(2) = (1 + zx^2)F(1),$$

$$F(1) = 1 + zx.$$

However, these equalities imply the desired result

$$F(n) = (1 + xz)(1 + x^2z)\cdots(1 + x^nz).$$

Remark. In a completely similar way one can show that

$$(1 + xz)(1 + x^3z)\cdots(1 + x^{2n-1}z)$$

$$= 1 + \sum_{k=1}^{n} \frac{(1 - x^{2n})(1 - x^{2n-2})\cdots(1 - x^{2n-2k+2})}{(1 - x^2)(1 - x^4)\cdots(1 - x^{2k})} x^{k^2} z^k.$$

PROBLEM 14. Let x and a be positive. Find the largest term in the expansion of $(x + a)^n$, where n is a positive integer.

Solution. Let the largest term be

$$T_k = \binom{n}{k} x^{n-k} a^k.$$

This term must not be less than the two neighbouring terms T_{k-1} and T_{k+1}; thus $T_k \geq T_{k-1}$ and $T_k \geq T_{k+1}$. Whence

$$\frac{k}{n - k + 1}\frac{x}{a} \leq 1 \quad \text{and} \quad \frac{n - k}{k + 1}\frac{a}{x} \leq 1.$$

The first of these inequalities yields

$$k \leq \frac{(n + 1)a}{x + a}$$

and from the second inequality we get

$$k \geq \frac{(n + 1)a}{x + a} - 1.$$

We assume first that $\frac{(n + 1)a}{x + a}$ is an integer. Then $\frac{(n + 1)a}{x + a} - 1$ will be an integer also, and since k is an integer satisfying

$$\frac{(n + 1)a}{x + a} - 1 \leq k \leq \frac{(n + 1)a}{x + a},$$

it can attain one of the two values

$$k = \frac{(n+1)a}{x+a}, \qquad k = \frac{(n+1)a}{x+a} - 1.$$

In this case there are two adjacent terms which are equal to each other but exceed all the remaining terms.

Now consider the case when

$$\frac{(n+1)a}{x+a}$$

is not an integer. We then have

$$\frac{(n+1)a}{x+a} = \left[\frac{(n+1)a}{x+a}\right] + \theta$$

where $0 < \theta < 1$ and the square brackets denoting the integer part of the number so enclosed; in other words, θ denotes the fractional part of the number $(n + 1)a/(x + a)$. In this case the inequalities take the form

$$k \le \left[\frac{(n+1)a}{x+a}\right] + \theta, \qquad k \ge \left[\frac{(n+1)a}{x+a}\right] - (1 - \theta).$$

It is clear that in this case there is only one value of k for which our inequalities are satisfied, namely,

$$k = \left[\frac{(n+1)a}{x+a}\right].$$

Hence, when $(n + 1)a/(x + a)$ is not an integer, there is only one largest term T_k.

PROBLEM 15. Let j and n be positive integers and put

$$S_j = 1^j + 2^j + 3^j + \cdots + n^j.$$

Show that

$$\binom{k+1}{1}S_1 + \binom{k+2}{2}S_2 + \cdots + \binom{k+1}{k}S_k = (n+1)^{k+1} - (n+1).$$

Solution. We get

$$\sum_{p=1}^{n} (p+1)^{k+1} = \sum_{p=1}^{n} p^{k+1} + \binom{k+1}{1}\sum_{p=1}^{n} p^k + \binom{k+1}{2}\sum_{p=1}^{n} p^{k-1}$$

$$+ \cdots + \binom{k+1}{k}\sum_{p=1}^{n} p + n$$

by summing p from 1 to n in the identity

$$(p + 1)^{k+1} = p^{k+1} + \binom{k+1}{1}p^k + \binom{k+1}{2}p^{k-1} + \cdots + \binom{k+1}{k}p + 1.$$

But

$$\sum_{p=1}^{n} (p + 1)^{k+1} = \sum_{p=1}^{n} p^{k+1} - 1 + (n + 1)^{k+1}$$

and

$$\binom{k+1}{m} = \binom{k+1}{k+1-m} \quad \text{for } m = 1,2,\ldots,k.$$

Remark. With the help of the recursion formula in Problem 15 and the elementary fact that $S_1 = n(n + 1)/2$ we can easily see that

$$S_2 = \frac{1}{6}n(n+1)(2n+1),$$

$$S_3 = \frac{1}{4}n^2(n+1)^2 = S_1^2,$$

$$S_4 = \frac{1}{30}n(n+1)(2n+1)(3n^2+3n-1),$$

$$S_5 = \frac{1}{12}n^2(n+1)^2(2n^2+2n-1),$$

$$S_6 = \frac{1}{42}n(n+1)(2n+1)(3n^4+6n^3-3n+1),$$

$$S_7 = \frac{1}{24}n^2(n+1)^2(3n^4+6n^3-n^2-4n+2),$$

$$S_8 = \frac{1}{90}n(n+1)(2n+1)(5n^6+15n^5+5n^4-15n^3-n^2+9n-3),$$

$$S_9 = \frac{1}{20}n^2(n+1)^2(2n^6+6n^5+n^4-8n^3+n^2+6n-3),$$

$$S_{10} = \frac{1}{66}n(n+1)(2n+1)(3n^8+12n^7+8n^6-18n^5-10n^4+24n^3+2n^2-15n+5).$$

PROBLEM 16. Prove that if k and m are positive integers, then

(a) $$k^m - k(k - 1)^m + \frac{k(k - 1)}{1 \cdot 2}(k - 2)^m + \cdots + (-1)^{k-1}\, k \cdot 1^m = 0$$

provided that $k > m$; if $k = m$, then

(b) $$m^m - m(m-1)^m + \frac{m(m-1)}{1\cdot 2}(m-2)^m + \cdots + (-1)^{m-1}\, m = m!.$$

Solution. We have

$$(x+1)^m - x^m = mx^{m-1} + \frac{m(m-1)}{1\cdot 2}\, x^{m-2} + \cdots + mx + 1.$$

Replacing x by x + 1, we obtain

$$(x+2)^m - (x+1)^m$$

$$= m(x+1)^{m-1} + \frac{m(m-1)}{1\cdot 2}(x+1)^{m-2} + \cdots + m(x+1) + 1.$$

Subtracting the preceding equality from the last one, we find

$$(x+2)^m - 2(x+1)^m + x^m = m(m-1)x^{m-2} + p_1 x^{m-3} + \cdots.$$

Analogously we obtain

$$(x+3)^m - 3(x+2)^m + 3(x+1)^m - x^m$$

$$= m(m-1)(m-2)x^{m-3} + p_2 x^{m-4} + \cdots.$$

Using the method of mathematical induction, we can prove the following general identity

$$(x+k)^m - \frac{k}{1}(x+k-1)^m + \frac{k(k-1)}{1\cdot 2}(x+k-2)^m - \cdots + (-1)^k\, x^m$$

$$= m(m-1)\cdots(m-k+1)x^{m-k} + px^{m-k-1} + \cdots,$$

from which it is easy to obtain that for k = m

$$(x+m)^m - \frac{m}{1}(x+m-1)^m + \cdots + (-1)^m\, x^m = m!.$$

If k > m, we get

$$(x+k)^m - \frac{k}{1}(x+k-1)^m + \frac{k(k-1)}{1\cdot 2}(x+k-2)^m - \cdots + (-1)^k\, x^m = 0.$$

Putting in the last two equalities x = 0, we get the required identities.

Remarks. It can be shown that there are exactly

$$k^m - \binom{k}{1}(k-1)^m + \binom{k}{2}(k-2)^m - \cdots + (-1)^{k-1}\binom{k}{k-1}1^m$$

m-digit numbers made up of and actually containing the digits 1, 2, 3, ..., k. Evidently, the identities in case $k = m$ and in case $k > m$ which we proved in Problem 16 are immediate consequences of this combinatorial result. In Problem 17 we shall take up a proposition which easily lends itself to prove the combinatorial result we invoked to interpret the identities discussed in Problem 16.

PROBLEM 17. *Principle of Inclusion and Exclusion.* Suppose that a set of N objects and a set of m properties $a_1, a_2, \ldots, a_m$ are given. Some of the N objects may have none of the m properties and some may have one or more of these properties. We use the symbol $N(a_i a_j \ldots a_k)$ to denote the number of objects which have at least the properties $a_i, a_j, \ldots, a_k$ (and possibly additional properties). If we wish to stress the fact that we are concerned with objects which lack a certain property, we prime the corresponding a. For example, $N(a_1 a_2 a_4')$ denotes the number of objects which have the properties a_1 and a_2 and do not have the property a_4 (the question of the remaining properties is left open). In line with this convention, $N(a_1' a_2' \ldots a_m')$ denotes the number of objects with none of the properties $a_1, a_2, \ldots, a_m$.

Prove the following relation

$$\begin{aligned} N(a_1' a_2' \ldots a_m') = {} & N - N(a_1) - N(a_2) - \cdots - N(a_m) \\ & + N(a_1 a_2) + N(a_1 a_3) + N(a_2 a_3) + \cdots + N(a_{m-2} a_{m-1}) \\ & - N(a_1 a_2 a_3) - \cdots - N(a_{m-2} a_{m-1} a_m) \\ & \cdots \\ & + (-1)^m N(a_1 a_2 \ldots a_m). \end{aligned} \tag{E.3}$$

The above sum is taken over all combinations of the properties $a_1, a_2, \ldots, a_m$ (without regard to their order). A summand involving an even number of properties enters with a plus sign, and a summand involving an odd number of properties enters with a minus sign. Relation (E.3) is refered to as the Principle of Inclusion and Exclusion. This name reflects the fact that we exclude all objects which have at least one of the properties $a_1, a_2, \ldots, a_m$, include all objects which have at least two of these properties, exclude all objects which have at least three of these properties, and so on.

Solution. To verify formula (E.3) we use induction on the number of properties. In the case of a single property a the formula is obviously true. Indeed, an object either has this property a or does not have it. Therefore,

$$N(a') = N - N(a).$$

Suppose formula (E.3) is true for m - 1 properties, that is, suppose

$$\begin{aligned} N(a_1'a_2'\ldots a_{m-1}') = N &- N(a_1) - \cdots - N(a_{m-1}) \\ &+ N(a_1a_2) + \cdots + N(a_{m-2}a_{m-1}) \\ &- N(a_1a_2a_3) - \cdots - N(a_{m-3}a_{m-2}a_{m-1}) \\ &\cdots \\ &+ (-1)^{m-1} N(a_1a_2\ldots a_{m-1}). \end{aligned} \tag{E.4}$$

We may use formula (E.4) with any number of objects. In particular, this formula holds for the set of $N(a_m)$ which have property a_m. If we replace N by $N(a_m)$, we obtain

$$\begin{aligned} N(a_1'a_2'\ldots a_{m-1}'a_m) = N(a_m) &- N(a_1a_m) - \cdots - N(a_{m-1}a_m) \\ &+ N(a_1a_2a_m) + \cdots + N(a_{m-2}a_{m-1}a_m) \\ &- N(a_1a_2a_3a_m) - \cdots - N(a_{m-3}a_{m-2}a_{m-1}a_m) \\ &\cdots \\ &+ (-1)^{m-1} N(a_1a_2\ldots a_{m-1}a_m) \end{aligned} \tag{E.5}$$

(to get (E.5) from (E.4) one takes in each set corresponding to a summand in (E.4) only those objects which have the property a_m). Now we subtract (E.5) from (E.4). The difference between the right-hand side of (E.5) and (E.4) is just the right-hand side of (E.3). The difference between the left-hand side of (E.5) and the left-hand side of (E.4) is

$$N(a_1'a_2'\ldots a_{m-1}') - N(a_1'a_2'\ldots a_{m-1}'a_m). \tag{E.6}$$

But $N(a_1'a_2'\ldots a_{m-1}')$ represents the number of objects which do not have the properties a_1, a_2, ..., a_{m-1} and possibly have the property a_m. However, $N(a_1'a_2',\ldots a_{m-1}'a_m)$ represents the number of objects which do not have the properties a_1, a_2, ..., a_{m-1}, but definitely have the property a_m. It follows that the difference in (E.6) is simply the number of objects which have none of the properties $a_1, a_2, \ldots, a_{m-1}, a_m$. In other words,

$$N(a_1'a_2'\ldots a_{m-1}') - N(a_1'a_2'\ldots a_{m-1}'a_m) = N(a_1'a_2'\ldots a_{m-1}'a_m').$$

This proves (E.3) for the case when the number of properties is m. Having proved the validity of (E.3) for m = 1 and the fact that the validity of (E.3) for m - 1 implies its validity for m, we conclude that (E.3) is true for any finite number of properties.

Remarks. Recalling the comments at the end of the Solution of Problem 16, we look at the following questions: How many positive integers of m digits exist such that each digit is 1, 2, and 3? How many of these contain all three digits 1, 2, and 3 at least once?

The answer to the first question is the number of permutations of three objects (where repetitions are allowed), m at a time, namely, 3^m. To answer the second question we let

a_1 signify the absence of 1's

a_2 signify the absence of 2's

a_3 signify the absence of 3's

and invoke the Principle of Inclusion and Exclusion. Then

$$N = 3^m, \quad N(a_1) = N(a_2) = N(a_3) = 2^m,$$

$$N(a_1a_2) = N(a_2a_3) = N(a_1a_3) = 1, \quad N(a_1a_2a_3) = 0;$$

thus (E.3) gives

$$N(a_1a_2a_3) = 3^m - 3 \cdot 2^m + 3.$$

PROBLEM 18. Let [t] denote the greatest integer less than or equal to t. If a is a positive integer, then [n/a] is the largest integer α such that $\alpha a \le n$. This definition is equivalent with saying that $[n/a] = \alpha$, where $n = \alpha a + r$ with $0 \le r < a$. Thus $[6/5] = 1$, and $[-6/5] = -2$.
Prove that for a and b greater than 0,

$$\left[\frac{\left[\frac{n}{a}\right]}{b}\right] = \left[\frac{n}{ab}\right].$$

Solution. Let $[n/a] = \alpha$ and $[\alpha/b] = \beta$. Then $n = \alpha a + r_1$, $0 \le r_1 < a$ and $\alpha = \beta b + r_2$, $0 \le r_2 < b$ Therefore

$$n = \beta ab + ar_2 + r_1 \quad \text{and} \quad [n/ab] = \beta + [(ar_2 + r_1)/ab].$$

However, r_2 is at most $b - 1$, and r_1 is at most $a - 1$, and thus $ar_2 + r_1$ is at most $a(b - 1) + a - 1 = ab - 1$. Thus $[n/ab] = \beta = [\alpha/b]$.

PROBLEM 19. Show:

(i) If p is a positive prime, then

$$[[n/p^s]/p^t] = [n/p^{s+t}].$$

(ii) If $n \ge a > 0$ and $b > 1$, then

$$[n/a] > [n/ab].$$

(iii) If m, n, and a are positive, then

$$[mn/a] \ge m[n/a].$$

(iv) If $n = n_1 + n_2 + \cdots + n_t$, where n_i, for $i = 1,2,\ldots,t$, are positive, then

$$[n/a] \ge [n_1/a] + [n_2/a] + \cdots + [n_t/a].$$

(v) For any real numbers a and b,

$$[2a] + [2b] \ge [a] + [a + b] + [b]. \tag{E.7}$$

Solution. Parts (i) to (iii) are simple to verify. Part (iv) can be shown as follows: Letting $[n_i/a] = \alpha_i$, we have $n_i = \alpha_i + r_i$, $0 \le r_i < a$. Therefore

$$n = (\alpha_1 + \alpha_2 + \cdots + \alpha_t)a + r_1 + r_2 + \cdots + r_t$$

and

$$[n/a] = \alpha_1 + \alpha_2 + \cdots + \alpha_t + [(r_1 + \cdots + r_t)/a].$$

Hence

$$[n/a] \ge [n_1/a] + [n_2/a] + \cdots + [n_t/a].$$

To see Part (v) we first note that $[x + n] = [x] + n$ for n an integer and x arbitrary. Thus both sides of (E.7) change by the same quantity if either a or b changes by an integer. It is thus sufficient to prove (E.7) only for the case $0 \le a < 1$, $0 \le b < 1$. It then reads as follows

$$[2a] + [2b] \ge [a + b].$$

If $[a + b] = 0$ we have nothing to prove. If $[a + b] = 1$, then $a + b \ge 1$, and hence at least one of the two numbers, say a, is $\ge 1/2$, and thus $[2a] + [2b] \ge 1$.

PROBLEM 20. If p is a positive prime, let $E_p(m)$ denote the exponent of the highest power of the prime p that is a divisor of m.
Show: If both n and the prime p are positive, the exponent of the highest power of p that divides n! is

$$E_p(n!) = [n/p] + [n/p^2] + \cdots + [n/p^s] \quad \text{with } [n/p^{s+1}] = 0.$$

Solution. Consider the set T of integers from 1 to n, that is,

$$T = \{1,2,\ldots,p,\ldots,2p,\ldots,p^k,\ldots,n\}.$$

The last integer of the set that is divisible by p is $[n/p]p$, and the coefficient of p shows that there are $[n/p]$ multiples of p in the set. All other integers of the set are prime to p. Hence

$$E_p(n!) = E_p(p \cdot 2p \cdots p^k \cdots [n/p]p).$$

Now we take out one factor p from each of these multiples of p that are in the set T, thereby obtaining the factor $p^{[n/p]}$. Therefore

$$E_p(n!) = [n/p] + E_p(1 \cdot 2 \cdots [n/p]).$$

But the last integer of the new set $\{1,2,\ldots,[n/p]\}$ that is a multiple of p is $[[n/p]/p] = [n/p^2]p$. We can, as before, remove the factor

$$p^{[n/p^2]}$$

from the product of the integers of the new set, showing that

$$E_p(n!) = [n/p] + [n/p^2] + E_p(1 \cdot 2 \cdots [n/p^2]).$$

Likewise, we remove the factors $p^{[n/p^3]}$, $p^{[n/p^4]}$, ... until we find that $p^s \le n < p^{s+1}$, so that $[n/p^s] \ne 0$, while $[n/p^{s+1}] = 0$. Hence all is proved.

Remark. If $n = mk$, then $E_p\{(mk)!\} \ge mE_p(k!)$.

PROBLEM 21. Let the positive integer n be written in the scale of the prime p so that

$$n = a_0p^s + a_1p^{s-1} + \cdots + a_s.$$

Show that

$$E_p(n!) = \frac{n - (a_0 + a_1 + \cdots + a_s)}{p - 1},$$

where $E_p(n!)$ is the exponent of the highest power of the prime p that is a divisor of n!.

Solution. Since $n = a_0p^s + a_1p^{s-1} + \cdots + a_s$ with $0 < a_0 < p$ and $0 \le a_i < p$ for $i = 1,2,\ldots,s$,

$$[n/p] = a_0p^{s-1} + a_1p^{s-2} + \cdots + a_{s-2}p + a_{s-1},$$

$$[n/p^2] = a_0p^{s-2} + a_1p^{s-3} + \cdots + a_{s-2},$$

$$\cdots$$

$$[n/p^s] = 0.$$

Thus

$$[n/p] + \cdots + [n/p^s] = a_0\frac{p^s - 1}{p - 1} + a_1\frac{p^{s-1} - 1}{p - 1} + \cdots + a_{s-1}$$

or

$$E_p(n!) = \frac{a_0p^s + a_1p^{s-1} + \cdots + a_{s-1}p - a_0 - a_1 - \cdots - a_{s-1}}{p - 1}$$

$$= \frac{a_0p^s + \cdots + a_{s-1}p + a_s - a_0 - \cdots - a_{s-1} - a_s}{p - 1}$$

$$= \frac{n - (a_0 + a_1 + \cdots + a_s)}{p - 1},$$

establishing the claim.

PROBLEM 22. Let $a_1, a_2, \ldots, a_t$, and n be positive integers. Show that

$$\frac{n!}{a_1!a_2!\cdots a_t!},$$

where $n = a_1 + a_2 + \cdots + a_t$, is an integer.

Solution. We shall establish that $n!/a_1!a_2!\cdots a_t!$ is an integer by proving that the highest power of any prime contained in the denominator is at least equaled by the highest power of that prime contained in the numerator. We know (see Problem 20) that

$$E_p(n!) = [n/p] + [n/p^2] + \cdots + [n/p^k] + \cdots + [n/p^s]$$

with $[n/p^{s+1}] = 0$. Since $a_1 + a_2 + \cdots + a_t = n$, if p^{s+1} exceeds each a_i, and therefore $[a_i/p^{s+1}] = 0$ for $i = 1,2,\ldots,t$. Hence

$$E_p(a_1!) = [a_1/p] + [a_1/p^2] + \cdots + [a_1/p^k] + \cdots + [a_1/p^s],$$

where, of course, some $[a_1/p^m]$, for $m < s + 1$, may be 0, in which case all integers that follow it in the sum are also 0. Likewise

$$E_p(a_2!) = [a_2/p] + \cdots + [a_2/p^k] + \cdots + [a_2/p^s],$$

$$\cdots$$

$$E_p(a_t!) = [a_t/p] + \cdots + [a_t/p^k] + \cdots + [a_t/p^s].$$

But by Part (iv) of Problem 19,

$$[n/p^k] \geq [a_1/p^k] + [a_2/p^k] + \cdots + [a_t/p^k].$$

Hence $E_p(n!) \geq E_p(a_1!) + E_p(a_2!) + \cdots + E_p(a_t!)$ and thus the given expression is an integer.

Remark. The product of n consecutive positive integers is divisible by n!. Equivalently, binomial coefficients are positive integers.

PROBLEM 23. Show that for any positive integer n, the number of odd binomial coefficients in the expansion of $(a + b)^n$ is 2^k, where k is the sum of the digits in the binary representation of n.

Solution. A typical binomial coefficient is

$$\binom{n}{r} = \frac{n!}{r!s!}, \tag{E.8}$$

where $0 \le r \le n$ and $s = n - r$. Let

$$n = \sum_{i=0}^{h} n_i 2^{h-i}, \quad n_i = 0 \text{ or } 1,\ n_0 = 1,$$

$$r = \sum_{i=0}^{h} r_i 2^{h-i}, \quad r_i = 0 \text{ or } 1,$$

$$s = \sum_{i=0}^{h} s_i 2^{h-i}, \quad s_i = 0 \text{ or } 1.$$

By Problem 21, the exponent of the highest power of 2 contained in $n!$, $r!$, and $s!$, respectively, is

$$n - \Sigma n_i, \quad r - \Sigma r_i, \quad s - \Sigma s_i.$$

Since (E.8) is an integer, we must have

$$n - \Sigma n_i \ge (r - \Sigma r_i) + (s - \Sigma s_i), \tag{E.9}$$

and (E.8) will be odd if and only if equality is attained in (E.9), that is, if and only if

$$\Sigma n_i = \Sigma r_i + \Sigma s_i. \tag{E.10}$$

Now, given n, the number of ways in which r can be chosen to satisfy (E.10) is easily determined by observing that, for $i = 0,1,\ldots,h$,

(i) if $n_i = 0$, r_i can be chosen in only one way, since we must have $r_i = 0$ (and $s_i = 0$);

(ii) if $n_i = 1$, r_i can be chosen in two ways, since we can have $r_i = 0$ or 1 (and $s_i = 1$ or 0).

The required number of ways is thus 2^k, where k is the number of 1's (or the sum of the digits) in the representation of n to the base 2.

PROBLEM 24. Let a and b be positive integers. Show that $\frac{(ab)!}{a!(b!)^a}$ is an integer.

Solution. We must show that

$$\sum_i [ab/p^i] - \sum_i [a/p^i] - a\sum_i [b/p^i] \geq 0$$

for every prime p. Let r and s denote the integers such that $p^r \leq a < p^{r+1}$ and $p^s \leq b < p^{s+1}$. Then

$$\sum_i [ab/p^i] - \sum_i [a/p^i] - a\sum_i [b/p^i]$$

$$= \sum_{i=1}^{s} [ab/p^i] + \sum_{i=s+1}^{r+s} [ab/p^i] + \sum_{i=r+s+1}^{\infty} [ab/p^i]$$

$$- \sum_{i=1}^{r} [a/p^i] - \sum_{i=1}^{s} a[b/p^i]$$

$$= \sum_{i=1}^{s} ([ab/p^i] - a[b/p^i]) + \sum_{i=1}^{r} ([ab/p^{s+i}] - [a/p^i]) + \sum_{i=r+s+1}^{\infty} [ab/p^i]$$

$$\geq \sum_{i=1}^{s} ([ab/p^i] - a[b/p^i]) + \sum_{i=1}^{r} ([ap^s/p^{s+i}] - [a/p^i])$$

$$= \sum_{i=1}^{s} ([ab/p^i] - a[b/p^i]) \geq 0$$

because $[ab/p^i] \geq a[b/p^i]$ by Part (ii) of Problem 19.

PROBLEM 25. Let a and b be positive integers. Show that $\frac{(2a)!(2b)!}{a!b!(a+b)!}$ is an integer.

Solution. Let p be a prime number, j a positive integer and put $ap^{-j} = a'$, $bp^{-j} = b'$. It suffices to show: $[2a'] + [2b'] \geq [a'] + [b'] + [a'+b']$. But we know this already from Part (v) of Problem 19.

PROBLEM 26. How many zeros are at the end of the number 1000! ?

Solution. The number of terminal zeros of a number depends on how often the factor $10 = 2\cdot 5$ occurs in its factorization. We must therefore find the exponents of the factors 2 and 5 in the prime factorization of 1000!. It is clear that the prime 2 occurs to a much higher power in the prime factorization of 1000! than the prime 5. By Problem 20,

$$E_5(1000!) = [1000/5] + [1000/5^2] + [1000/5^3] + [1000/5^4]$$
$$= 200 + 40 + 8 + 1 = 249.$$

Hence there are 249 zeros at the end of 1000!.

PROBLEM 27. Let x be a real number and n a positive integer. Show that

$$[x] + \left[x + \frac{1}{n}\right] + \left[x + \frac{2}{n}\right] + \cdots + \left[x + \frac{n-1}{n}\right] = [nx].$$

Solution. Let

$$f(x) = [nx] - [x] - [x + 1/n] - \cdots - [x + (n-1)/n].$$

Then

$$f(x + 1/n) = [nx + 1] - [x + 1/n] - [x + 2/n] - \cdots - [x + (n-1)/n] - [x + 1]$$
$$= [nx] - [x] - [x + 1/n] - \cdots - [x + (n-1)/n] = f(x).$$

On the other hand, if $0 \le x < 1/n$, we have $f(x) = 0$. Thus $f = 0$.

PROBLEM 28. For n even, show that

$$\frac{1}{1!(n-1)!} + \frac{1}{3!(n-3)!} + \frac{1}{5!(n-5)!} + \cdots + \frac{1}{(n-1)!1!} = 2^{n-1}/n!.$$

Solution. For n even,

$$\frac{n!}{1!(n-1)!} + \frac{n!}{3!(n-3)!} + \frac{n!}{5!(n-5)!} + \cdots + \frac{n!}{(n-1)!1!}$$
$$= \binom{n}{1} + \binom{n}{3} + \binom{n}{5} + \cdots + \binom{n}{n-1} = 2^{n-1}$$

because

$$1 + \binom{n}{1} + \binom{n}{2} + \cdots + \binom{n}{n-1} + \binom{n}{n} = (1 + 1)^n = 2^n,$$

$$1 - \binom{n}{1} + \binom{n}{2} - \cdots + (-1)^n \binom{n}{n} = (1 - 1)^n = 0.$$

PROBLEM 29. Let n be a positive integer. Show that

$$\frac{1}{2\sqrt{n}} < \frac{1}{4^n}\binom{2n}{n} < \frac{1}{\sqrt{2n+1}}.$$

Solution. Let

$$A = \frac{1}{4^n}\binom{2n}{n} = \frac{1}{2}\,\frac{3}{4}\,\frac{5}{6}\cdots\frac{2n-1}{2n}.$$

Then

$$A < \frac{2}{3}\,\frac{4}{5}\cdots\frac{2n}{2n+1} = \frac{2}{1}\,\frac{4}{3}\,\frac{6}{5}\cdots\frac{2n}{2n-1}\,\frac{1}{2n+1}$$

and so

$$A < \frac{1}{A}\,\frac{1}{2n+1} \quad \text{or} \quad A < \frac{1}{\sqrt{2n+1}}.$$

But, on the other hand,

$$A > \frac{1}{2}\,\frac{2}{3}\,\frac{4}{5}\cdots\frac{2n-2}{2n-1} \quad \text{and} \quad A = \frac{1}{2}\,\frac{3}{4}\,\frac{5}{6}\cdots\frac{2n-1}{2n}.$$

Multiplying these relationships, we find

$$A > \frac{1}{2\sqrt{n}}.$$

PROBLEM 30. Let $x_1, x_2, \ldots, x_n$ form an arithmetic progression. It is given that

$$x_1 + x_2 + \cdots + x_n = a \quad \text{and} \quad x_1^2 + x_2^2 + \cdots + x_n^2 = b^2.$$

Determine the progression.

Solution. Let d be the common difference of the given progression. Then $x_k = x_1 + d(k - 1)$. We have

$$\frac{x_1 + x_n}{2}\,n = a$$

and

$$nx_1 + d\,\frac{n(n-1)}{2} = a. \tag{E.11}$$

On the other hand

$$x_k^2 = x_1^2 + 2x_1 d(k - 1) + d^2(k - 1)^2$$

and so (see the Remark to Problem 15)

$$\sum_{k=1}^{n} x_k^2 = nx_1^2 + 2x_1^2 d \sum_{k=1}^{n} (k - 1) + d^2 \sum_{k=1}^{n} (k - 1)^2$$

$$= nx_1^2 + 2x_1 d \frac{n(n - 1)}{2} + d^2 \frac{(n - 1)n(2n - 1)}{6} = b^2. \qquad \text{(E.12)}$$

Squaring both sides of (E.11) and dividing by n, we find

$$nx_1^2 + 2x_1 d \frac{n(n - 1)}{2} + d^2 \frac{n(n - 1)^2}{4} = \frac{a^2}{n}. \qquad \text{(E.13)}$$

Hence subtracting (E.13) from (E.12) gives

$$\frac{d^2 n(n^2 - 1)}{12} = \frac{b^2 n - a^2}{n} \quad \text{or} \quad d = \pm \frac{2\sqrt{3(b^2 - a^2)}}{n\sqrt{n^2 - 1}}.$$

Substituting d into (E.11), we find x_1.

PROBLEM 31. Let n be a positive integer. Show that

$$\binom{n}{0}^2 + \binom{n}{1}^2 + \binom{n}{2}^2 + \cdots + \binom{n}{n}^2 = \binom{2n}{n} \quad \text{and}$$

$$\binom{2n}{0}^2 - \binom{2n}{1}^2 + \binom{2n}{2}^2 - \cdots - \binom{2n}{2n-1}^2 + \binom{2n}{2n}^2 = (-1)^n \binom{2n}{n}.$$

Solution. The equations follow from the identities

$$(1 + x)^n(1 + x)^n = (1 + x)^{2n} \quad \text{and} \quad (1 + x)^{2n}(1 - x)^{2n} = (1 - x^2)^{2n},$$

respectively.

PROBLEM 32. Let n be a positive integer. Show that

$$\binom{n}{1} - 2\binom{n}{2} + 3\binom{n}{3} - \cdots + (-1)^{n-1} n\binom{n}{n} = \begin{cases} 0 & \text{if } n \neq 1, \\ 1 & \text{if } n = 1. \end{cases}$$

Solution. Differentiate the identity

$$1 - (1 - x)^n = \binom{n}{1}x - \binom{n}{2}x^2 + \cdots + (-1)^{n-1}\binom{n}{n}x^n$$

and put $x = 1$.

PROBLEM 33. Let n be a positive integer. Show that

$$\binom{n}{1} - \frac{1}{2}\binom{n}{2} + \frac{1}{3}\binom{n}{3} - \cdots + (-1)^{n-1}\frac{1}{n}\binom{n}{n} = 1 + \frac{1}{2} + \frac{1}{3} + \cdots + \frac{1}{n}.$$

Solution. We have

$$\int_0^1 \frac{1 - (1 - x)^n}{x}\,dx = \int_0^1 \frac{1 - x^n}{1 - x}\,dx = \int_0^1 (1 + x + x^2 + \cdots + x^{n-1})\,dx$$

$$= 1 + \frac{1}{2} + \frac{1}{3} + \cdots + \frac{1}{n}.$$

On the other hand,

$$\frac{1 - (1 - x)^n}{x} = \frac{1 - 1 + \binom{n}{1}x - \binom{n}{2}x^2 + \cdots + (-1)^{n-1}\binom{n}{n}x^n}{x}$$

$$= \binom{n}{1} - \binom{n}{2}x + \cdots + (-1)^{n-1}\binom{n}{n}x^{n-1}.$$

PROBLEM 34. Let $s_n = 1 + q + q^2 + \cdots + q^n$ and

$$S_n = 1 + \frac{1 + q}{2} + \left(\frac{1 + q}{2}\right)^2 + \cdots + \left(\frac{1 + q}{2}\right)^n.$$

Show that

$$\binom{n+1}{1} + \binom{n+1}{2}s_1 + \binom{n+1}{3}s_2 + \cdots + \binom{n+1}{n+1}s_n = 2^n S_n. \tag{E.14}$$

Solution. We shall first prove (E.14) for $q \neq 1$ and then for $q = 1$. If $q \neq 1$,

$$s_n = \frac{1 - q^{n+1}}{1 - q} \tag{E.15}$$

and

$$S_n = \frac{1 - \left(\frac{1 + q}{2}\right)^{n+1}}{1 - \frac{1 + q}{2}} = 2\,\frac{1 - \left(\frac{1 + q}{2}\right)^{n+1}}{1 - q}. \tag{E.16}$$

When we multiply the left-hand side of (E.14) by $(1 - q)$ and observe, from (E.15), that $(1 - q)s_k = 1 - q^{k+1}$, we get

$$\binom{n+1}{1}(1 - q) + \binom{n+1}{2}(1 - q^2) + \binom{n+1}{3}(1 - q^3)$$

$$+ \cdots + \binom{n+1}{n+1}(1 - q^{n+1}). \tag{E.17}$$

Since $1 - q^0 = 0$, (E.17) will not change its value if we add $\binom{n+1}{0}(1 - q^0)$ as first term. After this modification, we write (E.17) as the difference

$$\left[\binom{n+1}{0}\cdot 1 + \binom{n+1}{1}\cdot 1 + \cdots + \binom{n+1}{n+1}\cdot 1\right]$$

$$- \left[\binom{n+1}{0}q^0 + \binom{n+1}{1}q + \cdots + \binom{n+1}{n+1}q^{n+1}\right]$$

and we recognize the expression inside the first bracket as the binomial expansion of $(1 + 1)^{n+1}$, and the expression in the second as the binomial expansion of $(1 + q)^{n+1}$. So far, we have shown that the left member of (E.14) multiplied by $(1 - q)$ is equal to

$$(1 + 1)^{n+1} - (1 + q)^{n+1} = 2^{n+1} - (1 + q)^{n+1} = 2^{n+1}\left\{1 - \left(\frac{1 + q}{2}\right)^{n+1}\right\}.$$

If we now multiply the right-hand member of (E.14) by $(1 - q)$ and use the expression derived in (E.16) for S_n, we get

$$2^n(1 - q)S_n = 2^n\cdot 2\left\{1 - \left(\frac{1 + q}{2}\right)^{n+1}\right\} = 2^{n+1}\left\{1 - \left(\frac{1 + q}{2}\right)^{n+1}\right\}.$$

The fact that the left and right members of (E.14), when multiplied by $1 - q$, are identical establishes the identity (E.14) for $q \neq 1$.

If $q = 1$, then $s_n = S_n = n + 1$. The left side of (E.14) becomes

$$\binom{n+1}{1} + \binom{n+1}{2}2 + \binom{n+1}{3}3 + \cdots + \binom{n+1}{n+1}(n + 1) \tag{E.18}$$

and may be written as

$$\frac{(n + 1)!}{n!} + \frac{2(n + 1)!}{2!(n - 1)!} + \frac{3(n + 1)!}{3!(n - 2)!} + \cdots + \frac{(n + 1)(n + 1)!}{(n + 1)!}$$

which in turn is equal to

$$(n + 1)\left[\binom{n}{0} + \binom{n}{1} + \cdots + \binom{n}{n}\right].$$

Thus the expression in (E.18) computes to $(n + 1)(1 + 1)^n = (n + 1)2^n$ which is precisely the value of $2^n \cdot S_n$ when $q = 1$, so the identity (E.14) holds also for $q = 1$.

PROBLEM 35. Let $a_1, a_2, \ldots, a_n$ represent an arbitrary arrangement of the numbers $1,2,\ldots,n$. Show that, if n is odd and the product

$$(a_1 - 1)(a_2 - 2)(a_3 - 3)\cdots(a_n - n)$$

is nonzero, then this product is an even integer.

Solution. The set $\{a_1, a_2, \ldots, a_n, 1, 2, \ldots, n\}$ contains exactly $n + 1$ odd numbers. However, there are only n factors. Hence at least one of the factors contains two odd numbers, say a_m and m so that $a_m - m$ is even.

PROBLEM 36. Show that in any set of ten different two digit numbers (in the decimal system) one can select two disjoint subsets such that the sum of the numbers in each of the subsets is the same.

Solution. In the set of ten numbers there are altogether $2^{10} = 1024$ subsets. Excluding the empty set, the set of all ten numbers and the ten sets of nine numbers (which could not give a solution) we are left with 1012 possible sets.

The smallest sum is 10, the largest is $92 + 93 + \cdots + 99 = 764$; hence there are not more than 755 possible sums. Thus there must be two subsets having the same sum.

PROBLEM 37. Suppose that $a_1, a_2, \ldots, a_n$ is a finite sequence of real numbers and that m is a positive integer, $m \le n$. A term a_k of the sequence will be called an m-*leader* if there is a positive integer p such that $1 \le p \le m$ and such that $a_k + \cdots + a_{k+p-1} \ge 0$. Thus, for instance, the 1-leaders are the nonnegative terms of the sequence; note, however, that if $m > 1$, an m-leader need not be nonnegative. Show the following result due to F. Riesz: The sum of the m-leaders is nonnegative.

Solution. If there are no m-leaders, the assertion is true. Otherwise, let a_k be the first m-leader and let $a_k + \cdots + a_{k+p-1}$ be the shortest nonnegative sum that it leads ($p \le m$). We assert that every a_h in this sum is itself an m-leader, and, in fact, that $a_h + \cdots + a_{k+p-1} \ge 0$. Indeed, if not, then $a_k + \cdots + a_{h-1} > 0$, contradicting the original choice of p. We proceed now inductively with the sequence $a_{k+p}, a_{k+p+1}, \ldots, a_n$; the sum of the shortest nonnegative sums so obtained is exactly the sum of the m-leaders.

PROBLEM 38. Discover integer solutions of the equation $n(n + 1)/2 = k^2$. Numbers of the form k^2 are of course square numbers and numbers of the form $n(n + 1)/2$ are sometimes called *triangular numbers*.

Solution. Setting $x = 2n + 1$ and $y = 2k$, the equation $n(n + 1)/2 = k^2$ becomes

$$x^2 - 2y^2 = 1. \tag{E.19}$$

If $x = u$ and $y = v$ are integers which satisfy (E.19), we say, for simplicity, that the number $u + v\sqrt{2}$ is a solution of (E.19); two solutions $u + v\sqrt{2}$ and $u' + v'\sqrt{2}$ are said to be equal if $u = u'$ and $v = v'$ and the first solution is said to be greater than the second if $u + v\sqrt{2} > u' + v'\sqrt{2}$.

Let us consider all solutions $x + y\sqrt{2}$ of (E.19) with positive x and y. among these there is a least solution $x_1 + y_1\sqrt{2}$, in which $x_1 + y_1\sqrt{2}$ is called the fundamental solution of the equation (E.19). In fact,

$$x_1 + y_1\sqrt{2} = 3 + 2\sqrt{2}.$$

Putting

$$x_j + y_j\sqrt{2} = (x_1 + y_1\sqrt{2})^j \quad \text{for } j = 2,3,\ldots,$$

we see that $x_j + y_j\sqrt{2}$ are also solutions of the equation (E.19); actually, all solutions of (E.19) can be obtained in this way. We leave it to the reader to check out that $(n,k) = (1,1)$, $(8,6)$, $(49,35)$, $(288,204)$, $(1681, 1189)$, $(9800,6930)$, $(57121,40391)$, $(332928,235416)$, and $(1940449,1372105)$ are the first nine solutions of $n(n + 1)/2 = k^2$.

Remarks. The equation $x^2 - Dy^2 = 1$ with D being a positive integer which is not a square number, is known as *Pell's Equation* in Diophantine analysis.

If $N\{i\} = i(i + 1)/2$, the i-th triangular number is a square number, then $N\{4i(i + 1)\} = 4N\{i\}(2i + 1)^2$ is also a square number. Since the first triangular number is a square number, there exist an infinite number of square triangular numbers.

PROBLEM 39. Let a_n be the n-th term of a sequence defined by

$$a_n = -a_{n-1} - 2a_{n-2}, \quad a_1 = 1, \quad a_2 = -1.$$

Show that $2^{n+1} - 7a_{n-1}^2$ is a square integer.

Solution. By induction on n we prove that

$$2^{n+1} - 7a_{n-1}^2 = (2a_n + a_{n-1})^2.$$

This clearly holds for n = 2, so we assume it correct for n, n ≥ 2. Then

$$(2a_{n+1} + a_n)^2 = (-a_n - 4a_{n-1})^2 = a_n^2 + 8a_na_{n-1} + 16a_{n-1}^2$$

$$= 2(4a_n^2 + 4a_na_{n-1} + a_{n-1}^2) + 14a_{n-1}^2 - 7a_n^2$$

$$= 2(2^{n+1} - 7a_{n-1}^2) + 14a_{n-1}^2 - 7a_n^2 = 2^{n+2} - 7a_n^2.$$

PROBLEM 40. Find the greatest common divisor of

$$\binom{2n}{1}, \binom{2n}{3}, \binom{2n}{5}, \ldots, \binom{2n}{2n-1}.$$

Solution. From the Solution of Problem 28 we see that

$$\binom{2n}{1} + \binom{2n}{3} + \binom{2n}{5} + \cdots + \binom{2n}{2n-1} = 2^{2n-1};$$

thus their common divisor must be of the form 2^p. If $n = 2^kq$, where q is an odd integer, then from $\binom{2n}{1} = 2^{k+1}q$ it follows that a common divisor of these coefficients cannot be larger than 2^{k+1}.

To show that 2^{k+1} divides all of them we write

$$\binom{2^{k+1}q}{p} = \frac{2^{k+1}q}{p}\binom{2^{k+1}q - 1}{p - 1}.$$

Since binomial coefficients are integers and p is odd, we get

$$\binom{2^{k+1}q}{p} = 2^{k+1}M$$

where M is an integer and $p = 1,3,\ldots,2n-1$. This proves that 2^{k+1} is the greatest common divisor of the given set of integers.

PROBLEM 41. Let a sequence $T_0, T_1, T_2, T_3, \ldots$ satisfy the recurrence relation

$$T_{n+2} = aT_{n+1} - bT_n.$$

Show that

$$\frac{T_{n+1}^2 - aT_nT_{n+1} + bT_n^2}{b^n}$$

is constant.

Solution. From the recurrence relation we have

$$T_{n+1}^2 - aT_nT_{n+1} + bT_n^2 = T_{n+1}^2 - T_nT_{n+2}$$

But

$$\begin{vmatrix} T_n & T_{n+1} \\ T_{n+1} & T_{n+2} \end{vmatrix} = \begin{vmatrix} T_n & T_{n+1} \\ aT_n - bT_{n-1} & aT_{n+1} - bT_n \end{vmatrix} = b\begin{vmatrix} T_{n-1} & T_n \\ T_n & T_{n+1} \end{vmatrix}$$

and it follows that

$$\begin{vmatrix} T_n & T_{n+1} \\ T_{n+1} & T_{n+2} \end{vmatrix} = b^n\begin{vmatrix} T_0 & T_1 \\ T_1 & T_2 \end{vmatrix}.$$

Hence

$$T_{n+1}^2 - aT_nT_{n+1} + bT_n^2 = C\,b^n,$$

where

$$C = -\begin{vmatrix} T_0 & T_1 \\ T_1 & T_2 \end{vmatrix}.$$

Remark. The result in Problem 41 is due to Euler and can be generalized as follows: If $T_0, T_1, T_2, T_3, \ldots$ is a sequence of numbers satisfying the recurrence relation

$$T_{n+k} = a_1 T_{n+k-1} + a_2 T_{n+k-2} + \cdots + a_k T_n$$

for some given nonzero numbers $a_1, a_2, \ldots, a_k$, then

$$\frac{1}{a_k^n} \begin{vmatrix} T_n & T_{n+1} & \cdots & T_{n+k-1} \\ T_{n+1} & T_{n+2} & \cdots & T_{n+k} \\ \cdots & & & \\ T_{n+k-1} & T_{n+k} & \cdots & T_{n+2k-2} \end{vmatrix}$$

is constant.

PROBLEM 42. Show the *Identity of Catalan*: If n is a positive integer, then

$$\frac{1}{n+1} + \frac{1}{n+2} + \frac{1}{n+3} + \cdots + \frac{1}{2n} = 1 - \frac{1}{2} + \frac{1}{3} - \frac{1}{4} + \cdots - \frac{1}{2n}$$

and generalize it.

Solution. Let $a_1, a_2, a_3, \ldots$ be a sequence such that

$$a_{pk} = b\, a_k, \tag{E.20}$$

where p and b are two constants. We wish to determine the sum

$$\frac{1}{a_{n+1}} + \frac{1}{a_{n+2}} + \cdots + \frac{1}{a_{np}} = S.$$

Clearly

$$S = \left(\frac{1}{a_1} + \frac{1}{a_2} + \cdots + \frac{1}{a_{np}}\right) - \left(\frac{1}{a_1} + \frac{1}{a_2} + \cdots + \frac{1}{a_n}\right),$$

and so, using (E.20),

$$S = \left(\frac{1}{a_1} + \frac{1}{a_2} + \cdots + \frac{1}{a_{np}}\right) - \left(\frac{b}{a_p} + \frac{b}{a_{2p}} + \cdots + \frac{b}{a_{np}}\right).$$

Thus

$$S = \left(\frac{1}{a_1} + \frac{1}{a_2} + \cdots + \frac{1}{a_{p-1}} + \frac{1-b}{a_p}\right)$$
$$+ \left(\frac{1}{a_{p+1}} + \cdots + \frac{1}{a_{2p-1}} + \frac{1-b}{a_{2p}}\right)$$
$$+ \cdots \tag{E.21}$$
$$+ \left(\frac{1}{a_{(n-1)p+1}} + \cdots + \frac{1-b}{a_{np}}\right).$$

If we take for a_1, a_2, a_3, ... the succession of positive integers 1, 2, 3, ..., we see that (E.20) is satisfied with $p = 2$ and $b = 2$. Hence $1 - b = -1$ and (E.21) gives

$$S = \frac{1}{1} - \frac{1}{2} + \frac{1}{3} - \frac{1}{4} + \cdots + \frac{1}{2n-1} - \frac{1}{2n}$$

and so the Identity of Catalan is established.

If $a_k = k$ $(k = 1,2,3,\ldots)$, but $p = 3$ and $b = 3$, we obtain

$$\frac{1}{n+1} + \frac{1}{n+2} + \frac{1}{n+3} + \cdots + \frac{1}{2n}$$
$$= \left(\frac{1}{1} + \frac{1}{2} - \frac{2}{3}\right) + \left(\frac{1}{4} + \frac{1}{5} - \frac{2}{6}\right) + \cdots + \left(\frac{1}{3n-2} + \frac{1}{3n-1} - \frac{1}{3n}\right).$$

If we take $a_k = k^2$ $(k = 1,2,3,\ldots)$, $p = 2$, and $b = 4$, we get

$$\frac{1}{(n+1)^2} + \frac{1}{(n+2)^2} + \cdots + \frac{1}{(2n)^2}$$
$$= \left(\frac{1}{1} - \frac{3}{4}\right) + \left(\frac{1}{9} - \frac{3}{16}\right) + \cdots + \left(\frac{1}{(2n-1)^2} - \frac{3}{(2n)^2}\right).$$

In a similar way we get the identity

$$\frac{1}{(n+1)^3} + \frac{1}{(n+2)^3} + \cdots + \frac{1}{(2n)^3}$$
$$= \left(\frac{1}{1} - \frac{7}{8}\right) + \left(\frac{1}{27} - \frac{7}{64}\right) + \cdots + \left(\frac{1}{(2n-1)^3} - \frac{7}{(2n)^3}\right),$$

and so forth.

PROBLEM 43. Show the identity

$$\frac{a_1 + a_2 + a_3 + \cdots + a_n}{a_0(a_0 + a_1 + a_2 + \cdots + a_n)}$$

$$= \frac{a_1}{a_0(a_0 + a_1)} + \frac{a_2}{(a_0 + a_1)(a_0 + a_1 + a_2)}$$

$$+ \frac{a_3}{(a_0 + a_1 + a_2)(a_0 + a_1 + a_2 + a_3)}$$

$$+ \cdots + \frac{a_n}{(a_0 + a_1 + \cdots + a_{n-1})(a_0 + a_1 + \cdots + a_n)}$$

Solution. The claim can easily be established by induction.

PROBLEM 44. Consider the two finite seqences

$A_0, A_1, A_2, \ldots, A_n,$

$B_0, B_1, B_2, \ldots, B_n,$

where in the first sequence the terms equidistant from the extremes are equal (that is, $A_0 = A_n$, $A_1 = A_{n-1}$ and so forth) and in the second sequence the terms equidistant from the extremes have constant sum 2G (that is, $B_0 + B_n = 2G$, $B_1 + B_{n-1} = 2G$ and so forth). Putting $S = A_0 + A_1 + A_2 + \cdots + A_n$, show that

$$\sum_{k=0}^{n} A_k B_k = GS.$$

Solution. Clearly,

$$\sum_{k=0}^{n} A_k B_k = A_0 B_0 + A_1 B_1 + A_2 B_2 + \cdots + A_n B_n$$

$$= A_n B_n + A_{n-1} B_{n-1} + A_{n-2} B_{n-2} + \cdots + A_0 B_0;$$

thus, using the assumption that $A_{n-k} = A_k$, we obtain

$$2\sum_{k=0}^{n} A_k B_k = A_0(B_0 + B_n) + A_1(B_1 + B_{n-1}) + \cdots + A_n(B_n + B_0)$$

$$= 2G(A_0 + A_1 + \cdots + A_n) = 2GS.$$

Remarks. We note some interesting applications of Problem 44.

Application 1: If A_k is the binomial coefficient $\binom{n}{k}$ and B_k stands for the k-th term a + kd in an arithmetical progression, then $S = 2^n$ and $2G = 2a + nd$; hence, by Problem 44,

$$\sum_{k=0}^{n} \binom{n}{k}(a + kd) = 2^{n-1}(2a + nd).$$

Application 2: If $A_k = \binom{n}{k}$, $B_k = \sin kx \cos(n - k)x$, then $S = 2^n$ and $2G = \sin nx$; hence we get

$$\sum_{k=0}^{n} \binom{n}{k} \sin kx \cos(n - k)x = 2^{n-1} \sin nx.$$

PROBLEM 45. Let n be a positive integer with $n \geq 2$. Show that the number

$$\frac{(\sqrt{2} + 1)^{2n-1} + (\sqrt{2} - 1)^{2n-1}}{2\sqrt{2}}$$

is the sum of two square integers.

Solution. Put

$$(\sqrt{2} + 1)^{2n-1} = x\sqrt{2} + y, \quad (\sqrt{2} - 1)^{2n-1} = x\sqrt{2} - y;$$

y is an odd number and x the proposed number. By multiplying these two equalities member by member we obtain the equation

$$2x^2 - y^2 = 1$$

or also

$$x^2 = \left(\frac{y - 1}{2}\right)^2 + \left(\frac{y - 1}{2} + 1\right)^2,$$

an equation of the form

$$X^2 = Y^2 + Z^2$$

whose solutions in integers are given by the formulas

$$X = a^2 + b^2, \quad Y = a^2 - b^2, \quad Z = 2ab,$$

where a and b represent integers. Thus the proposed number is indeed the sum of two square integers.

PROBLEM 46. Let a and b be integers and n be an integer larger than or equal to 2. Show that the number

$$\frac{(a + \sqrt{a^2 + b^2})^{2n-1} + (-a + \sqrt{a^2 + b^2})^{2n-1}}{2\sqrt{a^2 + b^2}}$$

is the sum of two square integers and also the sum of three square integers.

Solution. Let

$$A = a + \sqrt{a^2 + b^2}, \qquad B = -a + \sqrt{a^2 + b^2}.$$

We have, identically,

$$\frac{A^{2n-1} + B^{2n-1}}{A + B} = \left(\frac{A^n \pm B^n}{A + B}\right)^2 + \left(b\frac{A^{n-1} \mp B^{n-1}}{A + B}\right)^2, \tag{E.22}$$

where the upper sign is taken in case n is odd; note that

$$(A^{2n-1} + B^{2n-1})(A + B)$$

$$= A^{2n} \pm 2A^nB^n + B^{2n} + AB(A^{2n-2} \mp 2A^{n-1}B^{n-1} + B^{2n-2})$$

$$= A^{2n} + B^{2n} + A^{2n-1}B + AB^{2n-1}.$$

Next we verify that each term in (E.22) is an integer. We consider, for example, the fraction

$$\frac{A^n + B^n}{A + B} = \frac{(a + \sqrt{a^2 + b^2})^n + (-a + \sqrt{a^2 + b^2})^n}{2\sqrt{a^2 + b^2}}.$$

If, as we have supposed, n is odd, this number reduces to

$$\frac{n}{1} a^{n-1} + \frac{n(n-1)(n-2)}{1 \cdot 2 \cdot 3} a^{n-3}(a^2 + b^2) + \cdots + (a^2 + b^2)^{\frac{1}{2}(n-1)}$$

and this is an integer. So much for the verification of the first part of the claim.

We now turn to the verification of the second part of the claim; according to (E.22):

(i) If n is odd,

$$\frac{A^n + B^n}{A + B} = f^2 + g^2.$$

Thus

$$\left(\frac{A^n + B^n}{A + B}\right)^2 = (f^2 + g^2)^2 = (f^2 - g^2)^2 + (2fg)^2.$$

(ii) If n is even and larger than 2,

$$\frac{A^{n-1} + B^{n-1}}{A + B} = f^2 + g^2;$$

then

$$\left(b\frac{A^{n-1} + B^{n-1}}{A + B}\right)^2 = b^2(f^2 - g^2)^2 + (2bfg)^2.$$

In both cases the number

$$\frac{A^{2n-1} + B^{2n-1}}{A + B}$$

is the sum of three square numbers.

Remarks. We note the following application: $a = 2$, $b = 3$, $A = 2 + \sqrt{13}$, $B = -2 + \sqrt{13}$, $n = 3$. We find:

$$\frac{A^5 + B^5}{A + B} = 769 = \left(\frac{A^3 + B^3}{A + B}\right)^2 + \left(b\frac{A^2 - B^2}{A + B}\right)^2 = 35^2 + 12^2;$$

$$\left(\frac{A^3 + B^3}{A + B}\right)^2 = 7^2 + 24^2; \qquad 769 = 7^2 + 24^2 + 12^2.$$

If A, B are integers, taken arbitrarily, the first member of (E.22) can not be a sum of two squares; example: $(3^3 + 1)/(3 + 1) = 7$.

PROBLEM 47. Determine the value of the sum

$$\frac{q}{1 + q} + \frac{q^2}{(1 + q)(1 + q^2)} + \frac{q^3}{(1 + q)(1 + q^2)(1 + q^3)}$$

$$+ \cdots + \frac{q^n}{(1 + q)(1 + q^2)(1 + q^3)\cdots(1 + q^n)}.$$

Solution. We have

$$1 = \frac{a-1}{a} + \frac{1}{a}, \qquad 1 = \frac{a-1}{a} + \frac{b-1}{ab} + \frac{1}{ab},$$

$$1 = \frac{a-1}{a} + \frac{b-1}{ab} + \frac{c-1}{abc} + \frac{1}{abc}, \qquad \cdots$$

Therefore, setting

$$a = 1 + q, \quad b = 1 + q^2, \quad c = 1 + q^3, \quad \ldots,$$

we see that the desired answer is

$$1 - \frac{1}{(1 - q)(1 + q^2)(1 + q^3)\cdots(1 + q^n)}.$$

PROBLEM 48. Show that the product $(p + 2)(p + 3)\cdots(p + q)$ is divisible by $q!$ whenever $p + 1$ is prime to q (that is, $p + 1$ and q have unity as their greatest common divisor). What can be said if $p + 1$ is not prime to q?

Solution. We know that $(p + 1)(p + 2)\cdots(p + q)$ is divisible by $q!$ and $(p + 2)(p + 3)\cdots(p + q)$ is divisible by $(q - 1)!$; see Problem 22. We put

$$\frac{(p + 2)(p + 3)\cdots(p + q)}{(q - 1)!} = A$$

and have

$$\frac{(p + 1)(p + 2)\cdots(p + q)}{q!} = \frac{(p + 1)A}{q}.$$

But $(p + 1)A/q$ is an integer and $p + 1$ is prime to q; hence q divides A and so

$$\frac{A}{q} = \frac{(p + 2)(p + 3)\cdots(p + q)}{q!}$$

is an integer and this proves the claim

We now look at two examples:

$$\frac{5\cdot6\cdot7\cdot8\cdot9}{2\cdot3\cdot4\cdot5\cdot6} = 3\cdot7; \qquad \frac{9\cdot10\cdot11\cdot12\cdot13\cdot14\cdot15\cdot16\cdot17}{2\cdot3\cdot4\cdot5\cdot6\cdot7\cdot8\cdot9\cdot10} = 11\cdot13\cdot17.$$

In the first example $p = 3$, $p + 1 = 4$, and $q = 6$; but 4 is not prime to 6. In the second example $p = 7$, $p + 1 = 8$, and $q = 10$; but 8 is not prime to 10. Therefore the condition that $p + 1$ be prime to q is only a sufficient condition that $(p + 2)(p + 3)\cdots(p + q)/q!$ be an integer.

PROBLEM 49. Consider the *Fibonacci Sequence*

1, 1, 2, 3, 5, 8, 13, 21, 34, 55, 89, 144, 233, 377, ...;

it satisfies the recurrence relation

$$x_1 = 1, \quad x_2 = 1, \quad x_{k+2} = x_{k+1} + x_k \quad \text{for } k = 1,2,3,\ldots$$

Evaluate the sum

$$S_n = \frac{1}{1\cdot 2} + \frac{2}{1\cdot 3} + \frac{3}{2\cdot 5} + \frac{5}{3\cdot 8} + \frac{8}{5\cdot 13} + \cdots + \frac{x_{n+1}}{x_n x_{n+2}}.$$

Solution. Evidently

$$S_n = \frac{2-1}{1\cdot 2} + \frac{3-1}{1\cdot 3} + \frac{5-2}{2\cdot 5} + \frac{8-3}{3\cdot 8} + \frac{13-5}{5\cdot 13} + \cdots + \frac{x_{n+2} - x_n}{x_n x_{n+2}}$$

$$= \left(\frac{1}{1} - \frac{1}{2}\right) + \left(\frac{1}{1} - \frac{1}{3}\right) + \left(\frac{1}{2} - \frac{1}{5}\right) + \left(\frac{1}{3} - \frac{1}{8}\right) + \left(\frac{1}{5} - \frac{1}{13}\right)$$

$$+ \cdots + \left(\frac{1}{x_n} - \frac{1}{x_{n+2}}\right)$$

$$= \frac{1}{1} + \frac{1}{1} - \frac{1}{x_{n+1}} - \frac{1}{x_{n+2}} = 2 - \frac{x_n}{x_{n+1}x_{n+2}}.$$

PROBLEM 50. Show that for n = 1,2,3,... the number

$$A_n = 5^n + 2\cdot 3^{n-1} + 1$$

is a multiple of 8.

Solution. We have $A_1 = 8$ and

$$A_{k+1} - A_k = (5-1)5^k + (6-2)3^{k-1} = 4(5^k + 3^{k-1}).$$

But 5 and 3 are odd and so $5^k + 3^{k-1}$ is even; hence $A_{k+1} - A_k$ is a multiple of 8. Since A_1 and $A_2 - A_1$ are multiples of 8, A_2 must be a multiple of 8, etc.

PROBLEM 51. Show that if x and y are positive irrational numbers such that $1/x + 1/y = 1$, then the sequences

$$[x], \quad [2x], \quad \ldots, \quad [nx], \quad \ldots \qquad \text{and} \qquad [y], \quad [2y], \quad \ldots, \quad [ny], \quad \ldots$$

together include every positive integer exactly once. (The notation $[t]$ means the largest integer contained in t.)

Solution. Since x and y are both positive, $1/x$ and $1/y$ are both less than 1, so that both x and y are greater than 1. Thus no two multiples of x have the same integral part, and no multiples of y have the same integral part. Therefore no integer appears more than once in either of the two sequences. Suppose that an integer K appeared in both sequences. Then we could find integers p and q such that $K < px < K + 1$ and $K < qy < K + 1$; no equality is possible because x and y are irrational numbers. Solving these inequalities for $1/x$ and $1/y$, we find

$$\frac{p}{K + 1} < \frac{1}{x} < \frac{p}{K}, \qquad \frac{q}{K + 1} < \frac{1}{y} < \frac{q}{K}.$$

Adding we find

$$\frac{p + q}{K + 1} < 1 < \frac{p + q}{K} \qquad \text{or} \qquad K < p + q < K + 1,$$

which is impossible since K, p, and q are all integers. Finally, suppose that an integer M is missing from both sequences. Then we can find integers p and q such that

$$px < M, \qquad (p + 1)x < M + 1, \qquad (q + 1)y < M + 1.$$

Solving for $1/x$ and $1/y$ as before, we are led to $M - 1 < p + q < M$, which is again impossible. Therefore every positive integer is present in one sequence or the other, and each positive integer occurs exactly once.

PROBLEM 52. Let 2n points be distributed in space. Show that one may draw at least n^2 line segments connecting these points without obtaining a triangle. (Only the given points are to be considered as vertices of a triangle).

Solution. Let $P_1, P_2, P_3, P_4, \ldots, P_{2n}$ be the 2n given points. Connect P_1 with every point of even index, that is, $P_2, P_4, \ldots, P_{2n}$. Next, connect P_3 with every point of even index. In general, connect each point of odd index with every point of even index. In this manner we get n^2 line segments without obtaining a triangle.

PROBLEM 53. Give the prime factorization of 104060401.

Solution. Observe that

$$104060401 = x^4 + 4x^3 + 6x^2 + 4x + 1 = (x + 1)^4$$

for $x = 100$. But 101 is prime.

PROBLEM 54. Given an infinite number of points in a plane, show that if all distances determined between them are integers then the points are all in a straight line.

Solution. Assume that P_1, P_2, P_3 are not collinear and are in the set. Any point P_4 in the set lies on the line P_iP_j or on one of the hyperbolas

$$|d(P,P_i) - d(P,P_j)| = 1, 2, \ldots, d(P_i,P_j) - 1,$$

where $i \neq j$ and $i, j = 1, 2, 3$; $d(P,P_i)$ denotes distance between P and P_i. Since any two such loci have at most four intersections, the possible positions for P_4 are finite in number.

PROBLEM 55. Show that after deleting the square numbers from the list of positive integers the number we find in the n-th position is equal to $n + \{\sqrt{n}\}$, where $\{\sqrt{n}\}$ denotes the integer closest to $\sqrt{n}$.

Solution. To prove the formula by induction, it suffices to show that the difference $D = n + \{\sqrt{n}\} - (n - 1 + \{\sqrt{n-1}\}) = 1$ or 2, with the value 2 occuring if and only if the number $n + \{\sqrt{n-1}\}$ is a square number. For convenience, let $\{\sqrt{n-1}\} = q$. Then of course $q - 1/2 < \sqrt{n-1} < q + 1/2$ or better

$$q^2 - q + \frac{1}{4} < n - 1 < q^2 + q + \frac{1}{4}.$$

This gives

$$q^2 + \frac{5}{4} < n + \{\sqrt{n-1}\} < (q + 1)^2 + \frac{1}{4}.$$

Therefore the number $n + \{\sqrt{n-1}\}$ is a square number if and only if n equals $(q + 1)^2 - q$. However, then and only then

$$\sqrt{n} > q + \frac{1}{2} > \sqrt{n - 1}.$$

In other words, then and only then $\{\sqrt{n}\} - \{\sqrt{n - 1}\} = 1$, because this difference is never greater than 1.

PROBLEM 56. Let a, b, and c be positive integers and consider all polynomials of the form $ax^2 - bx + c$ which have two distinct zeros in the open interval $0 < x < 1$. Find the least positive integer a for which such a polynomial exists.

Solution. Let $f(x) = ax^2 - bx + c = a(x - r)(x - s)$. Then $f(0)f(1) = a^2 r(r - 1)s(s - 1)$. The graph of $r(r - 1)$ shows that $0 < r < 1$ implies $0 < r(r - 1) \le 1/4$, with equality if and only if $r = 1/2$. Similarly we get $0 < s(s - 1) \le 1/4$. Since $r \ne s$,

$$r(r - 1)s(s - 1) < \frac{1}{16} \quad \text{and} \quad 0 < f(0)f(1) < \frac{a^2}{16}.$$

The coefficients a, b, c are positive integers, and therefore $1 \le f(0)f(1)$. Consequently $a^2 > 16$, that is, $a \ge 5$. The discriminant $b^2 - 4ac$ shows that the minimum possible value for b is 5. Furthermore, $5x^2 - 5x + 1$ has two distinct roots between 0 and 1.

PROBLEM 57. Change the sum $S = 1 + A_1 + A_2 + A_3 + \cdots + A_n$ into a product and change the product

$$\frac{a_1}{b_1}\frac{a_2}{b_2}\frac{a_3}{b_3}\cdots\frac{a_n}{b_n}$$

into a sum.

Solution. Clearly

$$S = (1 + A_1)\frac{(1 + A_1 + A_2)}{1 + A_1}\frac{(1 + A_1 + A_2 + A_3)}{1 + A_1 + A_2}\cdots\frac{(1 + A_1 + \cdots + A_n)}{1 + A_1 + \cdots + A_{n-1}}$$

is valid. Conversely, setting

$$\frac{a_1}{b_1} = 1 + A_1, \quad \frac{a_2}{b_2} = \frac{1 + A_1 + A_2}{1 + A_1}, \quad \text{etc.}$$

we get

$$A_1 = \frac{a_1 - b_1}{b_1}, \qquad A_2 = \frac{a_1}{b_1}\frac{a_2 - b_2}{b_2}, \qquad \text{etc.}$$

and in general

$$A_n = \frac{a_1 a_2 \cdots a_{n-1}}{b_1 b_2 \cdots b_{n-1}} \frac{a_n - b_n}{b_n}.$$

PROBLEM 58. Show that if m and n are odd positive integers, then

$$\frac{(mn)!}{(m!)^{(n+1)/2}(n!)^{(m+1)/2}}$$

is an integer.

Solution. The numbers

$$\frac{(n^2)!}{(n!)^{n+1}}, \qquad \frac{(mn)!}{(m!)^n n!}, \qquad \frac{(mn)!}{(n!)^m m!}$$

are integers by Problem 24. As a product of integers, the number

$$\left(\frac{(mn)!}{(m!)^{(n+1)/2}(n!)^{(m+1)/2}}\right)^2$$

must be an integer. Since m and n are odd, the number

$$\frac{(mn)!}{(m!)^{(n+1)/2}(n!)^{(m+1)/2}}$$

is a rational number whose square is an integer. This implies that this number itself must be an integer.

Remark. Alternately, $(n^2)!/(n!)^{n+1}$ is an integer because it is the number of ways of dividing n^2 (different) objects into a collection of n unordered batches of n elements each;

$$\frac{(mn)!}{(m!)^n n!}$$

must be an integer because it is the number of ways of dividing mn (different) objects into n unordered batches of m elements each, etc.

PROBLEM 59. Compute the sum

$$\sum_{i_n=1}^{m} \sum_{i_{n-1}=1}^{i_n} \cdots \sum_{k=1}^{i_1} e_k \quad \text{with } e_k = 1.$$

Solution. Since

$$\sum_{k=1}^{i_1} e_k = i_1 = \binom{i_1}{1},$$

$$\sum_{i_1=1}^{i_2} \sum_{k=1}^{i_1} e_k = \sum_{i_1=1}^{i_2} \binom{i_1}{1} = \binom{i_2 + 1}{2},$$

$$\sum_{i_2=1}^{i_3} \sum_{i_1=1}^{i_2} \sum_{k=1}^{i_1} e_k = \sum_{i_2=1}^{i_3} \binom{i_2 + 1}{2} = \binom{i_3 + 2}{3},$$

...

we see that the value of our sum is $\binom{n + m}{n + 1}$.

PROBLEM 60. Find the coefficient of x^k in the expression

$(1 + x + x^2 + \cdots + x^{n-1})^2$.

Solution. We have

$$(1 + x + x^2 + \cdots + x^{n-1})^2 = \frac{(x^n - 1)^2}{(x - 1)^2}$$

$$= (x^{2n} - 2x^n + 1)(1 + 2x + 3x^2 + \cdots + mx^{m-1} + \cdots).$$

Hence the coefficient of x^k is $k + 1$, if $0 \le k \le n - 1$, and $2n - k - 1$, if $n \le k \le 2n - 2$. In either case, the coefficient in question is equal to $n - |n - k - 1|$.

PROBLEM 61. Find the coefficient of x^m in the expression

$(1 + x)^k + (1 + x)^{k+1} + \cdots + (1 + x)^n$

in the cases: (i) $m < k$ and (ii) $m \geq k$.

Solution. We have

$$(1 + x)^k + (1 + x)^{k+1} + \cdots + (1 + x)^n = \frac{(1 + x)^{n+1} - (1 + x)^k}{x}.$$

Hence the coefficient of x^m is $\binom{n+1}{m+1} - \binom{k}{m+1}$ for $m < k$ and $\binom{n+1}{m+1}$ for $m \geq k$.

PROBLEM 62. Show that in a plane (i) n straight lines, no two of which are parallel and no three of which meet in a point, divide the plane into $(n/2)(n + 1) + 1$ parts and (ii) n circles, each circle intersecting all others and no three meeting in a point, divide the plane into $n^2 - n + 2$ parts.

Solution. (i) Suppose that k of the lines have already been drawn in the plane; the $(k + 1)$st line meets each of the k lines which have already been drawn and the k points of intersection divide the $(k + 1)$st line into $k + 1$ parts. Consequently the $(k + 1)$st line cuts exactly $k + 1$ of all regions into which the plane has already been divided. Since it splits each of these regions into two parts, drawing the $(k + 1)$st line increases the number of pieces by $k + 1$. But if only one line is drawn, it will divide the plane into two pieces. It follows from this that after n lines have been drawn the plane will have been divided into $2 + 2 + 3 + 4 + \cdots + n$ parts (drawing the second line increases the number of parts by 2, drawing the third line increases it by 3 more, drawing the fourth line increases it by 4 more, etc.). But $2 + 2 + 3 + 4 + \cdots + n = (1 + 2 + 3 + \cdots + n) + 1 = (n/2)(n + 1) + 1$.

(ii) By reasoning as in Part (i), we can show that the $(k + 1)$st circle increases by 2k the number of parts into which the plane is divided. For the $(k + 1)$st circle intersects each of the first k circles in two points; these 2k points divide the $(k + 1)$st circle into 2k arcs. Each of these arcs divides in two one of the regions formed by the first k circles. Since one circle divides the plane into two parts, the total number of parts after drawing the n-th circle is $2 + 2 + 4 + 6 + 8 + \cdots + 2(n-1) = n^2 - n + 2$.

PROBLEM 63. Show that

$$\binom{m}{1} + \binom{m+1}{2} + \cdots + \binom{m+n-1}{n} = \binom{n}{1} + \binom{n+1}{2} + \cdots + \binom{n+m-1}{m}.$$

Solution. Using induction, we can see that the sum on the left-hand side reduces to

$$\sum_{k=1}^{n} \binom{m+k-1}{k} = \binom{m+n}{m} - 1.$$

The sum on the right-hand side reduces to

$$\sum_{k=1}^{m} \binom{n+k-1}{k} = \binom{n+m}{n} - 1.$$

But $\binom{m+n}{m} = \binom{m+n}{n}$.

PROBLEM 64. In Problem 49 we considered the Fibonacci sequence

1, 1, 2, 3, 5, 8, 13, 21, 34, ...

satisfying the recurrence relation $x_1 = 1$, $x_2 = 1$,

$$x_{k+2} = x_{k+1} + x_k \quad \text{for } k = 1,2,3,\ldots$$

Show that every positive integer K can be written as the sum of different terms of the Fibonacci sequence such that no two summands are neighbors in the Fibonacci sequence.

Solution. Let $x_n \le K < x_{n+1}$. Then $0 \le K - x_n < x_{n-1}$. Hence there is an $s < n - 1$ such that $x_s \le K - x_n < x_{s+1}$. But then $0 \le K - x_n - x_s < x_{s-1}$, and $s - 1 < n - 2$. After a sequence of such steps, we find that $K = x_n + x_s + x_p + \cdots + x_r$, where the neighboring indices n, s, p, ..., r differ by at least 2.

PROBLEM 65. The length of the sides CB and CA of a triangle ΔABC are a and b and the angle between them is $\gamma = 2\pi/3$. Show that the length v of the bisector of γ is

$$v = \frac{ab}{a + b} \quad \text{or} \quad \frac{1}{v} = \frac{1}{a} + \frac{1}{b}.$$

Solution. Let CD be the bisector of $\sphericalangle ACB = \gamma$. Since the area of ΔABC is the sum of the areas of the triangles ΔACD and ΔCDB, we have

$$ab \sin \gamma = av \sin \frac{\gamma}{2} + bv \sin \frac{\gamma}{2} = v(a + b)\sin \frac{\gamma}{2},$$

and since $\sin \gamma = 2 \sin(\gamma/2)\cos(\gamma/2)$, it follows that

$$2ab \sin \frac{\gamma}{2} \cos \frac{\gamma}{2} = v(a + b)\sin \frac{\gamma}{2};$$

thus

$$\frac{1}{v} \cos \frac{\gamma}{2} = \frac{1}{2} \frac{a + b}{ab} = \frac{1}{2}\left(\frac{1}{a} + \frac{1}{b}\right).$$

But $\gamma = 2\pi/3$ and so $\cos(\gamma/2) = \cos(\pi/3) = 1/2$.

Remarks. The equation $1/a + 1/b = 1/v$ has the same form as $1/d + 1/d' = 1/f$ which describes the relationship in optics between the distance d of an object from a lense, the distance d' of the image from the lense, and the focal length f of the lens. The solution of Problem 65 therefore affords the following simple construction of a diagram with the help of which one of the quantities d, d', f can be found if the other two are given: From a point 0, draw two lines that form an angle of 120 degrees at 0, draw the bisector of that angle, and mark off equal distances along these three lines using 0 as the initial point. The quantities d and d' will be represented by points on the sides of the 120 degrees angle, and f by a point on the bisector. If d and d' are given values, the straight line going though d and d' intersect the bisector at f.

PROBLEM 66. Let $m + n$ points on a circle divide it into $m + n$ arcs. We mark m of the points with an A and the remaining n points with a B. If both end points of an arc are marked with an A, then we associate with it the number 2. If both end points of an arc are marked with a B, then we associate with it the number 1/2. Finally, if the end points of an arc are marked with different letters, then we associate with it the number 1. Show that the value of the product of these numbers is 2^{n-m}.

Solution. We can easily verify the fact that permuting two neighboring letters A and B has no effect on the product (it is enough to consider arrangements AABA, BABB, and AABB). This means that we may suppose all letters A to be grouped together and all letters B to be grouped together. But in this case the claim is trivially correct.

PROBLEM 67. Show that $\binom{a}{0} + \binom{a+1}{1} + \cdots + \binom{a+n}{n} = \binom{a+1+n}{n}$.

Solution. The claim is valid for $n = 0$. We assume that the claim is true for $n = k$, that is,

$$\binom{a}{0} + \binom{a+1}{1} + \cdots + \binom{a+k}{k} = \binom{a+k+1}{k}$$

and show that it is true for $n = k + 1$. But

$$\binom{a}{0} + \binom{a+1}{1} + \cdots + \binom{a+k}{k} + \binom{a+k+1}{k+1} = \binom{a+k+1}{k} + \binom{a+k+1}{k+1} = \binom{a+k+2}{k+1}$$

because

$$\binom{m}{j} + \binom{m}{j+1} = \binom{m+1}{j+1}.$$

Remark. Since $\binom{a}{j} = \binom{a}{a-j}$, we get from the result in Problem 67 that

$$\binom{a}{a} + \binom{a+1}{a} + \cdots + \binom{a+n}{a} = \binom{a+1+n}{a+1}.$$

Taking, say $a = 3$, the foregoing equation yields the formula

$$1\cdot2\cdot3 + 2\cdot3\cdot4 + \cdots + n(n + 1)(n + 2) = \frac{n(n + 1)(n + 2)(n + 3)}{4}.$$

PROBLEM 68. Let n be an integer larger than or equal to 1. Evaluate the sums

$$\binom{n}{0} + \frac{1}{2}\binom{n}{1} + \frac{1}{3}\binom{n}{2} + \cdots + \frac{1}{n + 1}\binom{n}{n}$$

and

$$\binom{n}{1} + 2\binom{n}{2} + 3\binom{n}{3} + \cdots + n\binom{n}{n}.$$

Solution. Since

$$\frac{n + 1}{k + 1}\binom{n}{k} = \binom{n+1}{k+1},$$

it follows that

$$(n + 1)\left\{\binom{n}{0} + \frac{1}{2}\binom{n}{1} + \frac{1}{3}\binom{n}{2} + \cdots + \frac{1}{n + 1}\binom{n}{n}\right\}$$

$$= \binom{n+1}{1} + \binom{n+1}{2} + \binom{n+1}{3} + \cdots + \binom{n+1}{n+1} = 2^{n+1} - 1,$$

so that

$$\binom{n}{0} + \frac{1}{2}\binom{n}{1} + \frac{1}{3}\binom{n}{2} + \cdots + \frac{1}{n+1}\binom{n}{n} = \frac{2^{n+1} - 1}{n+1}.$$

To evaluate the second sum we note that

$$k\binom{n}{k} = n\binom{n-1}{k-1} \qquad \text{for } n \geq 1$$

and so, for $n \geq 1$,

$$\binom{n}{1} + 2\binom{n}{2} + 3\binom{n}{3} + \cdots + n\binom{n}{n}$$

$$= n\left\{\binom{n-1}{0} + \binom{n-1}{1} + \binom{n-1}{2} + \cdots + \binom{n-1}{n-1}\right\} = n\cdot 2^{n-1}.$$

Remark. It is easy to see from the foregoing that

$$\binom{n}{1} - 2\binom{n}{2} + 3\binom{n}{3} - \cdots + (-1)^{n-1}\, n\binom{n}{n}$$

$$= n\left\{\binom{n-1}{0} - \binom{n-1}{1} + \binom{n-1}{2} - \cdots + (-1)^{n-1}\binom{n-1}{n-1}\right\}$$

$$= 0 \quad \text{if } n > 1.$$

If $n = 1$ the expression in question reduces to the single term $\binom{1}{1} = 1$. (The foregoing result is already known to us from Problem 32.)

PROBLEM 69. Show that for $m > n$,

$$\sum_{k=0}^{n} \frac{n(n-1)\cdots(n-k+1)}{m(m-1)\cdots(m-k+1)} = \frac{m+1}{m-n+1}.$$

Solution. By the Remark to Problem 67, $\Sigma_{k=0}^{n} \binom{m-k}{m-n} = \binom{m+1}{m-n+1}$. Thus

$$\sum_{k=0}^{n} \frac{n!\,(m-k)!}{m!\,(n-k)!} = \frac{1}{\binom{m}{n}} \sum_{k=0}^{n} \binom{m-k}{m-n} = \frac{\binom{m+1}{m-n+1}}{\binom{m}{n}} = \frac{m+1}{m-n+1}.$$

PROBLEM 70. Let k be a positive integer, but not a square integer. Let U denote the set of all positive rational numbers whose square is larger than k, and let L denote all rational numbers not belonging to the set U. Show that U has no smallest member, and L has no largest member.

Solution. Suppose that L contained a largest member, say a, and U contained a smallest member, say b. Thus, by assumption, $a^2 < k$ and $b^2 > k$. We now select two positive integers m and n such that $m^2 > kn^2$ and put

$$\frac{ma + nk}{na + m} = a' \quad \text{and} \quad \frac{mb + nk}{nb + m} = b'.$$

Then

$$a' - a = \frac{n(k - a^2)}{na + m} > 0 \quad \text{and} \quad b' - b = \frac{n(k - b^2)}{nb + m} < 0,$$

$$(a')^2 - k = \frac{(ma + nk)^2 - k(na + m)^2}{(na + m)^2} = \frac{(m^2 - kn^2)(a^2 - k)}{(na + m)^2} < 0,$$

$$(b')^2 - k = \frac{(mb + nk)^2 - k(nb + m)^2}{(nb + m)^2} = \frac{(m^2 - kn^2)(b^2 - k)}{(nb + m)^2} > 0.$$

Thus $a' > a$, $(a')^2 < k$, and $0 < b' < b$, $(b')^2 > k$; this means that a' is in L and b' is in U, in violation of the assumption that a is the largest member of L and that b is the smallest member of U.

PROBLEM 71. Let the points $z_1, z_2, \ldots, z_n$ lie on one side of a straight line passing through the origin of the complex plane. Show that the points $1/z_1, 1/z_2, \ldots, 1/z_n$ are situated on the same side of the straight line and

$$z_1 + z_2 + \cdots + z_n \neq 0, \quad \frac{1}{z_1} + \frac{1}{z_2} + \cdots + \frac{1}{z_n} \neq 0.$$

Solution. We may assume without loss of generality that the straight line is the imaginary axis and that all points $z_1, z_2, \ldots, z_n$ are situated to the right of it (otherwise all z_k, $k = 1,2,\ldots,n$, should be multiplied by a certain complex number of the form $\cos t + i \sin t$ to perform a suitable rotation). But the transformation $w = 1/z$ is an inversion with respect to the unit circle followed by a reflection about the real axis; in other words, the mapping $z \to 1/z$ can be viewed as $z \to z/|z|^2$ followed by $z \to \overline{z}$. Therefore it is clear that if a point z is on the right-hand side of the imaginary axis, then so is the point $1/z$; hence the real part of z and the real part of $1/z$ are both positive.

Remarks. If $z_1 + z_2 + \cdots + z_n = 0$, then any straight line passing through the origin separates the points $z_1, z_2, \ldots, z_n$, provided only that they do not lie on this line. Any straight line passing through the center of gravity of a system of material points $z_1, z_2, \ldots, z_n$ in a plane having masses $m_1, m_2, \ldots, m_n$ separates these points, provided only that they do not lie on this line.

PROBLEM 72. Show the following result of Cauchy:

$$\begin{vmatrix} \frac{1}{a_1 + b_1} & \frac{1}{a_1 + b_2} & \cdots & \frac{1}{a_1 + b_n} \\ \frac{1}{a_2 + b_1} & \frac{1}{a_2 + b_2} & \cdots & \frac{1}{a_2 + b_n} \\ \cdots & & & \\ \frac{1}{a_n + b_1} & \frac{1}{a_n + b_2} & \cdots & \frac{1}{a_n + b_n} \end{vmatrix} = \frac{\prod_{i>k} (a_i - a_k)(b_i - b_k)}{\prod_{i,k} (a_i + b_k)}.$$

Solution. We denote the determinant in question by D_n. Subtracting the last row of the determinant from all other rows, we can pull out the factor

$$\frac{\prod_{i=1}^{n-1} (a_n - a_i)}{(a_n + b_1)(a_n + b_2)\cdots(a_n + b_n)};$$

in the resulting determinant, subtracting the last column from all other columns, we can pull out the factor

$$\frac{\prod_{i=1}^{n-1} (b_n - b_i)}{(a_1 + b_n)(a_2 + b_n)\cdots(a_{n-1} + b_n)}.$$

Observing that the remaining determinant equals D_{n-1}, we see that repetition of the foregoing process yields the desired result.

PROBLEM 73. Let $(1 + x)^n = a_0 + a_1x + a_2x^2 + a_3x^3 + \cdots + a_nx^n$, where n is a positive integer. Evaluate the product

$$T_n = \left(1 + \frac{a_0}{a_1}\right)\left(1 + \frac{a_1}{a_2}\right)\left(1 + \frac{a_2}{a_3}\right)\cdots\left(1 + \frac{a_{n-1}}{a_n}\right).$$

Solution. Let $k = 0,1,\ldots,n$. Since

$$a_k = \binom{n}{k} = \frac{n!}{k!(n-k)!} \quad \text{and} \quad \binom{n}{k} + \binom{n}{k-1} = \binom{n+1}{k},$$

we see that

$$1 + \frac{a_{k-1}}{a_k} = \frac{a_k + a_{k-1}}{a_k} = \frac{\binom{n+1}{k}}{\binom{n}{k}} = \frac{n+1}{n+1-k}$$

and

$$T_n = \frac{n+1}{n}\,\frac{n+1}{n-1}\,\frac{n+1}{n-2}\cdots\frac{n+1}{1} = \frac{(n+1)^n}{n!}.$$

PROBLEM 74. Show that if n is a positive integer and

$$Q(n) = \prod_{k=1}^{n-1} k^{2k-n-1},$$

then $Q(n)$ is an integer whenever n is prime.

Solution. We have

$$\prod_{k=1}^{n-1} k! = \prod_{k=1}^{n-1} (n-k)! = \prod_{k=1}^{n-1} k^{n-k}.$$

Thus

$$Q(n) = \prod_{k=1}^{n-1} \frac{k^{n-1}}{k^{2n-2k}} = \frac{\left\{\prod_{k=1}^{n-1} k\right\}^{n-1}}{\left\{\prod_{k=1}^{n-1} k^{n-k}\right\}^2} = \frac{\{(n-1)!\}^{n-1}}{\prod_{k=1}^{n-1} k!(n-k)!} = \prod_{k=1}^{n-1}\left\{\frac{\binom{n}{k}}{n}\right\}.$$

But if n is prime, then n divides $\binom{n}{k}$, $k = 1,2,\ldots,n-1$. Hence $Q(n)$ is an integer whenever n is prime.

PROBLEM 75. Let $0 < a_1 < a_2 < \cdots < a_n$ and $e_i = \pm 1$. Prove that the sum

$\Sigma_{i=1}^{n} e_i a_i$ assume at least $\binom{n+1}{2} + 1$ distinct values as the e_i range over the 2^n possible combinations of signs.

Solution. We put $c = \Sigma_{i=1}^{n} (-a_i)$ and observe that

$$c < c + 2a_1 < c + 2a_2 < \cdots < c + 2a_n$$

$$< c + 2a_n + 2a_1 < \cdots < c + 2a_n + 2a_{n-1}$$

$$< c + 2a_n + 2a_{n-1} + 2a_1 < \cdots < c + 2(\Sigma_{i=1}^{n} a_i) = \Sigma_{i=1}^{n} a_i,$$

so that there are at least

$$1 + n + (n - 1) + (n - 2) + \cdots + 2 + 1 = 1 + \frac{n(n + 1)}{2}$$

distinct values in the list. Since each value is one of the given sums, we have shown that the expression assumes at least $\binom{n+1}{2} + 1$ values.

PROBLEM 76. The *Well-Ordering Principle* asserts that eny nonempty set of positive integers has a smallest element; the Well-ordering Principle is equivalent with the Principle of Mathematical Induction.
Using the Well-Ordering Principle, show that $\sqrt{2}$ is an irrational number.

Solution. Suppose that $\sqrt{2} = n/m$, where n and m are positive integers. Then $n > m$, and there is an integer $p > 0$ such that $n = m + p$, and $2m^2$ is equal to $m^2 + 2pm + p^2$. This implies $m > p$. Thus for some integer $a > 0$, $m = p + a$, $n = 2p + a$ and $2(p + a)^2 = (2p + a)^2$. The last equality implies $a^2 = 2p^2$ so that the entire process may be repeated indefinitely giving

$$n > m > a > p > \cdots,$$

but since every nonempty set of positive integers has a smallest element, this is a contradiction and $\sqrt{2}$ is not a rational number.

PROBLEM 77. Let n be a positive integer and m be any integer with the same parity as n. Show that the product mn is equal to the sum of n consecutive odd integers. These odd integers are all positive if and only if $m \geq n$.

Solution. In all cases the required odd integers are those between the bounds $m \pm (n - 1)$ inclusive, for this set of $2n - 1$ consecutive integers contains exactly n odd integers, which have mean value m since they occur symmetrically with respect to m (the "middle" member of the set). Further, in order that the smallest odd integer of the set (namely, $m - n + 1$) be positive, $m \geq n$ is clearly necessary and sufficient.

As a particular consequence, if k is an integer satisfying $k > 1$, then n^k is the sum of the n consecutive odd integers between the bounds $m \pm (n - 1)$ inclusive. This follows if m is assigned the value n^{k-1}, since n^{k-1} has the same parity as n, and $n^{k-1} \geq n$. (Note that this particular case has already been taken up in Problem 8.)

PROBLEM 78. If $a_{n+1} = (1 + a_n a_{n-1})/a_{n-2}$ and $a_1 = a_2 = a_3 = 1$, show that a_n is an integer.

Solution. We define a sequence (b_n) of integers by

$$b_1 = b_2 = b_3 = 1, \quad b_4 = 2, \quad b_n = 4b_{n-2} - b_{n-4} \quad (n > 4).$$

Then

$$b_{n+1}b_{n-2} - b_n b_{n-2} = (4b_{n-1} - b_{n-3})b_{n-2} - (4b_{n-2} - b_{n-4})b_{n-1}$$
$$= b_{n-1}b_{n-4} - b_{n-2}b_{n-3},$$

so that, by induction,

$$b_{n+1}b_{n-2} - b_n b_{n-1} = 1, \quad \text{or} \quad b_{n+1} = (1 + b_n b_{n-1})/b_{n-2},$$

and hence $(a_n) = (b_n)$.

PROBLEM 79. Let $a_{n+1} = (k + a_n a_{n-1})/a_{n-2}$ and $a_1 = a_2 = 1$, $a_3 = p$, where k, p are positive integers such that $(k,p) = 1$ (that is, such that k and p have 1 as largest common divisor). Show that a necessary and sufficient condition that a_n be an integer is that $k = rp - 1$, where r is an integer.

Solution. Sufficiency: Since $a_{n+1}a_{n-2} = k + a_n a_{n-1}$ and $a_n a_{n-3}$ equals $k + a_{n-1}a_{n-2}$, we have $(a_{n+1} + a_{n-1})/a_n = (a_{n-1} + a_{n-3})/a_{n-2}$, and this ratio is equal to $(a_3 + a_1)/a_2 = p + 1$ if n is even and to $(a_4 + a_2)/a_3 = (k+p+1)/p$ $= r + 1$ if n is odd and $k = rp - 1$. Thus $(a_{n+1} + a_{n-1})/a_n$ is an integer, and by recurrence a_{n+1} is an integer if $a_1, a_2, \ldots, a_n$ are integers, whether n is even or odd.

Necessity: We have $a_5 = k + p(k + p)$, $a_6 = [k + \{k + p\}\{k + p(k+p)\}]/p$, and since a_6 is of the form $(k + k^2 + mp)/p$, where m is an integer, we have that p divides $k(k + 1)$. But if $p = 1$, k is of the form $rp - 1$; and if $p > 1$ then, since $(p,k) = 1$, p must divide $k + 1$. That is, $k = rp - 1$. This completes the proof.

PROBLEM 80. Show that

$$\sum_{n=1}^{r} \frac{1}{n} = \sum_{n=1}^{r} (-1)^{n+1} \frac{1}{n} \binom{r}{n}.$$

Solution. Evaluate

$$\int_0^1 \frac{1 - x^r}{1 - x} dx$$

directly to obtain the left-hand side of the given expression. Then evaluate the same integral after making the substitution $x = 1 - u$ to obtain the right -hand side of the given expression.

PROBLEM 81. Prove that every positive integer has a multiple whose decimal representation involves all ten digits.

Solution. Let N be a positive integer and k a positive integer such that $10^k > N$. Some multiple of N, say HN, satisfies

$$1230456789 \cdot 10^k \leq HN < 1230456789 \cdot 10^k + 10^k.$$

Clearly every integer in this range contains all ten digits and the proof is complete.

PROBLEM 82. Show that $x^2 - y^2 = a^3$ has integral solutions for x and y whenever a is a positive integer.

Solution. Putting $x + y = a^2$ and $x - y = a$, we get $x = (a^2 + a)/2$ and $y = (a^2 - a)/2$ which are clearly integers when a is an integer.

PROBLEM 83. Let n be a fixed positive integer. Two integers a and b are said to be *congruent modulo* n, symbolized by $a \equiv b \pmod{n}$, if n divides the difference $a - b$; that is, provided that $a - b = kn$ for some integer n. Let $P(x) = \Sigma_{k=0}^{m} c_k x^k$ be a polynomial function of x with integral coefficients c_k. Show that if $a \equiv b \pmod{n}$, then $P(a) \equiv P(b) \pmod{n}$.

Solution. Since $a \equiv b \pmod{n}$, we have

$$a^k \equiv b^k \pmod{n} \qquad \text{for } k = 0,1,\ldots,m;$$

note that $a^k - b^k = (a - b)(a^{k-1} + a^{k-2}b + \cdots + b^{k-1})$. Therefore

$$c_k a^k \equiv c_k b^k \pmod{n}$$

for all such k. Adding these $m + 1$ congruences, we conclude that

$$\sum_{k=0}^{m} c_k a^k \equiv \sum_{k=0}^{m} c_k b^k \pmod{n}$$

or, in different notation, $P(a) \equiv P(b) \pmod{n}$.

PROBLEM 84. Let $N = a_m 10^m + a_{m-1} 10^{m-1} + \cdots + a_1 10 + a_0$ be the decimal expansion of the positive integer N, $0 \le a_k < 10$, and let

$$S = a_0 + a_1 + \cdots + a_m \quad \text{and} \quad T = a_0 - a_1 + a_2 - \cdots + (-1)^m a_m.$$

Show that (i) 9 divides N if and only if 9 divides S and (ii) 11 divides N if and only if 11 divides T.

Solution. Part (i): Consider $P(x) = \Sigma_{k=0}^{m} a_k x^k$, a polynomial with integral coefficients. Since $10 \equiv 1 \pmod{9}$, by Problem 83, $P(10) \equiv P(1) \pmod{9}$. But $P(10) = N$ and $P(1) = S$, so that $N \equiv S \pmod{9}$, which is what we wanted to prove.

Part (ii): Put $P(x) = \Sigma_{k=0}^{m} a_k x^k$. Since $10 \equiv -1 \pmod{11}$, we get $P(10) \equiv P(-1) \pmod{11}$. But $P(10) = N$, whereas $P(-1) = T$, so that $N \equiv T \pmod{11}$. The implication is that both N and T are divisible by 11 or neither is.

PROBLEM 85. Show the following result, due to Fermat: If p is prime and p does not divide a, then $a^{p-1} \equiv 1 \pmod{p}$.

Solution. We begin by considering the first $p - 1$ positive multiples of a; that is, the integers

$$a, \quad 2a, \quad 3a, \quad \ldots, \quad (p - 1)a.$$

None of these numbers is congruent modulo p to any other, nor is any congruent to zero. Indeed, if it happened that

$$ra \equiv sa \pmod{p}, \qquad 1 \leq r < s \leq p - 1,$$

then a could be cancelled to give $r \equiv s \pmod{p}$, which is impossible. Therefore, the above set of integers must be congruent modulo p to $1, 2, 3, \ldots, p - 1$, taken in some order. Multiplying all these congruences together, we find that

$$a \cdot 2a \cdot 3a \cdots (p - 1)a \equiv 1 \cdot 2 \cdot 3 \cdots (p - 1) \pmod{p},$$

whence

$$a^{p-1}(p - 1)! \equiv (p - 1)! \pmod{p}.$$

Once $(p - 1)!$ is cancelled from both sides of the preceding congruence (this is possible since p does not divide $(p - 1)!$), we obtain the desired result $a^{p-1} \equiv 1 \pmod{p}$.

PROBLEM 86. Show that if p is a prime, then $a^p \equiv a \pmod{p}$ for any integer a.

Solution. When p divides a, the statement obviously holds; for, in this case, $a^p \equiv 0 \equiv a \pmod{p}$. If p does not divide a, then in accordance with the result in Problem 85, $a^{p-1} \equiv 1 \pmod{p}$. When this congruence is multiplied by a, the conclusion $a^p \equiv a \pmod{p}$ follows.

Remarks. There is a different proof of the fact that $a^p \equiv a \pmod{p}$, involving induction on a. If $a = 1$, the assertion is that $1^p \equiv 1 \pmod{p}$, which is clearly true, as is the case $a = 0$. Assuming that the result holds for a, we must confirm its validity for $a + 1$. By the binomial theorem,

$$(a + 1)^p = a^p + \binom{p}{1}a^{p-1} + \cdots + \binom{p}{k}a^{p-k} + \cdots + \binom{p}{p-1}a + 1.$$

Our argument hinges on the observation that $\binom{p}{k} \equiv 0 \pmod{p}$ for $1 \leq k \leq p-1$, a fact already used in the Solution of Problem 74. To see this, note that

$$k!\binom{p}{k} = p(p - 1)\cdots(p - k + 1) \equiv 0 \pmod{p},$$

by virtue of which p divides k! or p divides $\binom{p}{k}$. But p divides k! implies p divides j for some j satisfying $1 \le j \le k \le p - 1$, an absurdity. Therefore, p divides $\binom{p}{k}$ or, converting to a congruence statement, $\binom{p}{k} \equiv 0$ (mod p). The point which we wish to make is that $(a + 1)^p \equiv a^p + 1 \equiv a + 1$ (mod p), where the right-most congruence uses our inductive assumption. Thus the desired conclusion holds for a + 1 and, in consequence, for all a > 0. If a is a negative integer, there is no problem: since $a \equiv r$ (mod p) for some r, where $0 \le r \le p - 1$, we get

$$a^p \equiv r^p \equiv r \equiv a \pmod{p}.$$

PROBLEM 87. Show that the following quotient is integral for all integers n > 0 and m > 1,

$$\frac{\prod_{i=0}^{n-1} (m^n - m^i)}{n!}$$

Solution. The numerator may be written as

$$m^{n(n-1)/2}(m - 1)(m^2 - 1)\cdots(m^n - 1).$$

Let p be any prime $\le n$. Then p divides n! exactly α times, where

$$\alpha = \sum_{j \ge 1} [n/p^j]$$

(see Problem 20). If p divides m, then p divides the numerator $n(n - 1)/2$ times, and

$$\alpha < \sum_{j \ge 1} \frac{n}{p^j} = \frac{n}{p - 1} \le n \le \frac{n(n - 1)}{2},$$

provided that n > 2, as we may assume, since the cases n = 1 and n = 2 are trivial. If p does not divide m, then $m^{s(p-1)} - 1$ is divisible by p for all positive integers s; to see this, we use the result in Problem 85 together with the fact that $m^{p-1} - 1$ divides $(m^{p-1})^s - 1$ for all positive integers s. The number of multiples of (p - 1) up to n is $\beta = [n/(p - 1)]$, so the numerator is divisible by p^β. Using the inequality $[x + y] \ge [x] + [y]$, we have

$$\beta = \left[\frac{n}{p - 1}\right] + \left[\sum_{j\geq 1} \frac{n}{p^j}\right] \geq \sum_{j\geq 1} \left[\frac{n}{p^j}\right] = \alpha,$$

completing the proof.

PROBLEM 88. Given a set of distinct numbers $a_0, a_1, a_2, \ldots$, such that $a_j \neq a_k$ if $j \neq k$, show that for all positive integers n,

$$\sum_{j=0}^{n} \prod_{k=0,k\neq j}^{n} \frac{(a_k - a_{n+1})}{(a_k - a_j)} = 1.$$

Solution. Put

$$L_j(x) = \prod_{k=0,k\neq j}^{n} \frac{a_k - x}{a_k - a_j}.$$

Clearly $L_j(a_j) = 1$ while $L_j(a_k) = 0$ for $0 \leq k \leq n$, $k \neq j$. Consequently

$$\sum_{j=0}^{n} L_j(x) = 1$$

for $x = a_0, a_1, a_2, \ldots, a_n$. But $\sum_{j=0}^{n} L_j(x)$ is a polynomial of degree $\leq n$ which is equal to 1 for $n + 1$ distinct values of x. Therefore

$$\sum_{j=0}^{n} L_j(x) = 1$$

is an identity.

PROBLEM 89. Show that the numerator of the sum $1 - 1/2 + 1/3 - \cdots$ to $[2p/3]$ terms is divisible by p when p is a prime greater than 3.

Solution. The required sum is equal to

$$\sum_{1\leq k<2p/3} \frac{1}{k} - \sum_{1\leq 2k<2p/3} \frac{2}{2k} = \sum_{1\leq k<2p/3} \frac{1}{k} - \sum_{1\leq k<p/3} \frac{1}{k} = \sum_{p/3<k<2p/3} \frac{1}{k}$$

$$= \sum_{p/3<k<p/2} \frac{1}{k} + \sum_{p/3<k<p/2} \frac{1}{p - k} = \sum_{p/3<k<p/2} \left(\frac{1}{k} + \frac{1}{p - k}\right)$$

$$= p \sum_{p/3<k<p/2} \frac{1}{k(p - k)}.$$

PROBLEM 90. Show that the reciprocal of every integer greater than 1 is the sum of a finite number of consecutive terms of the infinite series

$$\sum_{j=1}^{\infty} \frac{1}{j(j + 1)}.$$

Solution. Since

$$\frac{1}{j(j + 1)} = \frac{1}{j} - \frac{1}{j + 1} \tag{E.23}$$

we see that

$$\sum_{j=a}^{b-1} \frac{1}{j(j + 1)} = \frac{1}{a} - \frac{1}{b}.$$

Thus the problem is equivalent to that of finding positive integers a, b such that

$$\frac{1}{a} - \frac{1}{b} = \frac{1}{m}$$

for fixed integer $m > 1$. From (E.23) it is obvious that a solution is

$$a = m - 1, \quad b = m(m - 1).$$

Therefore, if $m > 1$,

$$\frac{1}{m} = \sum_{j=m-1}^{m(m-1)-1} \frac{1}{j(j + 1)},$$

and we have what we set out to do.

PROBLEM 91. Find all positive integers x and y with $x \neq y$ such that $x^y = y^x$.

Solution. The only solutions are $x = 2$ and $y = 4$ and $x = 4$ and $y = 2$. To prove that these are the only solutions, since $x^y = y^x$, the prime factors of x and y must be the same. Write

$$x = p_1^{a_1} p_2^{a_2} \cdots p_n^{a_n} \quad \text{and} \quad y = p_1^{b_1} p_2^{b_2} \cdots p_n^{b_n},$$

where $p_1, p_2, \ldots, p_n$ are primes and the a's and b's are positive integers. Now

$$x^y = p_1^{a_1 y} p_2^{a_2 y} \cdots p_n^{a_n y} \quad \text{and} \quad y^x = p_1^{b_1 x} p_2^{b_2 x} \cdots p_n^{b_n x}$$

so that we have the list of equations

$$a_1 y = b_1 x, \quad a_2 y = b_2 x, \quad \ldots, \quad a_n y = b_n x.$$

Suppose that $x > y$, so that

$$a_1 > b_1, \quad a_2 > b_2, \quad \ldots, \quad a_n > b_n,$$

and x is divisible by y, that is, $x = ky$ for some integer k. Thus the original equation may be written as $(ky)^y = y^{ky}$. Hence $ky = y^k$, or $k = y^{k-1}$. Now, since $x > y$ and $x = ky$, then $k > 1$ and we must have $y > 1$. If $y = k = 2$, then we get $2 = 2^{2-1}$. If $y \geq 2$ and $k > 2$, then $k < 2^{k-1} \leq y^{k-1}$, and if $y > 2$ and $k = 2$, then $k = 2 < y = y^{k-1}$. Therefore $y = 2$ and $k = 2$, and $x = ky = 4$ is the only solution in positive integers if $x > y$. Similarly, if $y > x$, the only solution is $y = 4$ and $x = 2$.

PROBLEM 92. Show that the equation $m^{n^m} = n^{m^n}$ has no solutions in positive integers with $m \neq n$.

Solution. Without loss of generality assume $n > m \geq 1$. If $m = 1$, or if $m = 2$ and $n = 3$ or 4, we readily find that

$$n^{m^n} > m^{n^m}.$$

Suppose $m = 2$ and $n > 4$. Since $f(x) = x^{1/x}$ is a decreasing function for $x > e$, $4^{1/4} > n^{1/n}$, or $4^n > n^4$, or $2^n > n^2$, or $n^{2^n} > 2^{n^2}$, and $n^{m^n} > m^{n^m}$. Finally, suppose $n > m \geq 3$. Then $m^{1/m} > n^{1/n}$, or $m^n > n^m$, and again $n^{m^n} > m^{n^m}$. It follows that the equation $m^{n^m} = n^{m^n}$ has no solutions in positive integers if $m \neq n$.

PROBLEM 93. Find the product

$$(1 + 3^{-1})(1 + 3^{-2})(1 + 3^{-4})(1 + 3^{-8})\cdots(1 + 3^{-2^n}).$$

Solution. If the given product is multiplied by $1 - \frac{1}{3}$, we have

$$(1 - 3^{-1})(1 + 3^{-1})(1 + 3^{-2})(1 + 3^{-4})\cdots(1 + 3^{-2^n})$$

$$= (1 - 3^{-2})(1 + 3^{-2})(1 + 3^{-4})\cdots(1 + 3^{-2^n})$$

$$= (1 - 3^{-4})(1 + 3^{-4})\cdots(1 + 3^{-2^n})$$

$$\cdots$$

$$= (1 - 3^{-2^n})(1 + 3^{-2^n})$$

$$= 1 - 3^{-2^{n+1}}.$$

Thus

$$(1 + 3^{-1})(1 + 3^{-2})(1 + 3^{-4})(1 + 3^{-8})\cdots(1 + 3^{-2^n})$$

$$= \frac{1 - 3^{-2^{n+1}}}{1 - 3^{-1}} = \frac{3}{2}(1 - 3^{-2^{n+1}}).$$

PROBLEM 94. Show that

$$\frac{1}{2} + \frac{1}{3} + \frac{1}{4} + \cdots + \frac{1}{n}$$

is not an integer for any n.

Solution. Assume $1/2 + 1/3 + \cdots + 1/n = t$, an integer. Let 2^m be the largest power of 2 less than or equal to n, and consider

$$\frac{n!}{2} + \frac{n!}{3} + \cdots + \frac{n!}{n} = n!t.$$

Every term in this equation is divisible by a higher power of 2 than the term $n!/2^m$. (To see this, observe that 2^m is the only positive integer less than or equal to n which is divisible by 2^m, so 2^m "knocks out" more 2s from n! than any of the other denominators.) Let $n!/2^m = 2^c q$, with q odd. Divide both sides of the equation by 2^c. Then $n!/2^m 2^c$ is the only odd term. This is a contradiction.

PROBLEM 95. Show that any $n + 1$ integers taken from $1, 2, \ldots, 2n$ contain a pair a and b such that a divides b.

Solution. For any integer m, $m = 2^k t$, with t odd. Write each of the $n + 1$ given integers in this way. If a is one of the given integers and $a = 2^k t$, then call t its "odd part". Since $a \le 2n$, $t < 2n$, so t must be one of the n odd integers between 0 and 2n. Then, since we have $n + 1$ odd parts, some two of them must be equal. Hence for the given set of $n + 1$ integers there exist a and b with $a = 2^r t$ and $b = 2^s t$. Then either $r \le s$ and a divides b or vice versa.

Remark. Note that the set of n numbers $n + 1, n + 2, \ldots, 2n$ contains no pair such that one divides the other.

PROBLEM 96. Find the sum

$$\frac{1}{2!} + \frac{2}{3!} + \frac{3}{4!} + \cdots + \frac{n}{(n+1)!}.$$

Solution. Since

$$\frac{k}{(k+1)!} = \frac{1}{k!} - \frac{1}{(k+1)!}.$$

we see that

$$\frac{1}{2!} + \frac{2}{3!} + \frac{3}{4!} + \cdots + \frac{n}{(n+1)!}$$

$$= \left(1 - \frac{1}{2!}\right) + \left(\frac{1}{2!} - \frac{1}{3!}\right) + \cdots + \left(\frac{1}{n!} - \frac{1}{(n+1)!}\right)$$

$$= 1 - \frac{1}{(n+1)!}.$$

PROBLEM 97. Find the sum $1 \cdot 1! + 2 \cdot 2! + 3 \cdot 3! + \cdots + n \cdot n!$.

Solution. Since $k \cdot k! = (k + 1)! - k!$, we see that

$$1 \cdot 1! + 2 \cdot 2! + 3 \cdot 3! + \cdots + n \cdot n!$$

$$= 2! - 1! + 3! - 2! + 4! - 3! + \cdots + (n + 1)! - n!$$

$$= (n + 1)! - 1.$$

PROBLEM 98. Evaluate the sums

$$P_n = 1 + 2x + 3x^2 + \cdots + nx^{n-1}$$

and

$$Q_n = 1^2 + 2^2x + 3^2x^2 + \cdots + n^2x^{n-1}.$$

Solution. Since

$$S_n = 1 + x + x^2 + \cdots + x^n = \frac{1 - x^{n+1}}{1 - x},$$

we see that upon differentiation

$$\frac{d}{dx} S_n = P_n \quad \text{and} \quad \frac{d}{dx} (xP_n) = Q_n.$$

Thus

$$P_n = \frac{1 - (n + 1)x^n + nx^{n+1}}{(1 - x)^2}$$

and

$$Q_n = \frac{1 + x - (n + 1)^2x^n + (2n^2 + 2n - 1)x^{n+1} - n^2x^{n+2}}{(1 - x)^3}.$$

PROBLEM 99. Evaluate the sum

$$H_n = \frac{1}{2} \tan \frac{x}{2} + \frac{1}{4} \tan \frac{x}{4} + \cdots + \frac{1}{2^n} \tan \frac{x}{2^n}.$$

Solution. Let

$$T_n = \cos \frac{x}{2} \cos \frac{x}{4} \cdots \cos \frac{x}{2^n}.$$

Since $\sin 2t = 2 \sin t \cos t$, we see that $\{2^n \sin(x/2^n)\}T_n = \sin x$, that is, $T_n = (\sin x)/\{2^n \sin(x/2^n)\}$. But $-H_n = \frac{d}{dx}(\log T_n) = \cot x - 2^{-n} \cot(x/2^n)$ and thus $H_n = 2^{-n} \cot(x/2^n) - \cot x$.

PROBLEM 100. Give an example of ten consecute integers all of which are not prime.

Solution. The set of numbers

$$11! + 2, \quad 11! + 3, \quad \ldots, \quad 11! + 11$$

has the desired property.

PROBLEM 101. Show that

$$\frac{1}{x+1} + \frac{2}{x^2+1} + \cdots + \frac{2^n}{x^{2^n}+1} = \frac{1}{x-1} - \frac{2^{n+1}}{x^{2^{n+1}}-1}.$$

Solution. Since

$$\frac{1}{z-1} - \frac{1}{z+1} = \frac{2}{z^2-1},$$

we get

$$\frac{1}{x^{2^k}-1} - \frac{1}{x^{2^k}+1} = \frac{2}{x^{2^{k+1}}-1}$$

for $k = 0, 1, 2, \ldots, n$, or

$$\frac{2^k}{x^{2^k}-1} - \frac{2^k}{x^{2^k}+1} = \frac{2^{k+1}}{x^{2^{k+1}}-1} \qquad \text{for } k = 0, 1, 2, \ldots, n.$$

Adding the obtained results, we arrive at the validity of our claim.

PROBLEM 102. Prove the identity

$$\left(1 + \frac{1}{a-1}\right)\left(1 - \frac{1}{2a-1}\right)\left(1 + \frac{1}{3a-1}\right)\left(1 - \frac{1}{4a-1}\right)\times\cdots$$

$$\times\left(1 + \frac{1}{(2n-1)a-1}\right)\left(1 - \frac{1}{2na-1}\right)$$

$$= \frac{(n+1)a}{(n+1)a-1}\,\frac{(n+2)a}{(n+2)a-1}\cdots\frac{(n+n)a}{(n+n)a-1},$$

provided that $a \neq 0, 1, 1/2, \ldots, 1/2n$.

Solution. We have

$$\left(1 + \frac{1}{a-1}\right)\left(1 - \frac{1}{2a-1}\right)\left(1 + \frac{1}{3a-1}\right)\cdots\left(1 + \frac{1}{(2n-1)a-1}\right)\left(1 - \frac{1}{2na-1}\right)$$

$$= \frac{a(2a - 2)3a(4a - 2)\cdots(2n - 1)a(2na - 2)}{(a - 1)(2a - 1)(3a - 1)(4a - 1)\cdots(2na - 1)}$$

$$= \frac{1\cdot a\cdot 3\cdot a\cdot 5\cdot a\cdots(2n-1)a\cdot(a-1)(2a-1)\cdots(na-1)\cdot 2^n}{[(n+1)a-1][(n+2)a-1]\cdots[(n+n)a-1](a-1)(2a-1)\cdots(na-1)}$$

$$= \frac{1\cdot a\cdot 3\cdot a\cdot 5\cdot a\cdots(2n-1)a\cdot 2^n}{[(n+1)a-1][(n+2)a-1]\quad[(n+n)a-1]}$$

$$= \frac{(n+1)a}{(n+1)a - 1}\,\frac{(n+2)a}{(n+2)a - 1}\cdots\frac{(n+n)a}{(n+n)a - 1}$$

because

$$1\cdot 3\cdot 5\cdots(2n - 1)\cdot 2^n = \frac{1\cdot 2\cdot 3\cdot 4\cdot 5\cdots 2n}{2\cdot 4\cdot 6\cdots 2n}\cdot 2^n$$

$$= \frac{1\cdot 2\cdot 3\cdot 4\cdot 5\cdots 2n}{1\cdot 2\cdot 3\cdots n} = (n + 1)(n + 2)\cdots 2n.$$

PROBLEM 103. Find the sum of n numbers of the form

1, 11, 111, , 1111, ...

Solution. The sum in question is

$$\frac{10 - 1}{9} + \frac{10^2 - 1}{9} + \frac{10^3 - 1}{9} + \cdots + \frac{10^n - 1}{9}$$

$$= \frac{1}{9}\left\{10\,\frac{10^n - 1}{9} - n\right\}.$$

PROBLEM 104. Verify the identity

$$\left(x^{n-1} + \frac{1}{x^{n-1}}\right) + 2\left(x^{n-2} + \frac{1}{x^{n-2}}\right) + \cdots + (n - 1)\left(x + \frac{1}{x}\right) + n$$

$$= \frac{1}{x^{n-1}}\left(\frac{x^n - 1}{x - 1}\right)^2.$$

Solution. The sum considered may be written as

$$\left(\frac{1}{x^{n-1}} + \frac{2}{x^{n-2}} + \cdots + \frac{n - 1}{x}\right) + [x^{n-1} + 2x^{n-2} + \cdots + (n-1)x] + n.$$

The first bracketed expression equals

$$\frac{1}{x^n}[x + 2x^2 + \cdots + (n-1)x^{n-1}] = \frac{x[(n-1)x^n - nx^{n-1} + 1]}{x^n(x - 1)^2}$$

(see Problem 98). The second bracketed expression is obtained from the first one by replacing x by 1/x. Hence we get the desired result.

PROBLEM 105. Verify the identity

$$\frac{n}{2n + 1} + \frac{1}{2^3 - 2} + \frac{1}{4^3 - 4} + \cdots + \frac{1}{(2n)^3 - 2n}$$

$$= \frac{1}{n + 1} + \frac{1}{n + 2} + \cdots + \frac{1}{2n}.$$

Solution. We have

$$\frac{1}{(2k)^3 - 2k} = \frac{1}{2k}\cdot\frac{1}{(2k)^2 - 1} = \frac{1}{4k}\left(\frac{1}{2k - 1} - \frac{1}{2k + 1}\right)$$

$$= \frac{1}{2}\left(\frac{2k - (2k - 1)}{2k(2k - 1)} - \frac{(2k + 1) - 2k}{2k(2k + 1)}\right)$$

$$= \frac{1}{2}\left(\frac{1}{2k - 1} - \frac{1}{2k} - \frac{1}{2k} + \frac{1}{2k + 1}\right).$$

Therefore

$$\sum_{k=1}^{n} \frac{1}{(2k)^3 - 2k}$$

$$= \frac{1}{2}\left\{\left(1 + \frac{1}{3} + \cdots + \frac{1}{2n - 1}\right) + \left(\frac{1}{3} + \frac{1}{5} + \cdots + \frac{1}{2n - 1}\right) + \frac{1}{2n + 1}\right.$$

$$\left. - 2\left(\frac{1}{2} + \frac{1}{4} + \cdots + \frac{1}{2n}\right)\right\}$$

$$= \frac{1}{2}\left\{2\left(1 + \frac{1}{3} + \cdots + \frac{1}{2n - 1}\right) - 1 + \frac{1}{2n + 1} - 2\left(\frac{1}{2} + \frac{1}{4} + \cdots + \frac{1}{2n}\right)\right\}$$

$$= \left(1 + \frac{1}{3} + \frac{1}{5} + \cdots + \frac{1}{2n - 1}\right) - \left(\frac{1}{2} + \frac{1}{4} + \cdots + \frac{1}{2n}\right) - \frac{n}{2n + 1}.$$

Hence

$$\sum_{k=1}^{n} \frac{1}{(2k)^3 - 2k} + \frac{n}{2n + 1} = 1 - \frac{1}{2} + \frac{1}{3} - \frac{1}{4} + \cdots + \frac{1}{2n - 1} - \frac{1}{2n}$$

$$= \frac{1}{n + 1} + \frac{1}{n + 2} + \cdots + \frac{1}{2n} \quad \text{(see Problem 42).}$$

PROBLEM 106. Prove the identity

$$\frac{a}{a+1} + \frac{b}{(a+1)(b+1)} + \frac{c}{(a+1)(b+1)(c+1)}$$

$$\cdots + \frac{k}{(a+1)(b+1)\cdots(k+1)}$$

$$= 1 - \frac{1}{(a+1)(b+1)\cdots(k+1)}.$$

Solution. Since

$$\frac{a}{a+1} = 1 - \frac{1}{a+1},$$

$$\frac{b}{(a+1)(b+1)} = \frac{1}{a+1} - \frac{1}{(a+1)(b+1)},$$

$$\frac{c}{(a+1)(b+1)(c+1)} = \frac{1}{(a+1)(b+1)} - \frac{1}{(a+1)(b+1)(c+1)},$$

and, in general,

$$\frac{k}{(a+1)(b+1)\cdots(k+1)}$$

$$= \frac{1}{(a+1)(b+1)\cdots(j+1)} - \frac{1}{(a+1)(b+1)\cdots(j+1)(k+1)},$$

the desired result follows immediately.

PROBLEM 107. Let C be a continuous closed curve in a plane which does not cross itself and let Q be a point inside C. Show that there are points P_1 and P_2 on C such that Q is the midpoint of the line segment P_1P_2.

Solution. We introduce a rectangular coordinate system in the plane containing C with the point Q being at the origin, that is, having coordinates (0,0). We now map a point P on C with coordinates (x,y) into a point P' with coordinates (-x,-y); this mapping takes C into a curve C' with C' being congruent to C. The point Q is both in the interior of C and C' and so the interior of C and C' have nonempty intersection. On the other hand, the region enclosed by the curve C' cannot be properly contained in the region enclosed by the curve C because the areas of these two regions are the same. Therefore the curves C and C' must intersect. But if P_2 is a point common to C and C' then, being a point on C', it must be the image of a point P_1 on C under the mapping $(x,y) \to (-x,-y)$ and $Q = (0,0)$ is the midpoint of the segment P_1P_2.

PROBLEM 108. Show that every positive integral power of $\sqrt{2} - 1$ is of the form $\sqrt{m} - \sqrt{m-1}$, where m is a positive integer.

Solution. Let a and b, $a \geq b$, be two non-negative real numbers, and n a positive integer. Set

$$p = \frac{(a+b)^n + (a-b)^n}{2}.$$

Then

$$p^2 - (a^2 - b^2)^n = \frac{[(a+b)^n - (a-b)^n]^2}{4}$$

and it follows that

$$(a-b)^n = \sqrt{p^2} - \sqrt{p^2 - (a^2 - b^2)^n}.$$

Take $a = \sqrt{s}$, $b = \sqrt{s-t}$, s and t positive integers, $s \geq t$. Then

$$(\sqrt{s} - \sqrt{s-t})^n = \sqrt{p^2} - \sqrt{p^2 - t^n}.$$

Moreover, p^2 is a positive integer. Indeed,

$$p = \frac{(a+b)^n + (a-b)^n}{2}$$

$$= a^n + \binom{n}{2}a^{n-2}b^2 + \binom{n}{4}a^{n-4}b^4 + \cdots + [1 + (-1)^n]b^n.$$

Since $a = \sqrt{s}$, $b = \sqrt{s-t}$, s and t positive integers, and $s \geq t$, we see that p is the sum of positive integers if n is even; if n is odd, p is the sum of numbers of the form $k\sqrt{s}$, where k is a positive integer. Thus, p^2 is a positive integer for any positive integer n.

Finally, taking $s = 2$ and $t = 1$, we get

$$(\sqrt{2} - 1)^n = \sqrt{p^2} - \sqrt{p^2 - 1}.$$

PROBLEM 109. Show that the curve traced by a pointer which pulls taut an inextensible string passing round a given ellipse is a confocal ellipse.

Solution. The claim is a consequence of the following theorem, due to Graves: If two tangents are drawn to an ellipse from any point of a confocal ellipse, the excess of the sum of these two tangents over the intercepted arc is constant. - A proof of this theorem can be found in volume I, on page 166, of "A Course in Mathematical Analysis" by E. Goursat, Dover Publications, Inc., New York 1959.

PROBLEM 110. Let a and n denote integers greater than 1. Show that the quotient

$$\frac{n(n + 1)(n + 2)\cdots(na - 1)}{a^n}$$

is a fraction if a is prime, and is an integer if a is not prime.

Solution. To establish the claim we distinguish two cases, according as a is prime or is not.

I. The integer a is prime. Denote by x the exponent of the highest power of a that is a divisor of the numerator of the quotient. This number x, as we know from Problem 20, is given by the formula

$$x = \left[\frac{na - 1}{a}\right] + \left[\frac{na - 1}{a^2}\right] + \left[\frac{na - 1}{a^3}\right] + \cdots$$
$$- \left[\frac{n - 1}{a}\right] - \left[\frac{n - 1}{a^2}\right] - \left[\frac{n - 1}{a^3}\right] - \cdots,$$

where the square bracket denotes the largest integer function.

Clearly we have

$$\left[\frac{na - 1}{a^2}\right] = \left[\frac{n - 1}{a}\right],$$

$$\left[\frac{na - 1}{a^3}\right] = \left[\frac{n - 1}{a^a}\right],$$

$$\left[\frac{na - 1}{a^4}\right] = \left[\frac{n - 1}{a^3}\right],$$

and so forth.

Thus the formula given for x reduces to

$$x = \left[\frac{na - 1}{a}\right],$$

which gives $x = n - 1$.

Thus the factor a is contained $n - 1$ times in the numerator of the quotient. But the factor a is contained n times in the denominator of the quotient. Hence the quotient is a fraction, and the first part of the claim is verified.

II. The integer a is not prime. To show that, in this case, the quotient is an integer, we shall prove that all prime factors of a appear in the numerator at least as often as in the denominator.

Let b be any prime factor of a and B be its exponent. We have

$$a = b^B q,$$

and b divides Bn times the denominator.

In the numerator this factor b enters a number of times y given by the formula

$$y = \left[\frac{nb^Bq - 1}{b}\right] + \left[\frac{nb^Bq - 1}{b^2}\right] + \cdots + \left[\frac{nb^Bq - 1}{b^{B+1}}\right] + \left[\frac{nb^Bq - 1}{b^{B+2}}\right] + \cdots - \left[\frac{n - 1}{b}\right] - \left[\frac{n - 1}{b^2}\right] - \cdots.$$

But, clearly,

$$\left[\frac{nb^Bq - 1}{b^{B+1}}\right] \geq \left[\frac{n - 1}{b}\right],$$

$$\left[\frac{nb^Bq - 1}{b^{B+2}}\right] \geq \left[\frac{n - 1}{b^2}\right],$$

$$\left[\frac{nb^Bq - 1}{b^{B+3}}\right] \geq \left[\frac{n - 1}{b^3}\right],$$

and so forth.

Thus the formula giving y reduces to

$$y \geq \left[\frac{nb^Bq - 1}{b}\right] + \left[\frac{nb^Bq - 1}{b^2}\right] + \cdots + \left[\frac{nb^Bq - 1}{b^B}\right];$$

hence we can write

$$y \geq nb^{B-1}q - 1 + nb^{B-2} - 1 + \cdots + nq - 1,$$

and, consequently,

$$y \geq nq \frac{b^B - 1}{b - 1} - B.$$

It is sufficient, in order to establish the second claim, to verify the inequality

$$nq \frac{b^B - 1}{b - 1} - B \geq Bn$$

under the assumption that a and n are integers larger than 1.

To do this we distinguish two cases, according as B is equal to or larger than 1.

If B = 1, the inequality reduces to

$$nq - 1 \geq n \quad \text{or} \quad q \geq 1 + \frac{1}{n}.$$

It is sufficient that q be larger than 1. But this is so for if q is equal to 1 and at the same time B is 1, then a would be prime, contrary to the hypothesis.

If B > 1, the inequality can be written as

$$nq(b^B - 1) \geq (n + 1)(b - 1)B$$

or, putting b = 1 + c (with c being an integer greater than zero),

$$nq\{(1 + c)^B - 1\} \geq (n + 1)Bc.$$

We expand $(1 + c)^B$ and we divide the two numbers by c; we get

$$nq\{B + \frac{B(B - 1)}{1\cdot 2} c + \cdots\} \geq nB + B$$

or

$$nqB + nq \frac{B(B - 1)}{1\cdot 2} c + \cdots \geq nB + B.$$

But clearly we have

$$nqB \geq nB.$$

Moreover, Since B and n are both larger than 1, we also have

$$nq \frac{B(B - 1)}{1\cdot 2} c \geq B.$$

Thus the inequality is satisfied and the second claim is established as well.

Remarks. In the foregoing we restricted ourselves to the case $n > 1$. If $B > 1$, the inequality

$$nq \frac{B(B-1)}{1\cdot 2} c \geq B$$

becomes

$$nq(B-1)c \geq 2.$$

If we suppose $n = 1$, the last inequality takes on the form

$$q(B-1)c \geq 2$$

which is always true in the unique case

$$B = 2 \quad \text{and} \quad n = q = c = 1.$$

This unique case corresponds to the quotient

$$\frac{1\cdot 2\cdot 3}{4}.$$

Thus the quotient

$$\frac{1\cdot 2\cdot 3\cdots(a-1)}{a}$$

(which corresponds to $n = 1$) is a fraction when a is prime, and is an integer when a is not prime except in the unique case where $a = 4$. In this latter case we have the fraction

$$\frac{1\cdot 2\cdot 3}{4}.$$

PROBLEM 112. Let m and n be positive integers. Show that

$$\frac{\{\binom{2m}{m}\binom{2n}{n}\}^2}{\binom{m+n}{n}}$$

is an integer.

Solution. Since

$$\frac{\{\binom{2m}{m}\binom{2n}{n}\}^2}{\binom{m+n}{n}} = \frac{(2m)!\ (2m)!\ (2n)!\ (2n)!\ m!\ n!}{m!\ m!\ m!\ m!\ n!\ n!\ n!\ n!\ (m+n)!}$$

$$= \frac{(2m)!\ (2m)!\ (2n)!\ (2n)!}{m!\ m!\ m!\ n!\ n!\ n!\ (m+n)!},$$

it will be sufficient to show that

$$2[2m'] + 2[2n'] \geq 3[m'] + 3[n'] + [m' + n'] \qquad (E.24)$$

for all values of m', n' between 0 and 1 (inclusively).

(a) For m' = 1, n' = 1, we have in fact

$$4 + 4 = 3 + 3 + 2;$$

(b) For m' = 1, n' < 1, (E.24) takes on the form

$$4 + 2[2n'] \geq 3 + 1 = 4,$$

which is clear, since [2n'] is either 0 or 1;

(c) For m' < 1, n' = 1, (E.24) takes on the form

$$2[2m'] + 4 \geq 3 + 1 = 4,$$

similar to case (b);

(d) For m' < 1, n' < 1, we get

$$2[2m'] + 2[2n'] \geq [m' + n'].$$

Indeed, the larger of the two quantities 2m', 2n' is, by itself, larger than or equal to the arithmetic mean m' + n'.

Remark. We could also have used the result in Problem 25 to solve Problem 112.

PROBLEM 113. Solve in positive integers the equation $x^y = y^x + 1$.

Solution. Evidently this equation is satisfied for y = 0, whatever x is.

To obtain the solutions in finite numbers, we note at once that x and y have to differ by little from each other and that their difference has to be an odd number.

Suppose first that x > y and let x = y + n. Then

$$(y + n)^y - y^{y+n} = 1,$$

or, dividing by y^y,

$$\left(1 + \frac{n}{y}\right)^y - y^n = \frac{1}{y^y}.$$

But $\left(1 + \frac{n}{y}\right)^y$ is less than e^n which in turn is less than 3^n; hence it can not surpass 3^n. If we let y = 1, we have

$$n = 1, \quad x = 2.$$

For $y = 2$,

$$\left(1 + \frac{n}{2}\right)^2 - 2^n = \frac{1}{2^2}.$$

This equation is satisfied for $n = 1$, hence $x = 3$. For $n = 3$, the first member becomes negative, and so much more for $n > 3$. Hence we have the following two solutions:

$$y = 1, \quad x = 2$$

$$y = 2, \quad x = 3,$$

and there are no others with $x > y$.

Let $y > x$; putting $y = x + n$,

$$x^{x+n} - (x + n)^x = 1,$$

and, on dividing by x^x,

$$x^n - \left(1 + \frac{n}{x}\right)^x = \frac{1}{x^x}.$$

But x, after the preceding discussion, can not be smaller than 3; however, for $x = 3$, the first member becomes

$$3^n - \left(1 + \frac{n}{3}\right)^3,$$

a value which, for $n = 1$, exceeds already the second member, and so much more so for $n > 1$, because the positive term increases with n more rapidly than the negative term. Moreover, for $x > 3$,

$$x^n - \left(1 + \frac{n}{x}\right)^x > 4^n - e^n > 4^n - 3^n > 1,$$

while

$$\frac{1}{x^x} < 1.$$

Hence there is no solution for $y > n$, and the only solutions in positive integers are

$$y = 0, \quad x \text{ arbitrary},$$

$$y = 1, \quad x = 2,$$

$$y = 2, \quad x = 3.$$

PROBLEM 114. Solve the equation $3^x = 54x - 135$.

Solution. We note that $54 = 2 \cdot 3^3$ and $135 = 5 \cdot 3^3$; thus

$$3^x = 2 \cdot 3^3 x - 5 \cdot 3^3.$$

This shows that $x = 3$ and $x = 4$ are roots of the given equation. But these two roots are the only roots of the given equation since the curve $y = 3^x$ is strictly above the straight line $y = 54x - 135$ when $x < 3$ and when $x > 4$ and $y = 3^x$ is strictly below the straight line $y = 54x - 135$ for all x satisfying the inequality $3 < x < 4$.

PROBLEM 115. Show that

$$\frac{1}{5}x^5 + \frac{1}{3}x^3 + \frac{7}{15}x$$

is an integer for every integral value of x.

Solution. Since

$$3x^5 + 5x^3 + 7x = 3(x - 2)(x - 1)x(x + 1)(x + 2)$$
$$+ 4 \cdot 5(x - 1)x(x + 1) + 15x,$$

we need only to use the fact that the product of n consecutive integers is divisible by n, to see that the number

$$(3x^5 + 5x^3 + 7x)/15 = \frac{1}{5}x^5 + \frac{1}{3}x^3 + \frac{7}{15}x$$

is an integer.

PROBLEM 116. Denote $((n!)!)!$ by $n(!)^3$, etc., $n(!)^0 = n$. Show that for $k \geq 2$,

$$\frac{n(!)^k}{(n!)^{[n-1]![n!-1]![n(!)^2-1]!\cdots[n(!)^{k-2}-1]!}}$$

is an integer.

Solution. From the Remark to Problem 22 we know that the product of any n consecutive integers is divisible by n!. Now $n(!)^k$ is a product of

$n(!)^{k-1}$ consecutive integers. We divide these numbers into groups of n consecutive integers. Then we have

$$[n - 1]![n! - 1]![n(!)^2 - 1]!\cdots[n(!)^{k-2} - 1]!, \quad k \geq 2,$$

groups, since

$$\begin{aligned} n(!)^{k-1} &= n(!)^{k-2}[n(!)^{k-2} - 1]! \\ &= n(!)^{k-2}[n(!)^{k-3} - 1]![n(!)^{k-2} - 1]! \\ &= \cdots \\ &= n(n - 1)![n! - 1]!\cdots[n(!)^{k-2} - 1]!. \end{aligned}$$

Thus, $n(!)^k$ is divisible by

$$(n!)^{[n-1]![n!-1]!\cdots[n(!)^{k-2}-1]!}.$$

PROBLEM 117. Sum the series

$$\sum_{r=1}^{2n-1} \frac{(-1)^{r-1} r}{\binom{2n}{r}}.$$

Solution. Let S be the desired sum. Multiplying the identity

$$\frac{2n + 2}{2n + 1} \frac{1}{\binom{2n}{r}} = \frac{1}{\binom{2n + 1}{r}} + \frac{1}{\binom{2n + 1}{r + 1}}$$

by $(-1)^{r-1}r$ and sum from r = 1 to r = 2n. On the left-hand side we obtain $(S - 2n)(2n + 2)/(2n + 1)$, and on the right-hand side

$$\frac{1}{\binom{2n + 1}{1}} - \frac{1}{\binom{2n + 1}{2}} + \cdots + \frac{1}{\binom{2n + 1}{2n - 1}} - \frac{1}{\binom{2n + 1}{2n}} - 2n = - 2n,$$

the binomial coefficients cancelling in pairs. Solving for S we find that $S = n/(n + 1)$.

CHAPTER 2

INEQUALITIES

PROBLEM 1. Suppose that a real-valued function g, defined on a nonempty set T of real numbers, satisfies for arbitrary elements t_1, t_2, $t_1 \neq t_2$ of T the inequality

$$g\left(\frac{t_1 + t_2}{2}\right) < \frac{g(t_1) + g(t_2)}{2}.$$

Then the more general inequality

$$g\left(\frac{t_1 + t_2 + \cdots + t_n}{n}\right) < \frac{g(t_1) + g(t_2) + \cdots + g(t_n)}{n}$$

holds, where the t_i's are arbitrary elements of T but $t_i \neq t_j$ for at least one pair i, j.

Solution. The proof is carried out in two steps.

Step 1. Assume the validity of the claim for n = m and prove its validity for n = 2m. We have

$$g\left(\frac{t_1 + t_2 + \cdots + t_{2m}}{2m}\right) = g\left(\frac{\frac{t_1 + t_2}{2} + \cdots + \frac{t_{2m-1} + t_{2m}}{2}}{m}\right)$$

$$< \frac{g\left(\frac{t_1 + t_2}{2}\right) + \cdots + g\left(\frac{t_{2m-1} + t_{2m}}{2}\right)}{m}$$

$$< \frac{\frac{g(t_1) + g(t_2)}{2} + \cdots + \frac{g(t_{2m-1}) + g(t_{2m})}{2}}{m}$$

$$= \frac{g(t_1) + g(t_2) + \cdots + g(t_{2m-1}) + g(t_{2m})}{2m}$$

(since, by assumption, not all of the elements $t_1, t_2, \ldots, t_{2m}$ are equal to one another, they can be grouped so that, for example, $t_1 \neq t_2$). Thus the claim is valid when n is a power of 2.

Step 2. Let $n > 2$ and n not be a power of 2, that is, let $2^{m-1} < n < 2^m$. Then

$$g\left(\frac{t_1 + t_2 + \cdots + t_n + s_1 + s_2 + \cdots + s_p}{n + p}\right)$$

$$< \frac{g(t_1) + \cdots + g(t_n) + g(s_1) + \cdots + g(s_p)}{n + p}$$

(here $t_1, t_2, \ldots, t_n$ are not all equal to one another), by what we have already established in Step 1. Put

$$s_1 = s_2 = \cdots = s_p = \frac{t_1 + t_2 + \cdots + t_n}{n}.$$

Then

$$s_1 + s_2 + \cdots + s_p = \frac{t_1 + t_2 + \cdots + t_n}{n} p$$

and

$$g\left(\frac{t_1 + \cdots + t_n + s_1 + \cdots + s_p}{n + p}\right) = g\left(\frac{t_1 + \cdots + t_n}{n}\right).$$

On the other hand

$$\frac{g(t_1) + \cdots + g(t_n) + g(s_1) + \cdots + g(s_p)}{n + p}$$

$$= \frac{g(t_1) + \cdots + g(t_n) + p \cdot g\left(\frac{t_1 + \cdots + t_n}{n}\right)}{n + p}.$$

From the last inequality we obtain

$$g\left(\frac{t_1 + \cdots + t_n}{n}\right) < \frac{g(t_1) + \cdots + g(t_n)}{n}.$$

PROBLEM 2. Show that, for $x_i > 0$ with $i = 1,2,\ldots,n$,

$$\frac{n}{\frac{1}{x_1} + \frac{1}{x_2} + \cdots + \frac{1}{x_n}} \le \sqrt[n]{x_1x_2\cdots x_n} \le \frac{x_1 + x_2 + \cdots + x_n}{n}$$

holds with equality being obtained only in the case $x_1 = x_2 = \cdots = x_n$. In other words, the harmonic mean of n positive numbers is less than or equal to their geometric mean which in turn is less than or equal to their arithmetic mean.

Solution. Let $g(t) = -\log(1 + t)$. From the graph of the function g it is clear that

$$g\left(\frac{t_1 + t_2}{2}\right) < \frac{g(t_1) + g(t_2)}{2}$$

since $(t_1 + t_2)/2$ is the midpoint of the segment $[t_1,t_2]$. Hence, by Problem 1,

$$-\log\left(1 + \frac{t_1 + \cdots + t_n}{n}\right) < -\frac{\log(1 + t_1) + \cdots + \log(1 + t_n)}{n}$$

or

$$\log \sqrt[n]{(1 + t_1)\cdots(1 + t_n)} < \log\left(1 + \frac{t_1 + \cdots + t_n}{n}\right)$$

or

$$\sqrt[n]{(1 + t_1)\cdots(1 + t_n)} < 1 + \frac{t_1 + \cdots + t_n}{n}$$

$$= \frac{(1 + t_1) + \cdots + (1 + t_n)}{n}.$$

Putting $1 + t_i = x_i$ we get

$$\sqrt[n]{x_1\cdots x_n} < \frac{x_1 + \cdots + x_n}{n}.$$

Obviously, if we assume the possibility $x_1 = x_2 = \cdots = x_n$, then we will get

$$\sqrt[n]{x_1\cdots x_n} \le \frac{x_1 + \cdots + x_n}{n}.$$

Finally, replacing x_i by $1/x_i$ in the foregoing inequality we obtain that the harmonic mean is less than or equal to the geometric mean for the same set of positive numbers $x_1, \ldots, x_n$.

PROBLEM 3. Verify that, for $x_i > 0$ with $i = 1,2,\ldots,n$,

$$\left(\frac{x_1 + x_2 + \cdots + x_n}{n}\right)^k \le \frac{x_1^k + x_2^k + \cdots + x_n^k}{n}$$

when k is a positive integer.

Solution. Let $g(t) = t^k$. Then

$$g\left(\frac{t_1 + t_2}{2}\right) < \frac{g(t_1) + g(t_2)}{2} \qquad \text{for } t_1 \ne t_2.$$

The desired result now follows from Problem 1.

PROBLEM 4. Let $a_i > 0$, $b_i > 0$ for $i = 1,2,\ldots,n$. Prove that

$$\sqrt[n]{(a_1 + b_1)(a_2 + b_2)\cdots(a_n + b_n)} \le \sqrt[n]{a_1 a_2 \cdots a_n} + \sqrt[n]{b_1 b_2 \cdots b_n}.$$

Solution. Let $g(t) = \log\,(1 + e^t)$. Then

$$g\left(\frac{t_1 + t_2}{2}\right) < \frac{g(t_1) + g(t_2)}{2} \qquad \text{for } t_1 \ne t_2.$$

By Problem 1 we therefore get (changing the notation from e^t to $\exp(t)$)

$$\log\left(1 + \exp\frac{t_1 + t_2 + \cdots + t_n}{n}\right)$$

$$< \frac{\log\,(1 + \exp(t_1)) + \cdots + \log\,(1 + \exp(t_n))}{n}$$

or

$$1 + \exp\frac{1_1 + t_2 + \cdots + t_n}{n} < \sqrt[n]{\{1 + \exp(t_1)\}\cdots\{1 + \exp(t_n)\}}.$$

Putting $\exp(t) = s$, that is, $t = \log s$, we obtain

$$\sqrt[n]{\{1 + \exp(t_1)\}\cdots\{1 + \exp(t_n)\}}$$

$$= \sqrt[n]{(1 + s_1)\cdots(1 + s_n)}$$

$$> 1 + \exp\frac{\log s_1 + \cdots + \log s_n}{n}$$

or

$$\sqrt[n]{(1 + s_1)(1 + s_2)\cdots(1 + s_n)} > 1 + \sqrt[n]{s_1 s_2 \cdots s_n}.$$

For

$$\frac{b_i}{a_i} = s_i \quad (i = 1,2,\ldots,n)$$

this is the desired inequality.

PROBLEM 5. Let $a_1, a_2, \ldots, a_n$ form an arithmetic progression ($a_i > 0$ for $i = 1,2,\ldots,n$). Show that

$$\sqrt{a_1 a_n} \le \sqrt[n]{a_1 a_2 \cdots a_n} \le \frac{a_1 + a_n}{2}.$$

In particular

$$\sqrt{n} \le \sqrt[n]{n!} \le \frac{n + 1}{2}.$$

Solution. We have by Problem 2

$$\sqrt[n]{a_1 a_2 \cdots a_n} \le \frac{a_1 + a_2 + \cdots + a_n}{n}.$$

Since $a_1, a_2, \ldots, a_n$ form an arithmetic progression, the term on the right-hand side of the last inequality equals $(a_1 + a_n)/2$.

To prove the rest of the claim, consider

$$(a_1 a_2 \cdots a_n)^2 = (a_1 a_n)(a_2 a_{n-1})\cdots(a_n a_1).$$

But

$$a_k a_{n-k+1} \ge a_1 a_n$$

(because

$$a_k = a_1 + (k-1)d \quad \text{and} \quad a_{n-k+1} = a_n - (k-1)d$$

and hence

$$a_k a_{n-k+1} = a_1 a_n + d^2\{(k-1)(n-1) - (k-1)^2\};$$

indeed, in any arithmetic progression, whose common difference is not zero, the product of two terms equidistant from the extreme terms is the greater the closer these terms are to the middle term). Thus

$$(a_1a_2\cdots a_n)^2 \geq (a_1a_n)^n.$$

PROBLEM 6. Let $a > 1$ and n be a positive integer. Verify that

$$a^n - 1 \geq n\left(a^{\frac{n+1}{2}} - a^{\frac{n-1}{2}}\right).$$

Solution. Let $a = s^2$. It is required to prove that

$$s^{2n} - 1 \geq n(s^{n+1} - s^{n-1})$$

or, which is the same,

$$\frac{s^{2n} - 1}{s^2 - 1} \geq ns^{n-1}.$$

But

$$\frac{s^{2n} - 1}{s^2 - 1} = s^{2(n-1)} + s^{2(n-2)} + \cdots + s^2 + 1$$

$$\geq n \sqrt[n]{s^2s^4\cdots s^{2n-2}} = ns^{n-1}$$

because $2 + 4 + \cdots + (2n-2) = n(n-1)$.

PROBLEM 7. Let $x_i > 0$ for $i = 1,2,\ldots,n$. Show that

$$(x_1 + x_2 + \cdots + x_n)\left\{\frac{1}{x_1} + \frac{1}{x_2} + \cdots + \frac{1}{x_n}\right\} \geq n^2.$$

Solution. By Problem 2,

$$x_1 + \cdots + x_n \geq n \sqrt[n]{x_1\cdots x_n}, \qquad \frac{1}{x_1} + \cdots + \frac{1}{x_n} \geq n \sqrt[n]{\frac{1}{x_1}\cdots\frac{1}{x_n}},$$

and the result follows.

Remarks. If we carry out the multiplication

$$(x_1 + x_2 + \cdots + x_n)\left(\frac{1}{x_1} + \frac{1}{x_2} + \cdots + \frac{1}{x_n}\right)$$

we obtain the sum of the following n^2 terms:

$$1, \quad \frac{x_2}{x_1}, \quad \ldots, \quad \frac{x_n}{x_1},$$

$$\frac{x_1}{x_2}, \quad 1, \quad \ldots, \quad \frac{x_n}{x_2},$$

$$\cdots$$

$$\frac{x_1}{x_n}, \quad \frac{x_2}{x_n}, \quad \ldots, \quad 1.$$

But $(x_i/x_k) + (x_k/x_i) \geq 2$, since $t + 1/t \geq 2$ for $t > 0$. This again shows the validity of the claim in Problem 7.

An interesting analogue from integration is the following result: Let f be a continuous strictly positive valued function on the interval [a,b]. Then $I = \int_a^b f(x)\,dx \cdot \int_a^b \frac{1}{f(x)}\,dx \geq (b - a)^2$.

Indeed,

$$I = \iint_S \frac{f(x)}{f(y)}\,dx\,dy = \iint_S \frac{f(y)}{f(x)}\,dx\,dy,$$

where S is the square $[a,b]\times[a,b]$. Therefore

$$I = \frac{1}{2}\iint_S \left[\frac{f(x)}{f(y)} + \frac{f(y)}{f(x)}\right]dx\,dy = \iint_S \frac{f^2(x) + f^2(y)}{2\,f(x)\,f(y)}\,dx\,dy \geq \iint_S 1\,dx\,dy$$

because of the trivial inequality $2AB \leq A^2 + B^2$. Hence $I \geq (b - a)^2$.

PROBLEM 8. Let f be a real-valued function defined on an interval (a,b). Then f is said to be *convex* if for each x_1, x_2 in (a,b) we have

$$f(q_1x_1 + q_2x_2) \leq q_1f(x_1) + q_2f(x_2) \tag{E.1}$$

regardless of how the positive numbers q_1 and $q_2 = 1 - q_1$ are chosen. The function f is said to be *concave* if the inequality in (E.1) is reversed. Prove the following *Inequality of Jensen*: If f is a convex function on (a,b), then

$$f(q_1x_1 + \cdots + q_nx_n) \le q_1f(x_1) + \cdots + q_nf(x_n)$$
$$(q_1, \ldots, q_n > 0;\ q_1 + \cdots + q_n = 1) \tag{E.2}$$

holds for any points $x_1, \ldots, x_n$ of the interval (a,b).

Solution. We note that in case $n = 2$, we are back to the definition of convexity. We therefore assume that the inequality in question is true for $n \ge 2$ and show that it will also be true for $n + 1$. In other words, we pick $n + 1$ points in (a,b), namely, $x_1, \ldots, x_n, x_{n+1}$, and we select $n + 1$ positive numbers $q_1, \ldots, q_n, q_{n+1}$ such that $q_1 + \cdots + q_n + q_{n+1} = 1$, and we seek to establish that

$$f(q_1x_1 + \cdots + q_nx_n + q_{n+1}x_{n+1})$$
$$\le q_1f(x_1) + \cdots + q_nf(x_n) + q_{n+1}f(x_{n+1}). \tag{E.3}$$

To this end we replace in the left-hand side of the above inequality the sum $q_nx_n + q_{n+1}x_{n+1}$ by the sum

$$(q_n + q_{n+1})\left(\frac{q_n}{q_n + q_{n+1}}\, x_n + \frac{q_{n+1}}{q_n + q_{n+1}}\, x_{n+1}\right).$$

In this way we can use inequality (E.2) and see that the expression on the left-hand side of (E.3) is smaller than or equal to

$$q_1f(x_1) + \cdots + (q_n + q_{n+1})\ f\left(\frac{q_n}{q_n + q_{n+1}}\, x_n + \frac{q_{n+1}}{q_n + q_{n+1}}\, x_{n+1}\right).$$

We now only have to apply to the values of the function in the last expression the basic inequality (E.1) in order to obtain (E.3). Hence (E.2) is proved completely.

PROBLEM 9. If $x_i > 0$, $q_i > 0$ for $i = 1,2,\ldots,n$, and $q_1 + \cdots + q_n = 1$, show that

$$x_1^{q_1} \cdots x_n^{q_n} \le q_1x_1 + \cdots + q_nx_n.$$

Solution. Since $x_i > 0$ for all i we may set $y_i = \log x_i$. Then we will have

$$x_i^{q_i} = \exp(q_i \log x_i) = \exp(q_i y_i).$$

But $f(t) = e^t$ is convex on the entire real line and we may appeal to Problem 8 to write

$$x_1^{q_1} \cdots x_n^{q_n} = \exp(\Sigma\, q_i y_i) = f(\Sigma\, q_i y_i)$$

$$\le \Sigma\, q_i f(y_i) = \Sigma\, q_i \exp(y_i) = \Sigma\, q_i x_i,$$

where the summation on i is from 1 to n.

PROBLEM 10. Let

$$a_1, \quad a_2, \quad \ldots, \quad a_n,$$

$$b_1, \quad b_2, \quad \ldots, \quad b_n,$$

$$\cdots$$

$$s_1, \quad s_2, \quad \ldots, \quad s_n,$$

$$\alpha, \quad \beta, \quad \ldots, \quad \sigma,$$

be positive and $\alpha + \beta + \cdots + \sigma = 1$. Show that

$$a_1^\alpha b_1^\beta \cdots s_1^\sigma + a_2^\alpha b_2^\beta \cdots s_2^\sigma + \cdots + a_n^\alpha b_n^\beta \cdots s_n^\sigma \le A^\alpha B^\beta \cdots S^\sigma,$$

where

$$A = \sum_{i=1}^n a_i, \quad B = \sum_{i=1}^n b_i, \quad \ldots, \quad S = \sum_{i=1}^n s_i.$$

Solution. From Problem 9 we see that

$$\left(\frac{a_i}{A}\right)^\alpha \left(\frac{b_i}{B}\right)^\beta \cdots \left(\frac{s_i}{S}\right)^\sigma \le \alpha\,\frac{a_i}{A} + \beta\,\frac{b_i}{B} + \cdots + \sigma\,\frac{s_i}{S}$$

and so

$$\sum_{i=1}^n \left(\frac{a_i}{A}\right)^\alpha \left(\frac{b_i}{B}\right)^\beta \cdots \left(\frac{s_i}{S}\right)^\sigma \le \sum_{i=1}^n \alpha\,\frac{a_i}{A} + \beta\,\frac{b_i}{B} + \cdots + \sigma\,\frac{s_i}{S}$$

$$= \alpha + \beta + \cdots + \sigma = 1$$

Thus

$$1 \geq \sum_{i=1}^{n} \left(\frac{a_i}{A}\right)^{\alpha}\left(\frac{b_i}{B}\right)^{\beta} \cdots \left(\frac{s_i}{S}\right)^{\sigma} = \frac{1}{A^{\alpha}B^{\beta} \cdots S^{\sigma}} \sum_{i=1}^{n} \left(a_i^{\alpha}b_i^{\beta} \cdots s_i^{\sigma}\right).$$

PROBLEM 11. For $i = 1,2,\ldots,n$, let

$$a_{i1}x_1 + a_{i2}x_2 + \cdots + a_{in}x_n = y_i,$$

where $a_{ij} > 0$ and $x_j > 0$ for $j = 1,2,\ldots,n$. Moreover, it is given that

$$a_{k1} + a_{k2} + \cdots + a_{kn} = 1, \quad a_{1k} + a_{2k} + \cdots + a_{nk} = 1$$

for $k = 1,2,\ldots,n$. Prove that

$$y_1y_2 \cdots y_n \geq x_1x_2 \cdots x_n.$$

Solution. Since $\log t$ is concave and $a_{i1} + a_{i2} + \cdots + a_{in} = 1$,

$$\log y_i \geq a_{i1} \log x_1 + a_{i2} \log x_2 + \cdots + a_{in} \log x_n$$

for $i = 1,2,\ldots,n$. Hence

$$\sum_{i=1}^{n} \log y_i \geq (\log x_1) \sum_{i=1}^{n} a_{i1} + \cdots + (\log x_n) \sum_{i=1}^{n} a_{in} = \sum_{i=1}^{n} \log x_i.$$

PROBLEM 12. Let $x > 0$, $y > 0$, $k > 1$, $k' > 1$, and $1/k + 1/k' = 1$. Show that

$$xy \leq \frac{1}{k} x^k + \frac{1}{k'} y^{k'}.$$

Solution. In Problem 9 consider the special case $n = 2$. Then put $q_1 = 1/k$, $q_2 = 1/k'$, $x_1 = x^k$, and $x_2 = y^{k'}$.

PROBLEM 13. Let $x_i > 0$ and $y_i > 0$ for $i = 1,2,\ldots,n$. Moreover, put

$$p_1 + p_2 + \cdots + p_n = P.$$

Show that

$$\left(x_1^{p_1} \cdots x_n^{p_n}\right)^{1/P} \le \frac{p_1x_1 + \cdots + p_nx_n}{P}.$$

In particular

$$\sqrt[n]{x_1x_2 \cdots x_n} \le \frac{x_1 + x_2 + \cdots + x_n}{n}$$

for $p_1 = p_2 = \cdots = p_n = 1$.

Solution. In Problem 9 take $q_i = p_i/P$ and all is clear. Moreover, note that the last claim has already been established in Problem 2.

PROBLEM 14. *Hölder's Inequality:* Let $a_i > 0$ and $b_i > 0$ for $i = 1, 2, \ldots, n$ and suppose that $k > 1$, $k' > 1$, and $1/k + 1/k' = 1$. Show that

$$\sum_{i=1}^{n} a_ib_i \le \left(\sum_{i=1}^{n} a_i^k\right)^{1/k}\left(\sum_{i=1}^{n} b_i^{k'}\right)^{1/k'}. \tag{E.4}$$

For the special case when $k = k' = 2$, we get the *Cauchy-Schwarz Inequality*.

Solution. We first assume that

$$\sum_{i=1}^{n} a_i^k = \sum_{i=1}^{n} b_i^{k'} = 1 \tag{E.5}$$

and observe that the inequality to be proved will be of the form

$$\sum_{i=1}^{n} a_ib_i \quad 1.$$

In Problem 12 we put successively $x = a_i$, $y = b_i$ for $i = 1,2,\ldots,n$ and then add up all inequalities obtained in this way. By (E.5) we get what we have set out to do.

The general case can be reduced to the foregoing special case if we take in place of the numbers a_i, b_i the numbers

$$a_i' = \frac{a_i}{(\sum_{j=1}^{n} a_j^k)^{1/k}} \quad \text{and} \quad b_i' = \frac{b_i}{(\sum_{j=1}^{n} b_j^{k'})^{1/k'}}$$

for which condition (E.5) is fulfilled. By what has been shown $\sum_{i=1}^{n} a_i'b_i' \le 1$ holds and this is equivalent to (E.4).

PROBLEM 15. *Minkowski's Inequality:* Let $a_i > 0$ and $b_i > 0$ for $i = 1,2,\ldots,n$ and suppose that $k > 1$. Show that

$$\left(\sum_{i=1}^{n} (a_i + b_i)^k\right)^{1/k} \le \left(\sum_{i=1}^{n} a_i^k\right)^{1/k} + \left(\sum_{i=1}^{n} b_i^k\right)^{1/k}. \tag{E.6}$$

Solution. We observe that

$$\sum_{i=1}^{n} (a_i + b_i)^k = \sum_{i=1}^{n} a_i(a_i + b_i)^{k-1} + \sum_{i=1}^{n} b_i(a_i + b_i)^{k-1}.$$

Applying to the last two sums inequality (E.4) in Problem 14, we get (since $1/k + 1/k' = 1$)

$$\sum_{i=1}^{n} (a_i + b_i)^k$$

$$\le \left(\sum_{i=1}^{n} a_i^k\right)^{1/k}\left(\sum_{i=1}^{n} (a_i + b_i)^{(k-1)k'}\right)^{1/k'}$$

$$+ \left(\sum_{i=1}^{n} b_i^k\right)^{1/k}\left(\sum_{i=1}^{n} (a_i + b_i)^{(k-1)k'}\right)^{1/k'}$$

$$= \left[\left(\sum_{i=1}^{n} a_i^k\right)^{1/k} + \left(\sum_{i=1}^{n} b_i^k\right)^{1/k}\right]\left(\sum_{i=1}^{n} (a_i + b_i)^k\right)^{1/k'}$$

and finally arrive at inequality (E.6) by division using the last factor as divisor.

PROBLEM 16. Let

$$A_n = \frac{a_1 + a_2 + \cdots + a_n}{n}$$

with $a_j > 0$ for $j = 1,2,\ldots,n$. Show that, for $p > 1$,

$$\sum_{n=1}^{m} A_n^p \le \frac{p}{p - 1} \sum_{n=1}^{m} A_n^{p-1} a_n.$$

Solution. By Problem 12,

$$A_n^{p-1} A_{n-1} \le \frac{(p - 1)A_n^p + A_{n-1}^p}{p}.$$

But

$$A_n^p - \frac{p}{p-1} A_n^{p-1} a_n = A_n^p - \frac{p}{p-1}[nA_n - (n-1)A_{n-1}]A_n^{p-1}$$

$$= A_n^p\left(1 - \frac{np}{p-1}\right) + \frac{(n-1)p}{p-1} A_n^{p-1} A_{n-1}$$

$$\le A_n^p\left(1 - \frac{np}{p-1}\right) + \frac{n-1}{p-1}\{(p-1)A_n^p + A_{n-1}^p\} = \frac{1}{p-1}\{(n-1)A_{n-1}^p - nA_n^p\}.$$

PROBLEM 17. Verify the *Inequality of Hardy and Landau*: For $p > 1$ and $a_1, a_2, \ldots, a_n$ positive,

$$\sum_{n=1}^{m} \left(\frac{a_1 + a_2 + \cdots + a_n}{n}\right)^p < \left\{\frac{p}{p-1}\right\}^p \sum_{n=1}^{m} a_n^p.$$

Solution. Using Hölder's Inequality (see Problem 14) we get from the result in Problem 16 the following:

$$\sum_{n=1}^{m} A_n < \frac{p}{p-1}\sum_{n=1}^{m} A_n^{p-1} a_n \le \frac{p}{p-1}\left(\sum_{n=1}^{m} a_n^p\right)^{1/p}\left(\sum_{n=1}^{m} A_n^p\right)^{1/p'},$$

where $p' = p/(p-1)$. Dividing by the last factor on the right-hand side (which is certainly positive) and raising the result to the power p, we get the desired inequality.

PROBLEM 18. Prove the *Inequality of Carleman*: If $a_n > 0$ for $n = 1,2,3,\ldots$, then

$$\sum_{n=1}^{\infty} \sqrt[n]{a_1 a_2 \cdots a_n} \le e \sum_{n=1}^{\infty} a_n$$

provided that $\Sigma_{n=1}^{\infty} a_n$ converges.

Solution. From Problem 17 we see that

$$\sum_{n=1}^{\infty} \left(\frac{a_1^{1/p} + a_2^{1/p} + \cdots + a_n^{1/p}}{n}\right)^p \le \left(\frac{p}{p-1}\right)^p \sum_{n=1}^{\infty} a_n$$

provided that $\Sigma_{n=1}^{\infty} a_n$ converges; the "<" turned into "≤" because of the limit process involved (as $m \to \infty$). If we now let $p \to \infty$, and note that

$$\sqrt[n]{a_1 a_2 \cdots a_n} \le \frac{a_1 + a_2 + \cdots + a_n}{n}$$

(see Problem 2), we obtain the desired result since

$$\left(1 + \frac{1}{p-1}\right)^p \to e \quad \text{as} \quad p \to \infty.$$

PROBLEM 19. Prove *Hadamard's Inequality*: For any determinant

$$A = |a_{ik}| = \begin{vmatrix} a_{11} & a_{12} & \cdots & a_{1n} \\ a_{21} & a_{22} & \cdots & a_{2n} \\ \cdots & & & \\ a_{n1} & a_{n2} & \cdots & a_{nn} \end{vmatrix}$$

with real elements a_{ik} we have the estimate

$$A^2 \le \prod_{i=1}^{n} \sum_{k=1}^{n} a_{ik}^2.$$

Solution. We commence with some observations. In the Cauchy-Schwarz Inequality (see Problem 14) equality occurs if and only if $a_1/b_1 = a_2/b_2 = \cdots = a_n/b_n$. From the theory of determinants we recall the following fact:

$$a_{h1}A_{j1} + a_{h2}A_{j2} + \cdots + a_{hn}A_{jn} = \begin{cases} A & \text{if } h = j, \\ 0 & \text{if } h \ne j, \end{cases}$$

where A_{hk} denotes the cofactor of the element a_{hk}. Finally, a continuous real-valued function on a closed bounded set in Euclidean space assumes its largest value on this set.

We now return to our problem and let the elements a_{ik} vary, but keep the sums of the squares

$$\sum_{k=1}^{n} a_{ik}^2 = c_i^2 \qquad (i = 1, \ldots, n)$$

fixed. If A_{max}^2 is the largest value which the function A^2 of the elements a_{ik} can assume under these n conditions (such a maximum exists because we are clearly dealing with a continuous function on a bounded closed set), then the elements of A_{max} in every row must be proportional to the corresponding cofactors. Indeed, for a fixed h

$$A = a_{h1}A_{h1} + \cdots + a_{hn}A_{hn},$$

where A_{hk} is the cofactor of a_{hk}, and by the Cauchy-Schwarz Inequality

$$A^2 \leq \sum_{k=1}^{n} a_{hk}^2 \cdot \sum_{k=1}^{n} A_{hk}^2 = c_h^2 \cdot \sum_{k=1}^{n} A_{hk}^2;$$

if a_{hk} is not proportional to A_{hk}, then A^2 can certainly not assume its maximal value because then the inequality sign holds while by changing the n values a_{hk} $(k = 1, \ldots, n)$ under keeping fixed c_h^2 and A_{hk} the square of the determinant can be made equal to the right-hand side expression.

Multiplying A_{max} with itself we get

$$A_{max}^2 = \prod_{i=1}^{n} c_i^2$$

because, for $h \neq j$, $0 = a_{h1}A_{j1} + \cdots + a_{hn}A_{jn}$ and therefore, for $h \neq j$,

$$0 = a_{h1}a_{j1} + \cdots + a_{hn}a_{jn}$$

as

$$\frac{a_{j1}}{A_{j1}} = \cdots = \frac{a_{jn}}{A_{jn}}.$$

Thus for the initial determinant A we certainly have

$$A^2 \leq \prod_{i=1}^{n} c_i^2 = \prod_{i=1}^{n} \sum_{k=1}^{n} a_{ik}^2.$$

Equality holds if and only if, for $h \neq j$, $a_{h1}a_{j1} + \cdots + a_{hn}a_{jn} = 0$.

Remarks. Hadamard's Inequality has the following geometrical interpretation: The volume of a parallelepiped obtained by n vectors of given length in an n-dimensional space is the largest when the vectors are mutually perpendicular.

It is clear that if all elements a_{ij} of the determinant A are bounded in absolute value by the constant M, then

$$A \leq (n)^{n/2}M^n.$$

PROBLEM 20. A *trigonometric polynomial* is an expression of the form

$$T(x) = \frac{1}{2} a_0 + \sum_{k=1}^{n} (a_k \cos kx + b_k \sin kx).$$

If $|a_n| + |b_n| > 0$, then the number n is called the *order of* T. It can be shown that a trigonometric polynomial of order n cannot have more than 2n real roots in $[0,2\pi)$, even if each multiple root is counted the number of times it occurs (see e.g., A Zygmund, Trigonometric Series, vol. 2, p. 2, Cambridge University Press, 1959).

Prove the *Inequality of Bernstein*: If

$$T(x) = \frac{1}{2} a_0 + \sum_{k=1}^{n} (a_k \cos kx + b_k \sin kx)$$

and $|T(x)| \leq M$ for all $x \in [0,2\pi]$, then the derivative T' satisfies $|T'(x)| \leq nM$ for all $x \in [0,2\pi]$. Moreover, $T(x) = \sin nx$ shows that the result is the best possible.

Solution. Suppose, on the contrary,

$$\sup_{0 \leq x \leq 2\pi} |T'(x)| = n K,$$

where $K > M$. Since T' is continuous, it attains its bounds and so for some c, $T'(c) = \pm n K$. Let us suppose that $T'(c) = nK$. Since nK is a maximum value of T', $T''(c) = 0$. Define

$$S(x) = K \sin n(x-c) - T(x).$$

Then

$$R(x) = S'(x) = nK \cos n(x-c) - T'(x)$$

and S and R both have order n.

Consider the points

$$u_0 = c + \frac{\pi}{2n}, \quad u_k = u_0 + k\frac{\pi}{n} \quad (1 \leq k \leq 2n).$$

Then

$$S(u_0) = 1 - T(u_0) > 0,$$

$$S(u_1) = -1 - T(u_1) < 0,$$

$$\cdots$$

$$S(u_{2n}) = 1 - T(u_{2n}) > 0.$$

Each of the 2n intervals

$$(u_0,u_1),\quad (u_1,u_2),\quad \ldots,\quad (u_{2n-1},u_{2n})$$

then contains a root of S, say $S(y_i) = 0$, where

$$u_i < y_i < u_{i+1} \quad \text{with } 0 \le i \le 2n - 1.$$

Clearly $y_{2n-1} < y_0 + 2\pi$. Put $y_{2n} = y_0 + 2\pi$. Then $S(y_{2n}) = S(y_0) = 0$. By Rolle's Theorem, there is a root x_i of R inside each interval (y_i,y_{i+1}), where $0 \le i \le 2n - 1$. Evidently $x_{2n-1} < x_0 + 2\pi$.

Now $R(c) = nK - T'(c) = 0$. Since the trigonometric polynomial R of order n has at most 2n real roots, it follows that, for some k, $c \equiv x_k \pmod{2\pi}$. But $R(c) = -T''(c) = 0$. Therefore c (and so x_k) is a double root (at least) of R. Therefore the x_i with $0 \le i \le 2n - 1$ provide at least $2n + 1$ real roots of R. This is only possible if $R = 0$ and so S is a constant function. But $S(u_0) > 0$ and $S(u_1) < 0$ and we have a contradiction.

PROBLEM 21. If P is an algebraic polynomial of degree n and $|P(x)| \le M$ for all $x \in (-1,1)$, show that

$$|P'(x)| \le \frac{nM}{\sqrt{1 - x^2}}$$

for all $x \in (-1,1)$.

Solution. This is clearly the algebraic equivalent of Bernstein's Inequality (see Problem 20) and is obtained by putting $T(\theta) = P(\cos\theta)$ and noting that $T'(\theta) = -P'(\cos\theta)\sin\theta$.

Remark. The bound for $P'(x)$ given in Problem 21 fails at the endpoints -1 and 1. A better result, due to Markoff, will be taken up in Problem 24.

PROBLEM 22. Setting $\cos\theta = x$, the expression

$$T_n(x) = \cos n\theta = \cos(n \arccos x)$$

are polynomials in x of degree n, called *Chebyshew polynomials*. The leading coefficient of $T_n(x)$ is equal to 2^{n-1}. The first five polynomials of this kind are

$$T_1(x) = x,$$

$$T_2(x) = 2x^2 - 1,$$

$$T_3(x) = 4x^3 - 3x,$$

$$T_4(x) = 8x^4 - 8x^2 + 1,$$

$$T_5(x) = 16x^5 - 20x^3 + 5x.$$

The roots of T_n are all real and distinct and lie in the interior of the interval (-1,1). The roots of T_n are

$$\cos \frac{(2k-1)\pi}{2n} \quad (k = 1, 2, \ldots, n).$$

Let $x_1, x_2, \ldots, x_n$ be arbitrary distinct real or complex numbers. Set

$$f(x) = a_0(x - x_1)(x - x_2)\cdots(x - x_n) \quad \text{with } a_0 \neq 0,$$

$$f_k(x) = \frac{1}{f'(x_k)} \frac{f(x)}{x - x_k}$$

$$= \frac{(x - x_1)\cdots(x - x_{k-1})(x - x_{k+1})\cdots(x - x_n)}{(x_k - x_1)\cdots(x_k - x_{k-1})(x_k - x_{k+1})\cdots(x_k - x_n)}.$$

Every polynomial P of degree n-1 may be represented by means of its values at the points $x_1, x_2, \ldots, x_n$ as follows

$$P(x) = P(x_1)f_1(x) + P(x_2)f_2(x) + \cdots + P(x_n)f_n(x);$$

this is the *Lagrange interpolation formula*. The polynomials f_k are called the *basis polynomials of the interpolation*.

Prove the following result: Let

$$x_k = \cos \frac{(2k-1)\pi}{2n} \quad \text{for } k = 1,2,\ldots,n$$

be the roots of the Chebyshew polynomial T_n. If Q is a polynomial of degree less than or equal to n-1, then

$$Q(x) = \frac{1}{n} \sum_{k=1}^{n} (-1)^{k-1} \sqrt{1 - x_k^2}\, Q(x_k) \frac{T_n(x)}{x - x_k}.$$

Solution. Since $T_n(x) = \cos(n \text{ arc cos } x)$, we get

$$T_n'(x) = \frac{n}{\sqrt{1 - x^2}} \sin(n \text{ arc cos } x).$$

From

$$\text{arc cos } x_k = \frac{(2k - 1)\pi}{2n}$$

we obtain

$$\sin(n \text{ arc cos } x_k) = \sin \frac{(2k - 1)\pi}{2} = (-1)^{k-1}$$

and so

$$T_n'(x_k) = \frac{(-1)^{k-1} n}{\sqrt{1 - x_k^2}}.$$

To prove the equality in question, we note that on both sides we have polynomials of degree $\leq n - 1$ and so it is sufficient to show that they agree for n values x_k. As $x \to x_k$,

$$\frac{T_n(x)}{x - x_k} \to T_n'(x_k) = \frac{n(-1)^{k-1}}{\sqrt{1 - x_k^2}}.$$

Also, for $x = x_k$, every term on the right-hand side except the k-th vanishes because, for $i = 1,2,\ldots,n$ and $k \neq 1$,

$$\frac{T_n(x_i)}{x_i - x_k} = 0$$

as the x_i are roots of T_n. Thus the right-hand side expression is a Lagrange interpolation polynomial for Q.

PROBLEM 23. Let Q be a polynomial of degree $\leq n - 1$ and for $x \in [-1,1]$

$$|Q(x)| \leq \frac{1}{\sqrt{1 - x^2}}.$$

Show that in $[-1,1]$ we then have the estimate $|Q(x)| \leq n$.

Solution. With the notation of Problem 22, if $-x_1 = x_n \leq x \leq x_1$,

$$\sqrt{1 - x^2} \geq \sqrt{1 - x_1^2} = \sin \frac{\pi}{2n} \geq \frac{1}{n}.$$

Hence the assertion is true for $x_n \leq x \leq x_1$. For the remaining points of [-1,1] we apply the Lagrange interpolation formula to the polynomial Q found in Problem 22:

$$Q(x) = \frac{1}{n} \sum_{k=1}^{n} (-1)^{k-1} \sqrt{1 - x_k^2}\, Q(x_k) \frac{T_n(x)}{x - x_k}.$$

Since either $x < x_n$ or $x > x_1$, the numbers $x - x_k$ have the same sign. Hence

$$|Q(x)| \leq \frac{1}{n} \sum_{k=1}^{n} \left| \frac{T_n(x)}{x - x_k} \right| .$$

But

$$T_n(x) = 2^{n-1} \sum_{k=1}^{n} (x - x_k)$$

and so

$$\frac{T_n'(x)}{T_n(x)} = \sum_{k=1}^{n} \frac{1}{x - x_k}.$$

Therefore

$$|Q(x)| \leq \frac{1}{n} |T_n'(x)|.$$

But, since $x = \cos \theta$,

$$T_n'(x) = \frac{n \sin n\theta}{\sin \theta},$$

which gives $|T_n'(x)| \leq n^2$ because $|\sin n\theta| \leq n|\sin \theta|$ for $-\infty < \theta < \infty$ (as can easily be verified by induction).

PROBLEM 24. Prove *Markoff's Inequality*: If P is any polynomial of degree $\leq n$, then

$$|P'(x)| \leq n^2 \left\{ \sup_{-1 \leq x \leq 1} |P(x)| \right\}$$

for all $x \in [-1,1]$.

Solution. If

$$\sup_{-1\le x\le 1} |P(x)| = M,$$

take

$$Q(x) = \frac{P'(x)}{Mn}$$

in Problem 23. The stipulations in Problem 23 are satisfied in view of Problem 21.

Remark. Taking $P(x) = T_n(x)$, where T_n is as defined in Problem 22, we see that Markoff's Inequality is the best possible. Indeed,

$$T_n'(x) = \frac{n \sin(n \arccos x)}{\sqrt{1 - x^2}} = n \frac{\sin n\theta}{\sin \theta}$$

and so $T_n'(1) = n^2$.

PROBLEM 25. Let $0 < x < \pi/2$. Show that

$$x - \sin x \le \frac{1}{6}x^3.$$

Solution. We have

$$2 \sin \tfrac{1}{2}x - \sin x = 2 \sin \tfrac{1}{2}x \,(1 - \cos \tfrac{1}{2}x) = 4 \sin \tfrac{1}{2}x \sin^2 \tfrac{1}{4}x.$$

Hence

$$2 \sin \tfrac{1}{2}x - \sin x < 4 \,\frac{x}{2}\left(\frac{x}{4}\right)^2,$$

since $\sin t < t$ for $t > 0$. Thus $2 \sin \frac{1}{2}x - \sin x < \frac{1}{8}x^3$. (E.7)

Replacing x by $x/2, x/4, \ldots, x/2^{n-1}$, we find

$$2 \sin \tfrac{1}{4}x - \sin \tfrac{1}{2}x < \tfrac{1}{8}\left(\frac{x}{2}\right)^3, \tag{E.8}$$

$$2 \sin \tfrac{1}{8}x - \sin \tfrac{1}{4}x < \tfrac{1}{8}\left(\frac{x}{4}\right)^3, \tag{E.9}$$

...

$$2 \sin \frac{1}{2^n}x - \sin \frac{1}{2^{n-1}}x < \frac{1}{8}\left(\frac{x}{2^{n-1}}\right)^3 \tag{E.10}$$

Multiplying inequalities (E.7), (E.8), ..., (E.10) by 1, 2, ..., 2^{n-1}, respectively, and adding them, we get

$$2^n \sin \frac{1}{2^n}x - \sin x < \frac{1}{8}x^3\left(1 + \frac{1}{2^2} + \frac{1}{2^4} + \cdots + \frac{1}{2^{2n-2}}\right).$$

Passing to the limit as $n \to \infty$, we find

$$\lim_{n\to\infty}\left(\frac{\sin \frac{x}{2^n}}{\frac{x}{2^n}}x - \sin x\right)$$

$$\le \frac{1}{8}x^3 \lim_{n\to\infty}\left\{1 + \frac{1}{4} + \frac{1}{4^2} + \cdots + \frac{1}{4^{n-1}}\right\}.$$

But

$$\lim_{n\to\infty}\left\{1 + \frac{1}{4} + \frac{1}{4^2} + \cdots + \frac{1}{4^{n-1}}\right\} = \frac{1}{1 - \frac{1}{4}} = \frac{4}{3}$$

and

$$\lim_{n\to\infty}\frac{\sin \frac{x}{2^n}}{\frac{x}{2^n}} = 1.$$

Consequently

$$x - \sin x \le \frac{1}{6}x^3.$$

PROBLEM 26. Show that

$$1 + \frac{1}{\sqrt{2}} + \frac{1}{\sqrt{3}} + \cdots + \frac{1}{\sqrt{n}} > 2\sqrt{n + 1} - 2.$$

Solution. Since

$$\sqrt{n + 1} - \sqrt{n} = \frac{1}{\sqrt{n + 1} + \sqrt{n}} < \frac{1}{2\sqrt{n}}.$$

we have

$$\frac{1}{\sqrt{n}} > 2\sqrt{n + 1} - 2\sqrt{n}.$$

Therefore

$$1 > 2\sqrt{2} - 2,$$

$$\frac{1}{\sqrt{2}} > 2\sqrt{3} - 2\sqrt{2},$$

$$\frac{1}{\sqrt{3}} > 2\sqrt{4} - 2\sqrt{3}$$

$$\cdots$$

$$\frac{1}{\sqrt{n}} > 2\sqrt{n+1} - 2\sqrt{n}.$$

Adding these inequalities, we obtain the required result.

PROBLEM 27. Let $\sqrt{A}$ be an irrational number and a be some rational number such that $0 < \sqrt{A} - a < 1$. Show that

$$a + \frac{A - a^2}{2a + 1} < \sqrt{A} < a + \frac{A - a^2}{2a + 1} + \frac{1}{4(2a + 1)}.$$

Solution. Since $a < \sqrt{A} < a + 1$, we have

$$\sqrt{A} + a < 2a + 1, \quad \frac{\sqrt{A} + a}{2a + 1} < 1, \quad \sqrt{A} - a > 0.$$

Hence

$$\frac{(\sqrt{A} + a)(\sqrt{A} - a)}{2a + 1} < \sqrt{A} - a, \quad \frac{A - a^2}{2a + 1} < \sqrt{A} - a, \quad \sqrt{A} > a + \frac{A - a^2}{2a + 1}.$$

To prove the second inequality, we first note that $x(1 - x) = x - x^2 < 1/4$ for any real number with equality only for $x = 1/2$; indeed,

$$x - x^2 - \tfrac{1}{4} = -(x - \tfrac{1}{2})^2 \le 0.$$

Since it is possible to assume that $\sqrt{A} - a \neq 1/2$, we have

$$[1 - (\sqrt{A} - a)](\sqrt{A} - a) < 1/4,$$

$$1 - (\sqrt{A} - a) < \frac{1}{4(\sqrt{A} - a)},$$

$$(2a + 1) - (\sqrt{A} + a) < \frac{1}{4(\sqrt{A} - a)}.$$

Multiplying both members of this inequality by $\sqrt{A} - a > 0$, we get

$$(2a + 1)(\sqrt{A} - a) - (A - a^2) < \frac{1}{4}.$$

Thus, finally,

$$\sqrt{A} < a + \frac{A - a^2}{2a + 1} + \frac{1}{4(2a + 1)}.$$

PROBLEM 28. Let

$$\frac{a_1}{b_1}, \frac{a_2}{b_2}, \ldots, \frac{a_n}{b_n}$$

be n fractions with $b_i > 0$ for $i = 1,2,\ldots,n$. Show that the fraction

$$\frac{a_1 + a_2 + \cdots + a_n}{b_1 + b_2 + \cdots + b_n}$$

is contained between the largest and the smallest of these fractions.

Solution. Let m denote the smallest and M the largest of the given fractions. Then

$$m \le \frac{a_i}{b_i} \le M \quad \text{or} \quad mb_i \le a_i \le Mb_i$$

for $i = 1,2,\ldots,n$. Summing these inequalities, we find

$$m \sum_{i=1}^{n} b_i \le \sum_{i=1}^{n} a_i \le M \sum_{i=1}^{n} b_i$$

or

$$m \le \frac{\sum_{i=1}^{n} a_i}{\sum_{i=1}^{n} b_i} \le M.$$

PROBLEM 29. Let a, b, ..., d be positive real numbers and m, n, ..., p be positive integers. Show that

$$\sqrt[m+n+\cdots+p]{ab\cdots d}$$

is contained between the largest and the smallest of the numbers

$$\sqrt[m]{a}, \quad \sqrt[n]{b}, \quad \ldots, \quad \sqrt[p]{d}$$

(where, of course, we assume that the principal value of the root is taken everywhere).

Solution. Consider the fractions

$$\frac{\log a}{m}, \quad \frac{\log b}{n}, \quad \ldots, \quad \frac{\log d}{p}$$

and let s be the smallest and S be the largest of these fractions. By Problem 28,

$$s < \frac{\log a + \log b + \cdots + \log d}{m + n + \cdots + p} < S$$

or

$$s < \log \sqrt[m+n+\cdots+p]{ab\cdots d} < S;$$

the claim now follows.

PROBLEM 30. Let $0 < \alpha < \beta < \delta < \cdots < \lambda < \pi/2$. Show that

$$\tan \alpha < \frac{\sin \alpha + \sin \beta + \sin \delta + \cdots + \sin \lambda}{\cos \alpha + \cos \beta + \cos \delta + \cdots + \cos \lambda} < \tan \lambda.$$

Solution. See Problem 28.

PROBLEM 31. Show that any finite sum of fractions of the form $1/n^2$, where n is positive integer larger than 1, is less than 1.

Solution. Clearly, any such sum is less than or equal to

$$\frac{1}{2^2} + \frac{1}{3^2} + \frac{1}{4^2} + \cdots + \frac{1}{n^2}$$

which in turn is less than

$$\int_1^n \frac{1}{x^2}\, dx = 1 - \frac{1}{n}.$$

PROBLEM 32. Let n be any positive integer. Show that

$$\frac{1}{2} < 1 + \frac{1}{2} + \frac{1}{3} + \cdots + \frac{1}{n} - \log n < 1.$$

Solution. Let $[t]$ denote the largest integer less than or equal to t. But it is clear that

$$I_n = \int_{1/n}^{1} \left(\frac{1}{x} - \left[\frac{1}{x}\right]\right) dx = \log n - \frac{1}{2} - \frac{1}{3} - \cdots - \frac{1}{n}$$

and that $0 < I_n < 1/2$. Sketch a figure of the integrand!

PROBLEM 33. Let a and b denote real numbers and $|a|$ the absolute value of a, that is, $|a|$ is the larger of the two numbers a and -a if $a \neq 0$ and $|0| = 0$. Using the fact that $||a| - |b|| \leq |a + b| \leq |a| + |b|$, show that

$$|x + x_1 + x_2 + \cdots + x_n| \geq |x| - (|x_1| + |x_2| + \cdots + |x_n|).$$

Solution. We have $|x + x_1| \geq |x| - |x_1|$ and, replacing x_1 by $x_1 + x_2$, this yields $|x + x_1 + x_2| \geq |x| - |x_1 + x_2|$, that is, $|x + x_1 + x_2| + |x_1 + x_2| \geq |x|$. But

$$|x + x_1 + x_2| + |x_1 + x_2| \leq |x + x_1 + x_2| + |x_1| + |x_2|.$$

Hence

$$|x + x_1 + x_2| + |x_1| + |x_2| \geq |x|$$

or

$$|x + x_1 + x_2| \geq |x| - (|x_1| + |x_2|).$$

It is clear that the general case can be proved in an entirely similar manner.

PROBLEM 34. For n = 1,2,3,..., let

$$x_n = \frac{1000^n}{n!}.$$

Find the largest term of the sequence.

Solution. It is clear that $x_{999} = x_{1000}$. Since $x_{n+1} = \frac{1000}{n+1}x_n$, we see that x_n is increasing when n goes from 1 to 999 and that x_n is decreasing as n goes from 1000 to ∞. Thus x_n is largest for $n = 999$ and $n = 1000$.

PROBLEM 35. Show *Bernoulli's Inequality*: If $a_1, a_2, \ldots, a_n$ with $n \geq 2$ are real numbers larger than -1 and, moreover, all a_j's with $j = 1,2,\ldots,n$ have the same sign, then

$$(1 + a_i)(1 + a_2)\cdots(1 + a_n) > 1 + a_1 + a_2 + \cdots + a_n.$$

Solution. The case when $a_1, a_2, \ldots, a_n$ are positive is trivial. It only remains to show that

$$(1 - a_1)(1 - a_2)\cdots(1 - a_n) > 1 - (a_1 + a_2 + \cdots + a_n)$$

when $n \geq 2$ and $a_1, a_2, \ldots, a_n$ are positive but less than 1. Since

$$(1 - a_1)(1 - a_2) = 1 - (a_1 + a_2) + a_1a_2 > 1 - (a_1 + a_2),$$

the claim is seen to be valid for $n = 2$. We proceed by induction. Suppose that $k \geq 2$ and assume that

$$(1 - a_1)\cdots(1 - a_k) > 1 - (a_1 + \cdots + a_k).$$

Multiplying by the positive number $(1 - a_{k+1})$, we get

$$(1 - a_1)\cdots(1 - a_k)(1 - a_{k+1})$$

$$> 1 - (a_1 + \cdots + a_k) - a_{k+1} + a_{k+1}(a_1 + \cdots + a_k)$$

$$> 1 - (a_1 + \cdots + a_k + a_{k+1}).$$

PROBLEM 36. For any positive integer n let $s_1, s_2, \ldots, s_n$ be arbitrary real numbers and $t_1, t_2, \ldots, t_n$ be any real numbers such that the sum $t_1 + t_2 + \cdots + t_n = 0$. Show that $\Sigma_{k=1}^{n} \Sigma_{j=1}^{n} t_k t_j |s_k - s_j| \leq 0$.

Solution. When $n = 2$, we can use the identity

$$(t_1 + t_2)^2 - (t_1 - t_2)^2 = 4t_1t_2;$$

but $t_1 + t_2 = 0$.

When $n = 3$, we have

$$0 = (t_1 + t_2 + t_3)^2 = t_1^2 + t_2^2 + t_3^2 + 2t_1t_2 + 2t_2t_3 + 2t_3t_1.$$

Thus

$$2t_1t_2|s_1 - s_2| + 2t_2t_3|s_1 - s_2| + 2t_3t_1|s_1 - s_2|$$
$$+ 2t_1t_2|s_2 - s_3| + 2t_2t_3|s_2 - s_3| + 2t_3t_1|s_2 - s_3|$$
$$+ 2t_1t_2|s_3 - s_1| + 2t_2t_3|s_3 - s_1| + 2t_3t_1|s_3 - s_1|$$
$$= -(t_1^2 + t_2^2 + t_3^2)(|s_1 - s_2| + |s_2 - s_3| + |s_3 - s_1|)$$
$$+ 2t_3(t_2 + t_1)|s_2 - s_1| + 2t_1(t_2 + t_3)|s_2 - s_3| + 2t_2(t_1 + t_3)|s_1 - s_3|$$
$$= -(t_1^2 + t_2^2 + t_3^2)(|s_1 - s_2| + |s_2 - s_3| + |s_3 - s_1|)$$
$$+ 2t_3^2|s_2 - s_1| + 2t_1^2|s_2 - s_3| + 2t_2^2|s_3 - s_1|$$
$$= -t_1^2(|s_2 - s_1| + |s_1 - s_3|) + t_1^2|s_2 - s_3|$$
$$-t_2^2(|s_1 - s_2| + |s_2 - s_3|) + t_2^2|s_1 - s_3|$$
$$-t_3^2(|s_2 - s_3| + |s_3 - s_1|) + t_3^2|s_2 - s_1|.$$

But $|A - B| \leq |A - C| + |C - B|$ for any three real numbers A, B, and C. Hence the claim follows for $n = 3$.

The general case can be proved in an entirely similar manner.

PROBLEM 37. Show that

$$\frac{1}{2}\cdot\frac{3}{4}\cdot\frac{5}{6} \cdots \frac{2n - 1}{2n} < \frac{1}{\sqrt{2n + 1}}.$$

Solution. It is clear that

$$1\cdot 3 < 2^2, \quad 3\cdot 5 < 4^2, \quad \ldots, \quad (2n - 1)(2n + 1) < (2n)^2.$$

Thus

$$1^2\cdot 3^2\cdot 5^2 \cdots (2n - 1)^2(2n + 1) < 2^2\cdot 4^2\cdot 6^2 \cdots (2n)^2.$$

PROBLEM 38. Let f and g be positive-valued functions defined on a common interval and suppose that for any x_1 and x_2 of this interval we have

$$\left\{f\left(\frac{x_1 + x_2}{2}\right)\right\}^2 \le f(x_1)f(x_2) \quad \text{and} \quad \left\{g\left(\frac{x_1 + x_2}{2}\right)\right\}^2 \le g(x_1)g(x_2).$$

Show that

$$\left\{f\left(\frac{x_1 + x_2}{2}\right) + g\left(\frac{x_1 + x_2}{2}\right)\right\}^2 \le \left(f(x_1) + g(x_1)\right)\left(f(x_2 + g(x_2)\right).$$

Solution. What we have to show is the following: If a_1, b_1, c_1, a_2, b_2, c_2 are positive real numbers with $a_1c_1 - b_1^2 \ge 0$ and $a_2c_2 - b_2^2 \ge 0$, then $(a_1 + a_2)(c_1 + c_2) - (b_1 + b_2)^2 \ge 0$.

However

$$T = a_1a_2\{(a_1 + a_2)(c_1 + c_2) - (b_1 + b_2)^2\}$$

$$= a_2(a_1 + a_2)(a_1c_1 - b_1^2) + a_1(a_1 + a_2)(a_2c_2 - b_2^2)$$

$$+ (a_1b_2 - a_2b_1)^2.$$

If

$$a_1 > 0, \quad a_2 > 0, \quad a_1c_1 - b_1^2 \ge 0, \quad \text{and} \quad a_2c_2 - b_2^2 \ge 0,$$

then $T \ge 0$ and the claim follows.

Remark. A positive-valued function f defined on the interval I satisfying the property that for any x_1 and x_2 of I the square of the value of f at the midpoint of the interval $[x_1,x_2]$ is less than or equal to the product of the values of f at the endpoints x_1 and x_2 of the interval $[x_1,x_2]$ is called a *weakly log convex function on the interval* I. What we have shown therefore is that the sum of two weakly log convex functions on an interval is a weakly log convex function on the same interval.

PROBLEM 39. Let $a_1 \le a_2 \le \cdots \le a_n$ and $b_1 \le b_2 \le \cdots \le b_n$. Demonstrate that

$$\left(\frac{1}{n}\sum_{k=1}^{n} a_k\right)\left(\frac{1}{n}\sum_{k=1}^{n} b_k\right) \le \frac{1}{n}\sum_{k=1}^{n} a_k b_k.$$

Solution. Let

$$\sum a = \sum_{k=1}^{n} a_k, \quad \sum b = \sum_{k=1}^{n} b_k, \quad \sum ab = \sum_{k=1}^{n} a_k b_k;$$

then

$$\sum_{\lambda}\sum_{\sigma} (a_\lambda b_\lambda - a_\lambda b_\sigma) = \sum_{\lambda} (n a_\lambda b_\lambda - a_\lambda \sum b) = n \cdot \sum ab - \sum a \sum b,$$

$$\sum_{\lambda}\sum_{\sigma} (a_\sigma b_\sigma - a_\sigma b_\lambda) = \sum (n a_\sigma b_\sigma - a_\sigma \sum b) = n \cdot \sum ab - \sum a \sum b.$$

Hence

$$n \cdot \sum ab - \sum a \sum b = \frac{1}{2}\sum_{\lambda}\sum_{\sigma} (a_\lambda b_\lambda - a_\lambda b_\sigma + a_\sigma b_\sigma - a_\sigma b_\lambda)$$

$$= \frac{1}{2}\sum_{\lambda}\sum_{\sigma} (a_\lambda - a_\sigma)(b_\lambda - b_\sigma).$$

But $(a_\lambda - a_\sigma)(b_\lambda - b_\sigma) \ge 0$ for $\lambda = 1,2,\ldots,n$ and $\sigma = 1,2,\ldots,n$.

Remarks. The inequality in Problem 39 is due to Chebyshew. This inequality can be generalized as follows:

If $0 \le a_1 \le a_2 \le \cdots \le a_n$, $0 \le b_1 \le b_2 \le \cdots \le b_n$, ..., $0 \le c_1 \le c_2 \le \cdots \le c_n$, then

$$\frac{\sum a}{n}\,\frac{\sum b}{n} \cdots \frac{\sum c}{n} \le \frac{\sum ab\cdots c}{n}.$$

The conditions $0 \le a_1$, $0 \le b_1$, ..., $0 \le c_1$ are essential; for example, if $a_1 = 1$, $a_2 = 3$, $b_1 = 1$, $b_2 = 3$, $c_1 = -4$, $c_2 = -3$, we have

$$\frac{a_1 + a_2}{2}\,\frac{b_1 + b_2}{2}\,\frac{c_1 + c_2}{2} = -14 > -\frac{31}{2} = \frac{a_1 b_1 c_1 + a_2 b_2 c_2}{2}.$$

As an example of the generalized inequality we mention: If a, b, and c are positive numbers and n is a positive integer, then

$$(a + b + c)^n \le 3^{n-1}(a^n + b^n + c^n).$$

PROBLEM 40. Let n be a positive integer larger than 1 and $a > 0$. Show that

$$\frac{1 + a + a^2 + \cdots + a^n}{a + a^2 + a^3 + \cdots + a^{n-1}} \geq \frac{n + 1}{n - 1}. \tag{E.11}$$

Solution. It is clear that (E.11) holds for $n = 2$. Suppose that (E.11) is valid for $n = k$, that is,

$$\frac{1 + a + a^2 + \cdots + a^k}{a + a^2 + a^3 + \cdots + a^{k-1}} \geq \frac{k + 1}{k - 1}. \tag{E.12}$$

Since $a > 0$, (E.12) can be written as

$$1 + a + \cdots + a^k \geq \frac{k + 1}{k - 1}(a + a^2 + \cdots + a^{k-1}) \tag{E.13}$$

or

$$1 + a + \cdots + a^k + a^{k+1} \geq \frac{k + 1}{k - 1}(a + a^2 + \cdots + a^{k-1}) + a^{k+1}.$$

We shall show that

$$\frac{k + 1}{k - 1}(a + a^2 + \cdots + a^{k-1}) + a^{k+1} \geq \frac{k + 2}{k}(a + a^2 + \cdots + a^{k-1} + a^k) \tag{E.14}$$

proving that (E.11) is also valid for $n = k + 1$.

Let us assume that (E.14) does not hold, but instead that

$$\frac{k + 1}{k - 1}(a + a^2 + \cdots + a^{k-1}) + a^{k+1} < \frac{k + 2}{k}(a + a^2 + \cdots + a^{k-1} + a^k).$$

Then we find that

$$2(a + a^2 + \cdots + a^{k-1} + a^k) + (k^2 + k)a^k(a - 1) < 0,$$

which is clearly impossible if $a \geq 1$. Thus, by induction, (E.11) is established if $a \geq 1$.

If we set $a = 1/b$ $(b > 0)$, the expression on the left-hand side of (E.11) becomes

$$f(b) = \frac{1 + b + b^2 + \cdots + b^n}{b + b^2 + b^3 + \cdots + b^{n-1}}.$$

But the first part of the proof shows that

$$f(b) \geq \frac{n+1}{n-1} \quad \text{for } n > 1 \text{ and } b \geq 1.$$

Hence (E.11) is now established for all real $a > 0$.

PROBLEM 41. Show that if $a > b > 0$, then $A < B$, where

$$A = \frac{1 + a + \cdots + a^{n-1}}{1 + a + \cdots + a^n}, \qquad B = \frac{1 + b + \cdots + b^{n-1}}{1 + b + \cdots + b^n}.$$

Solution. It is clear that

$$\frac{1}{A} = 1 + \frac{a^n}{1 + a + \cdots + a^{n-1}}, \qquad \frac{1}{B} = 1 + \frac{b^n}{1 + b + \cdots + b^{n-1}},$$

that is,

$$\frac{1}{A} = 1 + \frac{1}{\frac{1}{a^n} + \frac{1}{a^{n-1}} + \cdots + \frac{1}{a}}, \qquad \frac{1}{B} = 1 + \frac{1}{\frac{1}{b^n} + \frac{1}{b^{n-1}} + \cdots + \frac{1}{b}}.$$

Whence $1/A > 1/B$ and so $A < B$.

PROBLEM 42. Let $x > 0$ and n be a positive integer. Show that

$$\frac{x^n}{1 + x + x^2 + \cdots + x^{2n}} \leq \frac{1}{2n+1}.$$

Solution. For $t > 0$ we have $t + t^{-1} \geq 2$ because $(t-1)^2 \geq 0$. Setting $t = x^k$ for $k = 1,2,\ldots,n$ and $x > 0$, we therefore get

$$\frac{x^n}{\sum_{k=0}^{2n} x^k} = \frac{1}{1 + \sum_{k=1}^{n} (x^k + x^{-k})} \leq \frac{1}{1 + 2n}.$$

PROBLEM 43. Let $a,b > 0$, $a + b = 1$, and $q > 0$. Show that

$$\left(a + \frac{1}{a}\right)^q + \left(b + \frac{1}{b}\right)^q \geq \frac{5^q}{2^{q-1}}.$$

Solution. The function

$$f(x) = \left(x + \frac{1}{x}\right)^q \quad \text{for } q > 0$$

is convex for $0 < x < 1$ because

$$f''(x) = q(q - 1)(x + x^{-1})^{q-2}(1 - x^{-2})^2 + 2qx^{-3}(x + x^{-1})^{q-1}$$

$$= q(x + x^{-1})^{q-2}q(1 - x^{-2})^2 + x^{-4} - 1 + 4x^2 > 0$$

for $q > 0$ and $0 < x < 1$. Consequently, for $a,b > 0$ and $a + b = 1$,

$$\frac{f(a) + f(b)}{2} \geq f\left(\frac{a + b}{2}\right) = f(\tfrac{1}{2}) = (\tfrac{1}{2} + 2)^q = \frac{5^q}{2^q}.$$

PROBLEM 44. Let $x,y > 0$ with $x \neq y$ and m and n be positive integers. Show that

$$x^m y^n + x^n y^m < x^{m+n} + y^{m+n}.$$

Solution. Consider $x^{m+n} - x^m y^n - x^n y^m + y^{m+n} = (x^m - y^m)(x^n - y^n)$. But

$$x^m - y^m = (x - y)(x^{m-1} + x^{m-2}y + \cdots + xy^{m-2} + y^{m-1})$$

and

$$x^n - y^n = (x - y)(x^{n-1} + x^{n-2}y + \cdots + xy^{n-2} + y^{n-1}).$$

Hence $(x^m - y^m)(x^n - y^n) > 0$ for $x \neq y$.

PROBLEM 45. Let $x > 0$ but $x \neq 1$ and n be a positive integer. Show that

$$x^{2n-1} + x < x^{2n} + 1.$$

Note in particular that

$$x^{n-1} + \frac{1}{x^{n-1}} < x^n + \frac{1}{x^n} \quad \text{for } x \neq 1,\ x > 0, \text{ and } n \text{ a positive integer.}$$

Solution. Since $(x^{2n-1} - 1)(x - 1) = x^{2n-2} + x^{2n-3} + \cdots + x + 1 > 0$, we have that

$$x^{2n} - x - x^{2n-1} + 1 > 0.$$

PROBLEM 46. Let $a > b > 0$ and n be a positive integer larger than 1. Show that

$$\sqrt[n]{a} - \sqrt[n]{b} < \sqrt[n]{a - b}.$$

Solution. Let, for $n > 1$, $x \geq 1$,

$$f(x) = x^{1/n} - (x - 1)^{1/n}.$$

Then, for $x > 1$,

$$nf'(x) = x^{1/n-1} - (x - 1)^{1/n-1} < 0;$$

thus $f(x)$ decreases for $x > 1$.

Since $f(1) = 1$ and $f(x) < 1$ for $x > 1$, we have

$$x^{1/n} - 1 < (x - 1)^{1/n} \quad \text{for } x > 1.$$

Letting $x = a/b$ and noting that $a/b > 1$, we get

$$\left(\frac{a}{b}\right)^{1/n} - 1 < \left(\frac{a}{b} - 1\right)^{1/n} \quad \text{or} \quad a^{1/n} - b^{1/n} < (a - b)^{1/n}.$$

PROBLEM 47. Let $a,b,x > 0$ and $a \neq b$. Show that

$$\left(\frac{a + x}{b + x}\right)^{b+x} > \left(\frac{a}{b}\right)^{x}.$$

Solution. Let, for $x \geq 0$,

$$f(x) = \left(\frac{a + x}{b + x}\right)^{b+x}.$$

Then

$$f'(x) = \left(\frac{b - a}{a + x} + \log \frac{a + x}{b + x}\right) f(x).$$

The sign of the derivative is the same as the sign of the function

$$g(x) = \frac{b - a}{a + x} + \log \frac{a + x}{b + x}.$$

Since

$$g'(x) = \frac{(a - b)^2}{(a + x)^2(b + x)} < 0,$$

$g(x) > g(+\infty) = 0$. Thus f is seen to be an increasing function.

PROBLEM 48. Let $a > b > 0$ and n be a positive integer larger than 1. Show that, for $k \geq 0$,

$$\sqrt[n]{a^n + k^n} - \sqrt[n]{b^n + k^n} \leq a - b.$$

Solution. Since

$$(x - y)(x^{n-1} + x^{n-2}y + \cdots + xy^{n-2} + y^{n-1}) = x^n - y^n$$

and putting

$$x = \sqrt[n]{a^n + k^n}, \quad y = \sqrt[n]{b^n + k^n}$$

we get, since $x \geq a$ and $y \geq b$, $k \geq 0$, and $a > b > 0$,

$$\sqrt[n]{a^n + k^n} - \sqrt[n]{b^n + k^n} \; (a^{n-1} + a^{n-2}b + \cdots + b^{n-1}) \leq a^n - b^n.$$

But $a^{n-1} + a^{n-2}b + \cdots + ab^{n-2} + b^{n-1}$ is positive and so the last inequality yields

$$\sqrt[n]{a^n + k^n} - \sqrt[n]{b^n + k^n} \leq a - b.$$

Remark. If $n = 2$ and a, b, and k are arbitrary real numbers, then

$$\sqrt{a^2 + k^2} - \sqrt{b^2 + k^2} \leq ||a| - |b||$$

holds.

PROBLEM 49. Compare the magnitudes of

$$(\sqrt{n})^{\sqrt{n+1}} \quad \text{and} \quad (\sqrt{n + 1})^{\sqrt{n}}.$$

Solution. We note that

$$(\sqrt{n})^{\sqrt{n+1}} > (\sqrt{n + 1})^{\sqrt{n}} \leftrightarrow \frac{\log \sqrt{n}}{\sqrt{n}} > \frac{\log \sqrt{n + 1}}{\sqrt{n + 1}}$$

and

$$(\sqrt{n})^{\sqrt{n+1}} < (\sqrt{n + 1})^{\sqrt{n}} \leftrightarrow \frac{\log \sqrt{n}}{\sqrt{n}} < \frac{\log \sqrt{n + 1}}{\sqrt{n + 1}}.$$

But

$$f(x) = \frac{\log x}{x}$$

is increasing for $0 < x < e$, where e is the basis of the natural logarithms, and decreasing for $x > e$. Moreover $\sqrt{7} < e < \sqrt{8}$. Hence

$$(\sqrt{n})^{\sqrt{n+1}} < (\sqrt{n+1})^{\sqrt{n}} \quad \text{for } n = 1,2,3,4,5,6$$

and

$$(\sqrt{n})^{\sqrt{n+1}} > (\sqrt{n+1})^{\sqrt{n}} \quad \text{for } n \geq 7.$$

PROBLEM 50. Using the elementary inequality

$$x \log x \geq x - 1 \quad \text{for } x > 0,$$

show that

$$\sum_{i=1}^{n} p_i \log p_i \geq \sum_{i=1}^{n} p_i \log q_i$$

for $p_i > 0$, $q_i > 0$ $(i = 1,2,\ldots,n)$ and

$$\sum_{i=1}^{n} p_i = \sum_{i=1}^{n} q_i.$$

Solution. Since $p_i/q_i > 0$, we get that

$$\frac{p_i}{q_i} \log \frac{p_i}{q_i} \geq \frac{p_i}{q_i} - 1 \quad \text{or} \quad p_i \log \frac{p_i}{q_i} \geq p_i - q_i$$

(since $q_i > 0$). Summing over i, we obtain

$$\sum_{i=1}^{n} p_i \log \frac{p_i}{q_i} \geq \sum_{i=1}^{n} (p_i - q_i).$$

But

$$\sum_{i=1}^{n} p_i = \sum_{i=1}^{n} q_i$$

and so

$$\sum_{i=1}^{n} p_i \log \frac{p_i}{q_i} \geq 0$$

or

$$\sum_{i=1}^{n} (p_i \log p_i - p_i \log q_i) \geq 0$$

or

$$\sum_{i=1}^{n} p_i \log p_i \geq \sum_{i=1}^{n} p_i \log q_i.$$

Remark. There is equality if and only if $p_i = q_i$ $(i = 1,\ldots,n)$.

PROBLEM 51. Show that if $a_1 \geq a_2 \geq a_3 \geq \cdots \geq a_n \geq 0$ and

$$\sum_{i=1}^{k} a_i \leq \sum_{i=1}^{k} b_i \quad \text{for } k = 1,2,\ldots,n, \tag{E.15}$$

then

$$\sum_{i=1}^{n} a_i^2 \leq \sum_{i=1}^{n} b_i^2. \tag{E.16}$$

Solution. After multiplication by $a_k - a_{k+1}$, (E.15) becomes

$$(a_k - a_{k+1}) \sum_{i=1}^{k} a_i \leq (a_k - a_{k+1}) \sum_{i=1}^{k} b_i \quad \text{for } k = 1,\ldots,n, \tag{E.17}$$

where $a_{n+1} = 0$. Summing both sides of (E.17) from $k = 1$ to $k = n$, we get

$$\sum_{i=1}^{n} a_i^2 \leq \sum_{i=1}^{n} a_i b_i. \tag{E.18}$$

By the Cauchy-Schwarz Inequality (see Problem 14), (E.18) yields

$$\left(\sum_{i=1}^{n} a_i^2\right)^2 \leq \left(\sum_{i=1}^{n} a_i b_i\right)^2 \leq \left(\sum_{i=1}^{n} a_i^2\right)\left(\sum_{i=1}^{n} b_i^2\right).$$

PROBLEM 52. For a positive integer n, let P(n) be the proposition

$$P(n):\ f(n) = \sqrt{n + \sqrt{n - 1 + \sqrt{\cdots + \sqrt{2 + \sqrt{1}}}}} < \sqrt{n} + 1$$

Verify this proposition.

Solution. Since $f(1) = 1 < 2$, $P(1)$ is true. Suppose now that $P(n)$ is valid. Then

$$\begin{aligned} f(n+1) &= \sqrt{n+1+f(n)} \\ &\le \sqrt{n+1+\sqrt{n}+1} \quad \text{(by the induction hypothesis)} \\ &\le \sqrt{(\sqrt{n}+1)^2} \\ &= \sqrt{n}+1 \\ &< \sqrt{n+1}+1. \end{aligned}$$

Thus, $P(n)$ implies $P(n+1)$.

PROBLEM 53. Let $x \ge 0$ and let m and n be real numbers such that $m \ge n > 0$. Prove that

$$(m+n)(1+x^m) \ge 2n\,\frac{1-x^{m+n}}{1-x^n}.$$

Solution. The inequality is an equality if $m = n$, so suppose that $m > n$. Let

$$f(x) = (m-n)x^{m+n} + (m+n)(x^n - x^m) + (n-m);$$

we shall show that $f(x)$ is negative if $0 \le x < 1$ and positive if $x > 1$, and this will be sufficient to verify the inequality. Differentiating, we get

$$f'(x) = (m+n)x^{n-1}g(x), \quad g(x) = (m-n)x^m - mx^{m-n} + n.$$

From this, we see that $f'(x)$ and $g(x)$ have the same sign for positive values of x. Since

$$g'(x) = m(m-n)x^{m-n-1}(x^n - 1),$$

we see that $g(x)$ is strictly decreasing on $[0,1]$ and strictly increasing on $[1,\infty)$ with a minimum value $g(1) = 0$ at $x = 1$. Since $g(0) = n > 0$, it follows that $g(x) > 0$ for all positive $x \ne 1$. Since $f'(x)$ has the same sign as $g(x)$, we see that $f(x)$ is strictly increasing. The verification is complete when we note that $f(0) = n - m < 0$ and $f(1) = 0$.

Remarks. Assuming that $m > n$ are positive real numbers and $x \neq 1$ is nonnegative, we have shown that

$$(m - n) - (m + n)x^n + (m + n)x^m - (m - n)x^{m+n}$$

is positive if $0 \le x < 1$ and negative if $x > 1$. It is easy to see that this implies the two inequalities

$$(m + n)(1 + x^m) > \frac{2n(1 - x^{m+n})}{1 - x^n}$$

and

$$\frac{2m(1 - x^{m+n})}{1 - x^m} > (m + n)(1 + x^n).$$

Along the same lines of reasoning we can establish that

$$(m - n) + (m + n)x^n - (m + n)x^m - (m - n)x^{m+n}$$

is positive if $0 \le x < 1$ and negative if $x > 1$ whenever $m > n$ are positive real numbers and so the two inequalities

$$\frac{2m(1 - x^{m+n})}{1 - x^n} > (m + n)(1 + x^m)$$

and

$$(m + n)(1 + x^n) > \frac{2n(1 - x^{m+n})}{1 - x^m}$$

must hold.

PROBLEM 54. If $a_i \ge 0$, $\Sigma_i\, a_i = 1$, and $0 \le x_i \le 1$ for $i = 1,2,\ldots,n$, show that

$$\frac{a_1}{1 + x_1} + \frac{a_2}{1 + x_2} + \cdots + \frac{a_n}{1 + x_n} \le \frac{1}{1 + x_1^{a_1} x_2^{a_2} \cdots x_n^{a_n}}.$$

Solution. Assume without loss of generality that $a_i > 0$ for all i. The proposed inequality follows from Jensen's Inequality (see Problem 8): If f is convex on an interval I, then for all y_i in I, $\Sigma\, a_i f(y_i) \le f(\Sigma\, a_i y_i)$ with equality if and only if $y_1 = y_2 = \cdots = y_n$. To apply this to the proposed inequality, let

$$y_i = \log x_i \quad \text{and} \quad f(y) = \frac{1}{1 + e^y},$$

assuming for the moment that $x_i > 0$ for all i. Since $0 < x_i \le 1$, it follows that $-\infty < y_i \le 0$ and since, for $y < 0$,

$$f''(y) = e^y(e^y - 1)\frac{1}{(1 + e^y)^3},$$

we see that f is strictly convex on $(-\infty, 0]$. Then

$$\sum_{i=1}^{n} \frac{a_i}{1 + x_i} = \sum_{i=1}^{n} a_i(1 + \exp y_i)^{-1} \le \left\{1 + \exp \sum_{i=1}^{n} a_i y_i\right\}^{-1}$$

$$= \left\{1 + \prod_{i=1}^{n} \exp(a_i y_i)\right\}^{-1} = \left\{1 + \prod_{i=1}^{n} x_i^{a_i}\right\}^{-1},$$

with equality if and only if $y_1 = y_2 = \cdots = y_n$, that is, if and only if $x_1 = x_2 = \cdots = x_n$.

If some $x_i = 0$, the above proof breaks down, but this case is easily handled on its own merits. Again, equality holds if and only if $x_1 = x_2 = \cdots = x_n$, which in this case means they are all zero.

If we allow $a_i = 0$ (and assume $0^0 = 1$), then the condition for equality becomes x_i = constant for all i for which $a_i > 0$.

Finally we note that the inequality is reversed if $x_i \ge 1$ for all i. This is because $f''(y) > 0$ on $(0,\infty)$ and thus f is concave on $[0,\infty)$.

PROBLEM 55. If a_i with $i = 1,2,\ldots,n$ denote real numbers, show that

$$n \min(a_i) \le \Sigma a_i - S \le \Sigma a_i + S \le n \max(a_i),$$

where

$$(n - 1)S^2 = \sum_{1 \le i < j \le n} (a_i - a_j)^2 \quad \text{and } S \ge 0;$$

equality holds if and only if all a_i are equal.

Solution. We assume that $a_1 \le a_2 \le \cdots \le a_n$. Then

$$S^2 = \frac{1}{n-1} \sum_{i=2}^{n} \sum_{j=1}^{i-1} (a_i - a_j)^2 \le \frac{1}{n-1} \sum_{i=2}^{n} (i-1)(a_i - a_1)^2$$

$$\le \sum_{i=2}^{n} (a_i - a_1)^2 \le \left\{ \sum_{i=1}^{n} (a_i - a_1) \right\}^2 .$$

Taking square roots we obtain

$$na_1 \le \sum_{i=1}^{n} a_i - S.$$

Similarly,

$$S^2 \le \frac{1}{n-1} \sum_{j=1}^{n-1} (n-j)(a_n - a_j)^2 \le \left\{ \sum_{j=1}^{n} (a_n - a_j) \right\}^2$$

from which follows that

$$\sum_{j=1}^{n} a_j + S \le na_n .$$

It is clear that equality holds anywhere if and only if it holds throughout and this is true if and only if all a_i are equal.

PROBLEM 56. In the Solution of Problem 42 we encountered the inequality

$$m + \frac{1}{m} \ge 2 \quad \text{for } m > 0$$

and equality if and only if m = 1. Prove that

$$m + \frac{4}{m^2} \ge 3 \quad \text{for } m > 0$$

and equality if and only if m = 2.

Solution. Since $m > 0$, the proposed inequality is equivalent with the inequality $m^3 + 4 - 3m^2 \ge 0$. But $m^3 + 4 - 3m^2 = (m + 1)(m - 2)^2$, $m > 0$ and $(m - 2)^2 \ge 0$. Hence $(m + 1)(m - 2)^2 \ge 0$ and $= 0$ if and only if $m = 0$.

PROBLEM 57. Assume that $a_i > 0$, $(i = 1,\dots,n)$, with $a_{n+1} = a_1$. Show:

$$\sum_{i=1}^{n} \left(\frac{a_i}{a_{i+1}} \right)^n \ge \sum_{i=1}^{n} \frac{a_{i+1}}{a_i}.$$

Solution. Let $b_i = a_i/a_{i+1}$ for $i = 1,2,\ldots,n$ and $b_{n+1} = 1$. Then the product $b_1 b_2 \cdots b_{n+1} = 1$, and the inequality to be proved is equivalent with

$$\sum_{i=1}^{n+1} \frac{1}{b_i} \le \sum_{i=1}^{n+1} b_i^n.$$

The inequality of the arithmetic and geometric means (see Problem 2) give

$$\frac{1}{b_i} = \frac{\prod_{k=1}^{n+1} b_k}{b_i} \le \frac{\left(\sum_{k=1}^{n+1} b_k^n\right) - b_i^n}{n}.$$

Therefore

$$\sum_{i=1}^{n+1} \frac{1}{b_i} \le \frac{1}{n} \sum_{i=1}^{n+1} \left\{\left(\sum_{k=1}^{n+1} b_k^n\right) - b_i^n\right\} = \sum_{i=1}^{n+1} b_i^n.$$

PROBLEM 58. Show *Abel's Inequality*: Let $\{a_1, a_2, \ldots, a_n\}$ and $\{b_1, b_2, \ldots, b_n\}$ be two sets of real numbers with $b_1 \ge b_2 \ge \cdots \ge b_n \ge 0$ and put

$$s_k = a_1 + a_2 + \cdots + a_k \quad \text{for } k = 1,2,\ldots,n,$$

with M and m denoting, respectively, the largest and the smallest of the numbers $s_1, s_2, \ldots, s_n$. Then

$$mb_1 \le \sum_{i=1}^{n} a_i b_i \le Mb_1.$$

Solution. Clearly

$$\sum_{i=1}^{n} a_i b_i = s_1 b_1 + (s_2 - s_1)b_2 + \cdots + (s_n - s_{n-1})b_n$$

$$= s_1(b_1 - b_2) + s_2(b_2 - b_3) + \cdots + s_{n-1}(b_{n-1} - b_n) + s_n b_n.$$

But

$$s_1(b_1 - b_2) + s_2(b_2 - b_3) + \cdots + s_{n-1}(b_{n-1} - b_n) + s_n b_n \le Mb_1$$

because

$$s_j(b_j - b_{j+1}) \le M(b_j - b_{j+1}) \quad \text{for } j = 1,2,\ldots,n-1,$$

and $s_n b_n \le Mb_n$. Therefore $\Sigma_{i=1}^n a_i b_i \le Mb_1$. Similarly, since

$$m(b_j - b_{j+1}) \le s_j(b_j - b_{j+1}) \quad \text{for } j = 1,2,\dots,n-1$$

and $mb_n \le s_n b_n$, we obtain that

$$s_1(b_1 - b_2) + s_2(b_2 - b_3) + \cdots + s_{n-1}(b_{n-1} - b_n) + s_n b_n \ge mb_1$$

or $mb_1 \le \sum_{i=1}^n a_i b_i$.

PROBLEM 59. Determine all polynomials of the form $x^n + a_1 x^{n-1} + \cdots + a_n$ with all $a_k = \pm 1$ and which have all roots real.

Solution. Let the roots be $x_1, \dots, x_n$. Since the roots are real, the numbers $x_1^2, \dots, x_n^2$ are positive, and we know that their sum is

$$x_1^2 + \cdots + x_n^2 = a_1^2 - 2a_2 = 1 \pm 2 = 3 \quad \text{(since -1 is negative).}$$

The product

$$x_1^2 \cdots x_n^2 = a_n^2 = 1.$$

Hence, by Problem 2, we have $1 \le 3/n$, that is, $n \le 3$, with equality only if $x_1^2 = x_2^2 = x_3^2 = 1$. In case $n = 3$, all roots are ± 1 and $a_2 = -1$. The rest is trivial.

PROBLEM 60. Show that

$$\left(\sum_{i=1}^n \frac{a_i}{i}\right)^2 \le \sum_{i=1}^n \sum_{j=1}^n \frac{a_i a_j}{i + j - 1}.$$

Solution. Since

$$\sum_{i=1}^n \sum_{i=j}^n \frac{a_i a_j}{i + j - 1} - \left(\sum_{i=1}^n \frac{a_i}{i}\right)^2 = \sum_{i=1}^n \sum_{j=1}^n \frac{b_i b_j}{i + j - 1}$$

with

$$b_k = \frac{(k - 1)a_k}{k} \quad \text{for } k = 1,2,\dots,n$$

and

$$\sum_{i=1}^{n}\sum_{j=1}^{n}\frac{b_i b_j}{i+j-1} = \int_0^1 \left(\sum_{i=1}^{n}\sum_{j=1}^{n} b_i b_j x^{i+j-2}\right) dx$$

$$= \int_0^1 \left|\sum_{i=1}^{n} b_i x^{i-1}\right|^2 dx \geq 0,$$

the desired inequality follows. Moreover, equality holds if and only if $a_2 = a_3 = \cdots = a_n = 0$.

PROBLEM 61. Let f and g be real-valued functions defined on the set of real numbers. Show that there are numbers x and y such that $0 \leq x \leq 1$, $0 \leq y \leq 1$, and $|xy - f(x) - g(x)| \geq 1/4$.

Solution. If $|xy - f(x) - g(x)| > 1/4$ for all x and y in the unit interval [0,1], then, in particular,

$$|f(0) + g(1)| < 1/4, \quad |f(0) + g(0)| < 1/4, \quad \text{and} \quad |f(1) + g(0)| < 1/4.$$

But then by the triangle inequality

$$\begin{aligned} |1 - f(1) - g(1)| &\geq 1 - |f(1) + g(1)| \\ &\geq 1 - |f(1) + g(0)| - |-g(0) - f(0)| \\ &\quad - |g(1) + f(0)| \\ &> \frac{1}{4}. \end{aligned}$$

PROBLEM 62. Let $t > 0$. Show that

$$t^\alpha - \alpha t \leq 1 - \alpha \quad \text{if } 0 < \alpha < 1 \tag{E.19}$$

and

$$t^\alpha - \alpha t \geq 1 - \alpha \quad \text{if } \alpha > 1. \tag{E.20}$$

Solution. Differentiating the function $f(t) = t^\alpha - \alpha t$ with respect to t yields $f'(t) = \alpha(t^{\alpha-1} - 1)$. Clearly, if $0 < \alpha < 1$, then

$$\begin{aligned} f'(t) &> 0 \quad \text{for } 0 < t < 1, \\ &< 0 \quad \text{for } t > 1, \\ &= 0 \quad \text{for } t = 1, \end{aligned}$$

and f assumes its largest value at $t = 1$. On the other hand, if $\alpha > 1$, then

$$f'(t) \begin{cases} < 0 & \text{for } 0 < t < 1, \\ > 0 & \text{for } t > 1, \\ = 0 & \text{for } t = 0. \end{cases}$$

and f assumes its smallest value at $t = 1$.

Remarks. Let a and b be positive real numbers and α and β be such that $\alpha < 1$, $\beta < 1$, and $\alpha + \beta = 1$, then $a^\alpha b^\beta \leq \alpha a + \beta b$. Indeed, take $t = a/b$ and denote $1 - \alpha$ by β; then (E.19) gives the desired result.

Let $A > B > 0$. If we put $t = 1 + 1/A$ and $\alpha = A/B$ in (E.20), we obtain

$$\left(1 + \frac{1}{A}\right)^{A/B} > 1 + \frac{1}{B}$$

or

$$\left(1 + \frac{1}{A}\right)^{A} > \left(1 + \frac{1}{B}\right)^{B}. \tag{E.21}$$

If we put $t = 1 - 1/(A + 1)$ and $\alpha = (A + 1)/(B + 1)$ in (E.20), we obtain

$$\left(1 - \frac{1}{A + 1}\right)^{\frac{A+1}{B+1}} > 1 - \frac{1}{B + 1} \quad \text{or} \quad \left(\frac{A}{A + 1}\right)^{A+1} > \left(\frac{B}{B + 1}\right)^{B+1}$$

or

$$\left(1 + \frac{1}{A}\right)^{A+1} < \left(1 + \frac{1}{B}\right)^{B+1}. \tag{E.22}$$

Inequalities (E.21) and (E.22) show that for $x > 0$ the function

$$g(x) = \left(1 + \frac{1}{x}\right)^{x}$$

increases with increasing x and for $x > 0$ the function

$$G(x) = \left(1 + \frac{1}{x}\right)^{x+1}$$

decreases with increasing x; in this connection see Problem 2 of Chapter 1.

PROBLEM 63. Let $s \neq 2k\pi$, where k is an integer, and

$$A = \sin t + \sin (t + s) + \sin (t + 2s) + \cdots + \sin [t + (j - 1)s]$$

and

$$B = \cos t + \cos (t + s) + \cos (t + 2s) + \cdots + \cos [t + (j - 1)s].$$

Show that

$$A = \sin\left(t + \frac{j-1}{2} s\right) \cdot \frac{\sin j \frac{s}{2}}{\sin \frac{s}{2}}$$

and

$$B = \cos\left(t + \frac{j-1}{2} s\right) \cdot \frac{\sin j \frac{s}{2}}{\sin \frac{s}{2}}.$$

Solution. Since

$$2 \sin \frac{s}{2} \cdot \sin(t + ks) = \cos\left(t + \frac{2k-1}{2} s\right) - \cos\left(t + \frac{2k+1}{2} s\right)$$

we get that

$$\begin{aligned} A \cdot 2 \sin \frac{s}{2} &= \cos\left(t - \frac{s}{2}\right) - \cos\left(t + \frac{s}{2}\right) \\ &\quad + \cos\left(t + \frac{s}{2}\right) - \cos\left(t + 3 \frac{s}{2}\right) \\ &\quad + \cos\left(t + 3 \frac{s}{2}\right) - \cos\left(t + 5 \frac{s}{2}\right) \\ &\quad \cdots \\ &\quad + \cos\left(t + \frac{2j-3}{2} s\right) - \cos\left(t + \frac{2j-1}{2} s\right) \\ &= \cos\left(t - \frac{s}{2}\right) - \cos\left(t + \frac{2j-1}{2} s\right) \\ &= 2 \sin\left(t + \frac{j-1}{2} s\right) \cdot \sin j \frac{s}{2}. \end{aligned}$$

Thus

$$A = \sin\left(t + \frac{j-1}{2} s\right) \cdot \frac{\sin j \frac{s}{2}}{\sin \frac{s}{2}}.$$

To prove the second formula we note that

$$2 \sin \frac{s}{2} \cdot \cos(t + ks) = \sin\left(t + \frac{2k+1}{2} s\right) - \sin\left(t + \frac{2k-1}{2} s\right)$$

and so

$$B \cdot 2 \sin \frac{s}{2}$$

$$= \sin\left(t + \frac{s}{2}\right) - \sin\left(t - \frac{s}{2}\right)$$

$$+ \sin\left(t + 3\,\frac{s}{2}\right) - \sin\left(t + \frac{s}{2}\right)$$

$$+ \sin\left(t + 5\,\frac{s}{2}\right) - \sin\left(t + 3\,\frac{s}{2}\right)$$

$$\cdots$$

$$+ \sin\left(t + \frac{2j - 1}{2}\,s\right) - \sin\left(t + \frac{2j - 3}{2}\,s\right)$$

$$= \sin\left(t + \frac{2j - 1}{2}\,s\right) - \sin\left(t - \frac{s}{2}\right)$$

$$= 2 \cos\left(t + \frac{j - 1}{2}\,s\right) \cdot \sin j\,\frac{s}{2}$$

Thus

$$B = \cos\left(t + \frac{j - 1}{2}\,s\right) \cdot \frac{\sin j\,\frac{s}{2}}{\sin \frac{s}{2}}.$$

Remark. If $x \neq 2k\pi$, then

$$\left|\sum_{k=p}^{q} \sin kx\right| \leq \frac{1}{\left|\sin \frac{x}{2}\right|} \quad \text{and} \quad \left|\sum_{k=p}^{q} \cos kx\right| \leq \frac{1}{\left|\sin \frac{x}{2}\right|}.$$

PROBLEM 64. Show that for any real number x and any positive integer n we have

$$\left|\sum_{k=1}^{n} \frac{\sin kx}{k}\right| \leq 2\sqrt{\pi}. \tag{E.21}$$

Solution. Let $0 < x < \pi$ and let m be the integer such that

$$m \leq \frac{\sqrt{\pi}}{x} < m + 1. \tag{E.22}$$

Then

$$\left|\sum_{k=1}^{n} \frac{\sin kx}{k}\right| \le \sum_{k=1}^{m} \left|\frac{\sin kx}{k}\right| + \left|\sum_{k=m+1}^{n} \frac{\sin kx}{k}\right|. \tag{E.23}$$

(If $m = 0$ the first summand on the right-hand side vanishes, if $m \ge n$ the second summand on the right-hand side becomes zero.) Since $|\sin t| \le |t|$ we get

$$\sum_{k=1}^{m} \left|\frac{\sin kx}{k}\right| \le \sum_{k=1}^{m} \frac{kx}{k} = mx \le \sqrt{\pi}. \tag{E.24}$$

By Abel's Inequality (see Problem 58) and the Remark to Problem 63, we see that

$$\left|\sum_{k=m+1}^{n} \frac{\sin kx}{k}\right| \le \frac{1}{\left|\sin \frac{x}{2}\right|} \cdot \frac{1}{m+1}.$$

Evidently $\sin t \ge (2/\pi)t$ for $0 \le t \le \pi/2$. This, together with (E.22), gives

$$\sin \frac{x}{2} > \frac{x}{\pi}, \qquad m + 1 > \frac{\sqrt{\pi}}{x}.$$

We therefore have

$$\left|\sum_{k=m+1}^{n} \frac{\sin kx}{k}\right| \le \frac{1}{\frac{x}{\pi} \cdot \frac{\sqrt{\pi}}{x}} = \sqrt{\pi},$$

which together with (E.23) and (E.24) gives the desired estimate (E.21) for $x \in (0,\pi)$. But the function

$$f(x) = \left|\sum_{k=1}^{n} \frac{\sin kx}{k}\right|$$

is an even function, that is, $f(-x) = f(x)$; therefore (E.21) also holds for $x \in (-\pi,0)$. Since (E.21) trivially holds for $x = \pm\pi$, (E.21) is valid for all x such that $-\pi \le x \le \pi$. Since f is a periodic function with period 2π, (E.21) is seen to hold for any real number x.

PROBLEM 65. Show that

$$\frac{1}{2} \cdot \log(2n + 1) < 1 + \frac{1}{3} + \frac{1}{5} + \cdots + \frac{1}{2n - 1} < 1 + \frac{1}{2} \cdot \log(2n - 1).$$

Solution. For $k \le x \le k + 1$ we have

$$\frac{1}{2k + 1} \leq \frac{1}{2x - 1} \leq \frac{1}{2k - 1}.$$

Integration gives

$$\frac{1}{2k + 1} < \int_k^{k+1} \frac{dx}{2x - 1} < \frac{1}{2k - 1};$$

thus

$$\frac{1}{3} + \frac{1}{5} + \cdots + \frac{1}{2n - 1} < \int_1^n \frac{dx}{2x - 1}$$

and

$$\int_1^{n+1} \frac{dx}{2x - 1} < 1 + \frac{1}{3} + \frac{1}{5} + \cdots + \frac{1}{2n - 1}.$$

Therefore

$$\int_1^{n+1} \frac{dx}{2x - 1} < 1 + \frac{1}{3} + \frac{1}{5} + \cdots + \frac{1}{2n - 1} < 1 + \int_1^n \frac{dx}{2x - 1}$$

or

$$\frac{1}{2}\cdot\log(2n + 1) < 1 + \frac{1}{3} + \frac{1}{5} + \cdots + \frac{1}{2n - 1} < 1 + \frac{1}{2}\cdot\log(2n - 1).$$

PROBLEM 66. Show that

$$\frac{1}{2}\cdot\log(2n + 1) < \int_0^{\pi/2} \frac{\sin^2 nx}{\sin n}\, dx < 1 + \frac{1}{2}\cdot\log(2n - 1).$$

Solution. Since

$$\sin^2 \frac{nt}{2} = \frac{1 - \cos nt}{2}$$

$$= \frac{(1 - \cos t) + (\cos t - \cos 2t) + \cdots + [\cos(n - 1)t - \cos nt]}{2}$$

and

$$\cos A - \cos B = 2 \sin \frac{A + B}{2} \sin \frac{B - A}{2},$$

it follows that

$$\sin^2 \frac{nt}{2} = \left(\sin \frac{t}{2} + \sin \frac{3t}{2} + \cdots + \sin \frac{(2n - 1)t}{2}\right) \sin \frac{t}{2}.$$

Putting t = 2x, it follows that

$$\sin^2 nx = [\sin x + \sin 3x + \cdots + \sin(2n - 1)x] \sin x$$

and so

$$\int_0^{\pi/2} \frac{\sin^2 nx}{\sin x}\, dx = 1 + \frac{1}{3} + \frac{1}{5} + \cdots + \frac{1}{2n - 1}.$$

The desired result now follows by invoking the inequality established in Problem 65.

Remark. Since $\log(2n + 1) > \log 2n > \log(2n - 1)$, we see that

$$\frac{1}{2}\cdot\log 2n < \int_0^{\pi/2} \frac{\sin^2 nx}{\sin x}\, dx < 1 + \frac{1}{2}\cdot\log 2n.$$

This in turn shows that

$$\lim_{n\to\infty} \frac{2}{\log n} \int_0^{\pi/2} \frac{\sin^2 nx}{\sin x}\, dx = 1.$$

PROBLEM 67. Let $f(x) = a_1 \sin x + a_2 \sin 2x + \cdots + a_n \sin nx$ satisfy $|f(x)| \le 1$ for all real x. Show that $|a_1 + 2a_2 + \cdots + na_n| \le n$.

Solution. By the Inequality of Bernstein (see Problem 20) we have that $|f'(x)| \le n$ for all real x. Hence it follows that $|f'(0)| \le n$.

PROBLEM 68. Suppose that

$$-1 \le ax^2 + bx + c \le 1 \quad \text{for} \quad -1 \le x \le 1,$$

where a, b, and c are real numbers. Without using A. A. Markoff's Inequality (see Problem 24), show that

$$-4 \le 2ax + b \le 4 \quad \text{for} \quad -1 \le x \le 1.$$

Solution. The graph of the linear function

$$f'(x) = 2ax + b$$

derived from the function

$$f(x) = ax^2 + c$$

is a straight line; therefore, as x varies over the interval [-1,1], $f'(x)$ assumes its largest and smallest values at the endpoints of that interval. Hence, it is enough to show that the values of $f'(x)$ at $x = -1$ and $x = 1$, that is, the numbers

$$f'(-1) = -2a + b \quad \text{and} \quad f'(1) = 2a + b$$

are neither less than -4 nor larger than 4. To estimate these numbers, we substitute -1, 0, and 1 for x in $f(x) = ax^2 + bx + c$ and use the given inequalities. We get

$$-1 \le f(-1) = a - b + c \le 1, \tag{E.25}$$

$$-1 \le -f(0) = -c \le 1, \tag{E.26}$$

$$-1 \le f(1) = a + b + c \le 1. \tag{E.27}$$

The sum of (E.25) and (E.26) gives

$$-2 \le a - b \le 2, \tag{E.28}$$

the sum of (E.26) and (E.27) gives

$$-2 \le a + b \le 2, \tag{E.29}$$

and the sum of (E.28) and (E.29) gives

$$-4 \le (a + b) + (a - b) \le 4$$

or

$$-2 \le a \le 2. \tag{E.30}$$

Adding first (E.28) and (E.30), then (E.29) and (E.30), we obtain

$$-4 \le 2a - b \le 4, \quad \text{or} \quad -4 \le -2a + b \le 4$$

and

$$-4 \le 2a + b \le 4,$$

which shows that the extreme values of $f'(x)$ (and hence certainly all is values) in the interval $-1 \le x \le 1$ are situated within the specified range.

Remark. Note that the quadratic polynomial $f(x) = 2x^2 - 1$ has the properties

$$-1 \le f(x) \le 1 \quad \text{for} \quad -1 \le x \le 1$$

and $f'(-1) = -4$, $f'(1) = 4$.

PROBLEM 69. Show that if x is larger than any of the numbers a_1, a_2, ..., a_n then

$$\frac{1}{x - a_1} + \frac{1}{x - a_2} + \cdots + \frac{1}{x - a_n} \geq \frac{n}{x - (a_1 + a_2 + \cdots + a_n)/n}.$$

Solution. Let

$$x_1 = x - a_1, \quad x_2 = x - a_2, \quad \ldots, \quad x_n = x - a_n.$$

It is clear that $x_i > 0$ for $i = 1,2,\ldots,n$ and that we may use the result in Problem 2; thus

$$\frac{n}{\frac{1}{x_1} + \frac{1}{x_2} + \cdots + \frac{1}{x_n}} \leq \frac{x_1 + x_2 + \cdots + x_n}{n}$$

or

$$\frac{1}{x_1} + \frac{1}{x_2} + \cdots + \frac{1}{x_n} \geq \frac{n^2}{x_1 + x_2 + \cdots + x_n}.$$

But $x_1 + x_2 + \cdots + x_n = nx - (a_1 + a_2 + \cdots + a_n)$ and everything is clear.

PROBLEM 70. Let a, b, r, R, and x be positive, $r \leq R$, and $r^2x = aR + bR^3$. Show that $x \geq 2\sqrt{ab}$.

Solution. Since $R^2x \geq aR + bR^3$, we have that

$$x \geq \frac{a}{R} + bR.$$

But

$$\frac{\frac{a}{R} + bR}{2} \geq \sqrt{ab}$$

by Problem 2.

PROBLEM 71. Let a_1, a_2, ..., a_n and b_1, b_2, ..., b_n be positive numbers and suppose that M denotes the largest of the fractions

$$\frac{a_1}{b_1}, \frac{a_2}{b_2}, \dots, \frac{a_n}{b_n}.$$

Show that

$$\frac{a_1 + a_2^2 + \cdots + a_n^n}{b_1 + Mb_2^2 + \cdots + M^{n-1}b_n^n} \le M.$$

Solution. It is clear that

$$\frac{a_1}{b_1} \le M, \quad \frac{a_2^2}{Mb_2^2} \le M, \quad \dots, \quad \frac{a_n^n}{M^{n-1}b_n^n} \le M;$$

hence we may use the result in Problem 28.

PROBLEM 72. Let $x_i > 0$, $p_i > 0$ for $i = 1,2,\dots,n$. Letting $\Sigma_j\, p_j$ stand for the sum $p_1 + p_2 + \cdots + p_n$, show that

$$\left(a_1^{p_1} \cdots a_n^{p_n}\right)^{1/\Sigma_j p_j} \le \frac{p_1a_1 + \cdots + p_na_n}{p_1 + \cdots + p_n}. \tag{E.31}$$

Solution. Put, for $i = 1,2,\dots,n$,

$$q_i = \frac{p_i}{p_1 + \cdots + p_n}$$

in Problem 9.

Remark. Setting $p_1 = p_2 = \cdots = p_n = 1$ in (E.31), we obtain the familiar inequality (see Problem 2)

$$(a_1a_2 \cdots a_n)^{1/n} \le \frac{a_1 + a_2 + \cdots + a_n}{n}.$$

PROBLEM 73. Let

$$C_j = \frac{n!}{j!(n - j)!} \quad \text{for } j = 1,2,\dots,n.$$

Show that

$$\sqrt{C_1} + \sqrt{C_2} + \cdots + \sqrt{C_n} \le \sqrt{n(2^n - 1)}.$$

Solution. The Binomial Theorem gives

$$1 + C_1 + C_2 + \cdots + C_n = (1 + 1)^n = 2^n$$

or

$$C_1 + C_2 + \cdots + C_n = 2^n - 1.$$

Our claim now follows by setting $a_i = 1$ and $b_i = \sqrt{C_i}$ for $i = 1,2,\ldots,n$ in the Cauchy-Schwarz Inequality (see Problem 14).

PROBLEM 74. Let $y = f(x)$ be a continuous, strictly increasing function of x for $x \ge 0$, with $f(0) = 0$. Examining the areas represented by the integrals, we see that

$$ab \le \int_0^a f(x)\,dx + \int_0^b f^{-1}(y)\,dy,$$

where $f^{-1}(y)$ is the inverse function to $f(x)$ and a and b are nonnegative constants. It is easily seen that there is strict inequality unless $b = f(a)$. This is the *Inequality of Young*.

Show that, for $t \ge 1$ and $s \ge 0$,

$$ts \le t \log t - t + e^s.$$

Solution. Let $y = f(x) = \log(x + 1)$ in Young's Inequality, and put $a = t - 1$ and $b = s$.

Remark. Letting $y = f(x) = x^{p-1}$, p 1, in Young's Inequality, we obtain

$$ab \le \frac{a^p}{p} + \frac{b^q}{q},$$

where p and q are connected by the equation $1/p + 1/q = 1$.

PROBLEM 75. Let p, f, and g be continuous functions on the interval [a,b] and suppose that p is positive-valued and f and g are monotonically increasing on [a,b]. Verify the following inequality, due to Chebyšhew:

$$\left\{\int_a^b p(x)\ f(x)\ dx\right\}\left\{\int_a^b p(x)\ g(x)\ dx\right\}$$

$$\leq \left\{\int_a^b p(x)\ dx\right\}\left\{\int_a^b p(x)\ f(x)\ g(x)\ dx\right\}.$$

Solution. Consider the difference

$$D = \int_a^b p(x)f(x)g(x)\ dx \cdot \int_a^b p(x)\ dx\ - \int_a^b p(x)f(x)\ dx \cdot \int_a^b p(x)g(x)\ dx.$$

But

$$\int_a^b p(x)\ dx = \int_a^b p(y)\ dy \quad \text{and} \quad \int_a^b p(x)g(x)\ dx = \int_a^b p(y)g(y)\ dy$$

and so

$$D = \int_a^b\int_a^b p(x)p(y)f(x)[g(x) - g(y)]dx\,dy.$$

Exchanging the roles of x and y, we get

$$D = \int_a^b\int_a^b p(x)p(y)f(y)[g(y) - g(x)]dx\,dy.$$

Forming the arithmetical mean of the last two expressions for D yields

$$D = \frac{1}{2}\int_a^b\int_a^b p(x)p(y)[f(x) - f(y)][g(x) - g(y)]dx\,dy.$$

Since f and g are monotonically increasing, the expressions in the two square brackets have the same sign; hence the integrand is nonnegative. Thus $D \geq 0$, giving us the desired inequality.

As we can easily see, the inequality remains true if both functions f and g are monotonically decreasing on [a,b]. If, however, one of these two functions is increasing, while the other is decreasing, then the inequality reverses.

PROBLEM 76. Let $f(x) = x^n + a_1x^{n-1} + a_2x^{n-2} + \cdots + a_n$, where n is odd and the coefficients $a_1, a_2, \ldots, a_n$ are real numbers. Letting

$$M = \max\{1, |a_1| + |a_2| + \cdots + |a_n|\},$$

show that $f(t) > 0$ for $t > M$ and $f(t) < 0$ for $t < -M$.

Solution. For sufficiently large positive t we have

$$\begin{aligned} f(t) &\geq t^n - (|a_1|t^{n-1} + |a_2|t^{n-2} + \cdots + |a_n|) \\ &\geq t^n - t^{n-1}(|a_1| + |a_2| + \cdots + |a_n|) \\ &= t^{n-1}[t - (|a_1| + |a_2| + \cdots + |a_n|)] \\ &> 0; \end{aligned}$$

here the first inequality holds for all positive t, the second for all $t \geq 1$ and the third for all $t > M$. On the other hand, since n is odd (and so $(-t)^n = -t^n$),

$$\begin{aligned} f(-t) &\leq -t^n + |a_1|t^{n-1} + |a_2|t^{n-2} + \cdots + |a_n| \\ &\leq -t^n + t^{n-1}(|a_1| + |a_2| + \cdots + |a_n|) \\ &= -t^{n-1}[t - (|a_1| + |a_2| + \cdots + |a_n|)] \\ &< 0; \end{aligned}$$

here the first inequality holds for all positive t, the second for all $t \geq 1$ and the third for all $t > M$. But $f(-t) < 0$ for $t > M$ is equivalent to $f(t) < 0$ for $t < -M$.

Remark. Let f and M be as in Problem 76. Since f is continuous, f must have at least one real root in the interval $(-M,M)$. Moreover all real roots of f must be in this interval.

PROBLEM 77. Let f be a continuous function on $[a,b]$ and put $f(t) = 0$ for $t \notin [a,b]$. For $h > 0$, we define the *Steklow Function of* f, denoted by f_h, as follows:

$$f_h(x) = \frac{1}{2h}\int_{x-h}^{x+h} f(t)\, dt.$$

It is clear that

$$f_h(x) = \frac{1}{2h}[F(x + h) - F(x - h)],$$

where

$$F(x) = \int_{a-h}^{x} f(t)\, dt;$$

moreover, F is a continuous function.

Show that

$$\int_a^b |f_h(x)|\, dx \le \int_a^b |f(x)|\, dx.$$

Solution. We assume first that $f(x) \ge 0$ for all x in [a,b] and consider the function $f(z + t)$ on the rectangle $[a,b] \times [-h,h]$. We see that

$$\int_a^b \int_{-h}^{h} f(z + t)\, dz\, dt = \int_{-h}^{h} \int_a^b f(z + t)\, dt\, dz. \tag{E.32}$$

Since

$$\int_{-h}^{h} f(z + t)\, dz = \int_{t-h}^{t+h} f(x)\, dx = 2hf_h(t),$$

the integral on the left-hand side of (E.32) equals

$$2h \int_a^b f_h(t)\, dt.$$

The integral on the right-hand side of (E.32) can be put into the form

$$\int_{-h}^{h} \int_{a+z}^{b+z} f(x)\, dx$$

and we only need to observe that

$$\int_{a+z}^{b+z} f(x)\, dx \le \int_a^b f(x)\, dx \tag{E.33}$$

in order to obtain the desired result in this special case. Indeed, for $z = 0$, we have equality in (E.33). If, however, $z > 0$ (hence $f \ge 0$), then

$$\int_{a+z}^{b+z} f(x)\, dx = \int_{a+z}^{b} f(x)\, dx \le \int_a^b f(x)\, dx.$$

The case $z < 0$ is handled with equal ease.

We now drop the assumption $f(x) \geq 0$ and denote the Steklow function of $|f|$ by $\overline{f}_h$. We get

$$|\overline{f}_h(x)| = \left| \frac{1}{2h} \int_{x-h}^{x+h} f(t)\, dt \right| \leq \frac{1}{2h} \int_{x-h}^{x+h} |f(t)|\, dt = \overline{f}_h(x)$$

and

$$\int_a^b |f_h(x)|\, dx \leq \int_a^b \overline{f}_h(x)\, dx. \tag{E.34}$$

But, as shown already,

$$\int_a^b \overline{f}_h(x)\, dx \leq \int_a^b |f(x)|\, dx,$$

which together with (E.34) gives the desired result.

PROBLEM 78. Let $N(a)$ denote the number of times the positive integer a occurs as a binomial coefficient, $\binom{n}{k}$. We have $N(1) = \infty$, $N(2) = 1$, $N(3) = N(4) = N(5) = 2$, $N(6) = 3$, etc. Clearly, for $a > 1$, $N(a) < \infty$. Show that $N(a) \leq 2 + 2 \log_2 a$.

Solution. Let b be the first b such that $\binom{2b}{b} > a$. Now

$$\binom{i+j}{i} = \binom{i+j}{j}$$

is monotonically increasing in i and in j; hence

$$\binom{b+i+b+j}{b+i} \geq \binom{b+b+j}{b} \geq \binom{2b}{b} > a$$

for all $i, j \geq 0$. Thus $\binom{i+j}{j} = a$ implies $i < b$ or $j < b$. Again by monotonicity, for each value of i (or j),

$$\binom{i+j}{j} = a$$

has at most one solution. Hence $N(a) < 2b$. Now $\binom{2b}{b} \geq 2^b$, so we have

$$a \geq \binom{2(b-1)}{b-1} \geq 2^{b-1}, \quad \text{i.e., } b \leq \log_2 a + 1 \text{ and } N(a) \leq 2 + 2\log_2 a.$$

PROBLEM 79. Let a_1/b_1, a_2/b_2, ... with $b_1 > 0$, $b_2 > 0$, ... be a strictly increasing sequence. Denoting by

$$A_j = a_1 + a_2 + \cdots + a_j \quad \text{and} \quad B_j = b_1 + b_2 + \cdots + b_j,$$

show that A_1/B_1, A_2/B_2, ... is a strictly increasing sequence.

Solution. For $i = 1,2,\ldots,n$ we have

$$\frac{a_i}{b_i} < \frac{a_{n+1}}{b_{n+1}}, \quad \text{that is,} \quad b_{n+1}a_i < a_{n+1}b_i,$$

whence

$$b_{n+1}A_n < a_{n+1}B_n.$$

Thus

$$\frac{A_{n+1}}{B_{n+1}} - \frac{A_n}{B_n} = \frac{a_{n+1}B_n - b_{n+1}A_n}{B_{n+1}B_n} > 0.$$

PROBLEM 80. Let m, n be positive integers and $a_1, a_2, \ldots, a_n$ be positive real numbers. For $i = 1,2,3,\ldots,$ put $a_{n+i} = a_i$ and

$$b_i = a_{i+1} + a_{i+2} + \cdots + a_{i+m}.$$

Show that

$$m^n a_1 a_2 \cdots a_n < b_1 b_2 \cdots b_n,$$

except if all the a_i are equal.

Solution. The desired inequality is equivalent with

$$\prod_{i=1}^{n} \left(\frac{b_i}{m}\right) \geq a_1 a_2 \cdots a_n.$$

Noting that (see Problem 2)

$$\frac{b_i}{m} = \frac{1}{m} \sum_{k=1}^{m} a_{i+k} \geq \left(\prod_{k=1}^{m} a_{i+k}\right)^{1/m},$$

with equality only if all the a_{i+k} are equal, we have

$$\prod_{i=1}^{n}\left(\frac{b_i}{m}\right) > \prod_{i=1}^{n}\left(\prod_{k=1}^{m} a_{i+k}\right)^{1/m} = \left(\prod_{i=1}^{n} a_i^m\right)^{1/m} = a_1a_2\cdots a_n,$$

unless all the a_i are equal.

Note that for $m = 1$ and any set $\{a_1, a_2, \ldots, a_n\}$, the inequality must be replaced by equality.

PROBLEM 81. Let $a_1, a_2, \ldots, a_n$ be real numbers. Show that

$$\min_{a_i \neq a_j} (a_i - a_j)^2 = M^2(a_1^2 + \cdots + a_n^2), \quad M^2 = \frac{12}{n(n^2 - 1)}.$$

Solution. We may assume that $a_1 \le a_2 \le \cdots \le a_n$ and, by homogeneity, $\Sigma_{i=1}^n a_i^2 = 1$. Now

$$\sum_{1\le i<j\le n} (a_i - a_j)^2 = n \sum_{i=1}^{n} a_i^2 - \left(\sum_{i=1}^{n} a_i\right)^2. \tag{E.35}$$

Assume $a_{i+1} - a_i > M > 0$, $i = 1,2,\ldots,n-1$. Then

$$(a_j - a_j)^2 > (i - j)^2M^2, \quad i,j = 1,2,\ldots,n$$

and

$$\sum_{1\le i<j\le n} (a_i - a_j)^2 > M^2 \cdot \sum_{1\le i<j\le n} (i - j)^2 = M^2\,\frac{n^2(n^2 - 1)}{12} = n.$$

Inserting this in (E.35) we get

$$n < n \sum_{i=1}^{n} a_i^2 - \left(\sum_{i=1}^{n} a_i\right)^2 \le n \sum_{i=1}^{n} a_i^2,$$

or $\Sigma_{i=1}^n a_i^2 > 1$, a contradiction to our normalization. Hence

$$\min_{a_i \neq a_j} (a_i - a_j)^2 \le M^2,$$

as asserted.

PROBLEM 82. Let x and y be non-zero real numbers. Show that

$$|x - y| \geq \frac{1}{4}(|x| + |y|)\left|\frac{|x|}{x} - \frac{|y|}{y}\right|.$$

Solution. We have

$$|x|\left|\frac{|x|}{x} - \frac{|y|}{y}\right| \leq |x|\left|\frac{x}{|x|} - \frac{y}{|x|}\right| + |x|\left|\frac{y}{|x|} - \frac{y}{|y|}\right|$$

$$\leq |x - y| + \frac{|(|y| - |x|)y|}{|y|} \leq |x - y| + ||y| - |x|| \tag{E.36}$$

$$\leq 2|x - y|.$$

Similarly by adding and subtracting $x/|y|$ we obtain

$$|y|\left|\frac{x}{|x|} - \frac{y}{|y|}\right| \leq 2|x - y|. \tag{E.37}$$

The desired result follows from (E.36) and (E.37).

PROBLEM 83. Let x and a be real numbers and n be a nonnegative integer. Show that

$$|x - a|^n |x + na| \leq (x^2 + na^2)^{(n+1)/2}.$$

Solution. We have for $A = (x - a)^2$ and $B = (x + na)^2$ that

$$A^{n/(n+1)} B^{1/(n+1)} \leq \frac{nA + B}{n + 1} = x^2 + na^2$$

from Problem 2.

PROBLEM 84. Given an arbitrary finite set of n pairs of positive numbers $\{(a_i,b_i): i = 1,2,\ldots,n\}$, show that

$$\prod_{i=1}^{n} [xa_i + (1 - x)b_i] \leq \max\left\{\prod_{i=1}^{n} a_i, \prod_{i=1}^{n} b_i\right\}$$

for all $x \in [0,1]$, with equality attained only at $x = 0$ or $x = 1$, if and only if

$$\left(\sum_{i=1}^{n} \frac{a_i - b_i}{a_i}\right)\left(\sum_{i=1}^{n} \frac{a_i - b_i}{b_i}\right) \geq 0.$$

Solution. Let

$$f(x) = \prod_{i=1}^{n} [xa_i + (1 - x)b_i], \quad g(x) = \log f(x).$$

Then

$$g'(x) = \sum_{i=1}^{n} \frac{a_i - b_i}{xa_i + (1 - x)b_i}, \quad g''(x) = -\sum_{i=1}^{n} \left(\frac{a_i - b_i}{xa_i + (1 - x)b_i}\right)^2.$$

Since $g''(x) < 0$ for all $x \in [0,1]$, the maximum of $g(x)$ [hence of $f(x)$] is attained at $x = 0$ or $x = 1$ if and only if $g'(0)$ and $g'(1)$ do not differ in sign, that is, $g'(0)g'(1) \geq 0$. Since

$$f(0) = \prod_{i=1}^{n} a_i, \quad f(1) = \prod_{i=1}^{n} b_i,$$

$$g'(0) = \sum_{i=1}^{n} \frac{a_i - b_i}{a_i}, \quad \text{and} \quad g'(1) = \sum_{i=1}^{n} \frac{a_i - b_i}{b_i},$$

the assertion is proved.

PROBLEM 85. Show that if m and n are positive integers, the smaller of the numbers $\sqrt[n]{m}$ and $\sqrt[m]{n}$ cannot exceed $\sqrt[3]{3}$.

Solution. Let $f(x) = x^{1/x}$. Elementary calculus shows that $f(x) \to 0$ as $x \to 0$, $f(x) \to 1$ as $x \to \infty$, f is increasing in $[0,e]$, and f is decreasing in $[e,\infty)$. Hence

$$C = \sup\{f(k) : k = 1,2,3,\ldots\}$$

$$= \max\{f(2), f(3)\}.$$

Since $3^2 > 2^3$, $f(3) > f(2)$, whence $C = 3^{1/3}$. Then $f(m) \leq 3^{1/3}$ for all positive integers m. If $m \leq n$, then $m^{1/n} \leq m^{1/m} \leq 3^{1/3}$, whence $\min\{m^{1/n}, n^{1/m}\} \leq 3^{1/3}$.

PROBLEM 86. Show that if $a \geq 2$ and $x > 0$, then $a^x + a^{1/x} \leq a^{x+1/x}$, equality holding if, and only if, $a = 2$ and $x = 1$.

Solution. Let $f(x,a) = (a^x + a^{1/x})/a^{x+1/x}$ for $a \geq 2$ and $x > 0$. We observe that $f(x,a) = f(1/x,a)$ and hence the problem will be solved if we show that $f(x,a) \le 1$ for $x \ge 1$ with equality if and only if $a = 2$ and $x = 1$. Next we note that $f(x,a)$ is a strictly decreasing function of a for each x and, hence, that it is enough to show that $f(x,2) < 1$ for $x > 1$, it being clear that $f(1,2) = 1$. We will obtain this latter inequality by verifying that

$$F(x) = 2^x f(x,2) - 2^x < 0 \quad \text{for } x > 1.$$

It is easily seen that

$$F'(x) = 2^{x-1/x}(\log 2)(1 + 1/x^2 - 2^{1/x}) = 2^{x-1/x}(\log 2)g(1/x).$$

The proof will be complete if we can show that $g(1/x) < 0$ for $x > 1$. Now

$$g''(t) = 2 - 2^t(\log 2)^2 > 2 - 2(\log 2)^2 > 0$$

for $0 < t < 1$, so that $g(t)$ is strictly convex for these t. Since $g(t)$ is continuous and $g(0) = g(1) = 0$, it follows that $g(t) < 0$ for $0 < t < 1$ or that $g(1/x) < 0$ for $x > 1$.

PROBLEM 87. Show that if $x_i \ge 0$ for $i = 1,2,\ldots,n$ and $\Sigma_{i=1}^{n} 1/(1 + x_i) \le 1$, then $\Sigma_{i=1}^{n} 2^{-x_i} \le 1$.

Solution. The result is obvious for $n = 1$; while for $n = 2$, $x_1,x_2 \ge 0$ and $(1 + x_1)^{-1} + (1 + x_2)^{-1} \le 1$ imply $x_1x_2 \ge 1$ and $x_1,x_2 > 0$. But then

$$2^{-x_1} + 2^{-x_2} \le 2^{-x_1} + 2^{-1/x_1},$$

and the latter is no greater than 1 by Problem 86.

The proof proceeds by induction. $\Sigma_{i=1}^{n} (1 + x_i)^{-1} \le 1$ implies that

$$(1 + z)^{-1} + \sum_{i=3}^{n} (1 + x_i)^{-1} \le 1,$$

where z is defined by $(1 + z)^{-1} = (1 + x_1)^{-1} + (1 + x_2)^{-1} \le 1$, and hence $z \ge 0$. By the inductive assumption,

$$2^{-z} + \sum_{i=3}^{n} 2^{-x_i} \le 1.$$

It remains to show that

$$2^{-x_1} + 2^{-x_2} \le 2^{-z} \quad \text{or} \quad 2^{-y_1} + 2^{-y_2} \le 1,$$

where $y_i = x_i - z$ with $i = 1, 2$. This, however, follows from the case for $n = 2$ upon verifying that $y_1, y_2 \ge 0$ and $(1 + y_1)^{-1} + (1 + y_2)^{-1} \le 1$.

PROBLEM 88. If x and y are positive, show that $x^y + y^x > 1$.

Solution. The inequality is trivially true if either x or $y \ge 1$. Let $0 < x, y < 1$, and put $y = kx$. Because of the symmetry we consider only $0 < k \le 1$. Now

$$f(x) = x^{kx} + (kx)^x = (x^x)^k + k^x x^x \ge a^k + ka,$$

where $a = \exp(-1/e) = \min(x^x)$ and $k^x \ge k$. But $F(k) = a^k + ka$ has a unique minimum at $k_0 = 1 - e < 0$ and is increasing for $k > k_0$. Since $F(0) = 1$, $F(1) = 2a > 1$, $f(x) > 1$.

PROBLEM 89. Let $0 \le a_i < 1$, $i = 1,\dots,n$ and put $\Sigma_{i=1}^n a_i = A$. Show that

$$\sum_{i=1}^{n} \frac{a_i}{1 - a_i} \ge \frac{nA}{n - A},$$

equality occuring only if all the a_i are equal.

Solution. Let $b_i = 1 - a_i$, $B = \Sigma_{i=1}^n b_i$, $B_{ij} = b_i/b_j + b_j/b_i$. Then

$$B_{ij} = \frac{b_i^2 + b_j^2}{b_i b_j} \ge 2,$$

$$\sum_{i=1}^{n} \frac{B}{b_i} = n + \sum_{i=1}^{n-1} \sum_{j=i+1}^{n} B_{ij} \quad n + 2 \sum_{i=1}^{n-1} (n - i) = n^2,$$

whence, after dividing by B,

$$\sum_{i=1}^{n} \left(\frac{1}{b_i} - 1\right) \ge \frac{n(n - B)}{B},$$

which is the required inequality for $a_i < 1$. Equality holds only if every $B_{ij} = 2$, only if all the b_i are equal, only if all the a_i are equal.

PROBLEM 90. Show that for $n \geq 2$,

$$\prod_{i=0}^{n} \binom{n}{i} \leq \{(2^n - 2)/(n - 1)\}^{n-1}.$$

Solution. Since

$$\prod_{i=0}^{n} \binom{n}{i} = \prod_{i=1}^{n-1} \binom{n}{i}, \qquad 2^n - 2 = \sum_{i=1}^{n-1} \binom{n}{i},$$

the desired inequality is merely the arithmetic-geometric means inequality (see Problem 2) for the $n - 1$ numbers $\binom{n}{i}$, $i = 1,2,\ldots,n-1$.

PROBLEM 91. Let $b_1, \ldots, b_n$ be any rearrangement of the positive numbers $a_1, \ldots, a_n$. Show that $a_1/b_1 + \cdots + a_n/b_n \geq n$.

Solution. Since the arithmetic mean of n positive numbers is at least as large as their geometric mean (see Problem 2), we have

$$\frac{1}{n} \sum_{i=1}^{n} \frac{a_i}{b_i} \geq \left(\prod_{i=1}^{n} \frac{a_i}{b_i} \right)^{1/n}.$$

But, by the conditions of the problem, $\prod_{i=1}^{n} a_i = \prod_{i=1}^{n} b_i$. Thus

$$\sum_{i=1}^{n} \frac{a_i}{b_i} \geq n.$$

PROBLEM 92. Given that $\Sigma_{i=1}^{n} b_i = b$ with each b_i a nonnegative number. Show that

$$\sum_{j=1}^{n-1} b_j b_{j+1} \leq \frac{b^2}{4}.$$

Solution. Let b_k denote the largest of the numbers b_i, where $i = 1,2, \ldots,n$. Then

$$\sum_{i=1}^{n-1} b_i b_{i+1} = \sum_{i=1}^{k-1} b_i b_{i+1} + \sum_{i=1}^{n-1} b_i b_{i+1} \leq b_k \sum_{i=1}^{k-1} b_i + b_k \sum_{i=k}^{n-1} b_{i+1}$$

$$= b_k(b - b_k) = \frac{b^2}{4} - \left(\frac{b}{2} - b_k\right)^2 \leq \frac{b^2}{4}.$$

Note that there is equality if and only if $b_k = b/2$, in which case $n = 2$. Thus for $n \geq 3$, $\Sigma_{i=1}^{n-1} b_i b_{i+1} < b^2/4$.

PROBLEM 93. Let $n \geq 2$ and $0 < x_1 < x_2 < \cdots < x_n \leq 1$. Show that

$$\frac{n \sum_{k=1}^{n} x_k}{\sum_{k=1}^{n} x_k + n x_1 x_2 \cdots x_n} \geq \sum_{k=1}^{n} \frac{1}{1 + x_k}.$$

Solution. By Problem 7, if $a_k > 0$ for $i = 1,\ldots,n$, then

$$\sum_{k=1}^{n} a_k \cdot \sum_{k=1}^{n} \frac{1}{a_k} \geq n^2$$

and thus

$$\sum_{k=1}^{n} a_k \cdot \sum_{k=1}^{n} a_k^{-1}(1 - a_k) \geq n \sum_{k=1}^{n} (1 - a_k).$$

Setting $a_k = x_k(1 + x_k)^{-1}$, we obtain next

$$\sum_{k=1}^{n} x_k^{-1} \cdot \left(n - \sum_{k=1}^{n} (1 + x_k)^{-1}\right) \geq n \sum_{k=1}^{n} (1 + x_k)^{-1}.$$

Multiplying both sides of this result by the product $x_1 x_2 \cdots x_n$ and observing that for $n \geq 2$,

$$\sum_{k=1}^{n} x_k \geq x_1 x_2 \cdots x_n \cdot \sum_{k=1}^{n} x_k^{-1},$$

we obtain the proposed inequality, after rearrangement of terms.

PROBLEM 94. Show the following inequality, due to Kantorovich: Let f be a continuous function on the interval $[0,1]$ such that $0 < m \leq f(x) \leq M$. Then

$$\left(\int_0^1 \frac{dx}{f(x)}\right)\left(\int_0^1 f(x)\, dx\right) \leq \frac{(m + M)^2}{4mM}. \tag{E.38}$$

Solution. Since

$$\frac{\{f(x) - m\}\{f(x) - M\}}{f(x)} \le 0 \quad \text{for } 0 \le x \le 1,$$

we obtain by integrating $f - (m + M) + (mM)/f$ over $[0,1]$

$$\int_0^1 f(x)\,dx + mM\int_0^1 \frac{dx}{f(x)} \le m + M.$$

Putting

$$u = mM\int_0^1 \frac{dx}{f(x)},$$

we obtain

$$u\int_0^1 f(x)\,dx \le (m + M)u - u^2 \le \frac{(m + M)^2}{4}.$$

Remarks. In Problem 94 the unit interval $[0,1]$ can of course be replaced by any interval of length 1 and in place of requiring f to be continuous it would have been sufficient to assume that f be integrable. If in Problem 94 the unit interval $[0,1]$ is replaced by the interval $[a,b]$, then (E.38) changes into

$$\left(\int_a^b f(x)\,dx\right)\left(\int_a^b \frac{dx}{f(x)}\right) \le \frac{(m + M)^2}{4mM}(b - a)^2.$$

PROBLEM 95. Let $x > 0$ and $x \ne 1$. Show the following inequalities due to Karamata:

$$\text{(i)} \quad \frac{\log x}{x - 1} \le \frac{1}{\sqrt{x}}, \tag{E.39}$$

$$\text{(ii)} \quad \frac{\log x}{x - 1} \le \frac{1 + x^{1/3}}{x + x^{1/3}}. \tag{E.40}$$

Solution. We start with the expansion

$$\log\frac{1 + t}{1 - t} = 2\left(t + \frac{1}{3}t^3 + \frac{1}{5}t^5 + \frac{1}{7}t^7 + \cdots\right),$$

which holds if $-1 < t < 1$. To prove Part (i), we put

$$x = \frac{(1 + t)^2}{(1 - t)^2}$$

and note that (E.39) becomes

$$\frac{1}{t} \log \frac{1 + t}{1 - t} - \frac{1}{1 - t^2} \leq 0 \quad \text{for } 0 < |t| < 1.$$

Using the power series expansion, we get

$$\sum_{n=1}^{\infty} \left(1 - \frac{1}{2n + 1}\right) t^{2n} \geq 0 \quad \text{for } 0 \leq |t| < 1,$$

which is evidently true.

To prove Part (ii), we set

$$x = \frac{(1 + t)^3}{(1 - t)^3}$$

and observe that (E.40) becomes

$$\frac{3}{2t} \log \frac{1 + t}{1 - t} - \frac{t^2 + 3}{1 - t^4} \leq 0 \quad \text{for } 0 < |t| < 1.$$

Using power series expansions, we get

$$\sum_{n=0}^{\infty} t^{4n+2} + 3 \sum_{n=0}^{\infty} t^{4n} - 3 \sum_{n=0}^{\infty} \frac{t^{2n}}{2n + 1} \geq 0 \quad \text{for } |t| < 1,$$

that is,

$$3 \sum_{n=0}^{\infty} \left(1 - \frac{1}{4n + 1}\right) t^{4n} + \sum_{n=1}^{\infty} \left(1 - \frac{3}{4n + 3}\right) t^{4n+2} \geq 0,$$

which is obviously true.

Remarks. Another way of establishing the inequality (E.40) runs as follows: For $u \geq 1$ we have $0 \leq (u - 1)^3(u^3 - 1)$. Adding $3u^5 + 6u^3 + 3u$ on both sides we obtain $3u(1 + 2u^2 + u^4) \leq u^6 + 3u^4 + 4u^3 + 3u^2 + 1$. Dividing by the positive number $u^2(1 + u^2)^2$, we get

$$\frac{3}{u} \leq \frac{u^6 + 3u^4 + 4u^3 + 3u^2 + 1}{u^2(1 + u^2)^2}.$$

Letting

$$f(u) = 3 \log u, \quad g(u) = \frac{(u^2 - 1)(u^2 + u + 1)}{u(1 + u^2)},$$

we therefore see that $f'(u) \le g'(u)$. Since $f(1) = g(1)$, this leads to

$$f(u) \le g(u) \quad \text{for all } u \ge 1.$$

Hence, letting $u = x^{1/3}$ with $x > 1$, we obtain

$$\frac{\log x}{x - 1} \le \frac{1 + x^{1/3}}{x + x^{1/3}} \quad \text{for } x > 1.$$

Applying the transformation $x \to 1/x$, the proof is complete.

PROBLEM 96. Let $0 < y < x$. Show that

$$\frac{x + y}{2} > \frac{x - y}{\log x - \log y}. \tag{E.41}$$

Solution. We integrate the inequality

$$\frac{1}{2t} > \frac{2}{(1 + t)^2} \quad (t > 1),$$

and we get

$$\frac{1}{2}\int_1^{x/y} \frac{dt}{t} > 2\int_1^{x/y} \frac{dt}{(1 + t)^2},$$

that is,

$$\frac{1}{2} \log \frac{x}{y} > \frac{x - y}{x + y} \quad \text{for } 0 < y < x.$$

This inequality is equivalent with (E.41).

PROBLEM 97. Let $x > 0$. Show that

$$\frac{2}{2x + 1} < \log\left(1 + \frac{1}{x}\right) < \frac{1}{\sqrt{x^2 + x}}.$$

Solution. Let

$$f(x) = \frac{2}{2x + 1} - \log\left(1 + \frac{1}{x}\right) \quad \text{for } x > 0.$$

Since

$$f'(x) = \frac{1}{x(1 + x)(2x + 1)^2} > 0 \quad \text{for } x > 0,$$

f is an increasing function. From this fact and since $\lim_{x \to \infty} f(x) = 0$, we conclude that $f(x) < 0$ for $x > 0$.

Let

$$g(x) = \log\left(1 + \frac{1}{x}\right) - \frac{1}{\sqrt{x^2 + x}} \quad \text{for } x > 0.$$

Since

$$g'(x) = \frac{2x + 1}{2(x^2 + x)^{3/2}} - \frac{1}{x(1 + x)} > 0 \quad \text{for } x > 0,$$

we see that g is an increasing function. Since, moreover, $\lim_{x \to \infty} g(x) = 0$, we obtain that $g(x) < 0$ for $x > 0$.

Remarks. If in the second inequality of (E.42), for $x > 1$, we replace x by $1/(x - 1)$, and for $0 < x < 1$ we replace x by $x/(1 - x)$, we obtain (E.39) of Problem 95.

Replacing $1 + (1/x)$ by x/y with $x > y > 0$ in the first inequality of (E.42), we get (E.41) of Problem 96.

PROBLEM 98. Let F be a positive-valued, continuous, and decreasing function over the interval $[a,b]$. Show that

$$\frac{\int_a^b x\{F(x)\}^2\, dx}{\int_a^b x\, F(x)\, dx} \le \frac{\int_a^b \{F(x)\}^2\, dx}{\int_a^b F(x)\, dx}.$$

Solution. For $p(x) = \{F(x)\}^2$, $f(x) = x$, $g(x) = 1/F(x)$, the result in Problem 75 yields the desired inequality.

PROBLEM 99. Let $S_n = 1 + 1/2 + 1/3 + \cdots + 1/n$. Show that

$$n\{(1 + n)^{1/n} - 1\} < S_n < n\left\{1 - (n + 1)^{-1/n} - \frac{1}{n + 1}\right\}.$$

Solution. By Problem 2,

$$\frac{2}{1} + \frac{3}{2} + \frac{4}{3} + \cdots + \frac{n+1}{n} > (n+1)^{1/n}.$$

But

$$\frac{1}{1} + \frac{2}{2} + \frac{3}{3} + \cdots + \frac{n}{n} = n$$

and so

$$S_n > n\{(1+n)^{1/n} - 1\}.$$

Again by Problem 2,

$$\frac{1}{2} + \frac{2}{3} + \frac{3}{4} + \cdots + \frac{n}{n+1} > n(n+1)^{-1/n}.$$

Noting that

$$\frac{2}{2} + \frac{3}{3} + \frac{4}{4} + \cdots + \frac{n+1}{n+1} = n,$$

we see that

$$S_n < n\{1 - (n+1)^{-1/n} - \frac{1}{n+1}\}.$$

PROBLEM 100. Let $x > 0$ and $y > 0$. Show that

$$\frac{1 - e^{-x-y}}{(x+y)(1-e^{-x})(1-e^{-y})} - \frac{1}{xy} \le \frac{1}{12}.$$

Solution. The foregoing inequality is equivalent with

$$\frac{1}{1-e^{-x}} + \frac{1}{1-e^{-y}} - 1 \le \frac{1}{x} + \frac{1}{y} + \frac{x+y}{12},$$

that is,

$$f(x) + f(y) \le 1,$$

where

$$f(x) = \frac{1}{1-e^{-x}} - \frac{1}{x} - \frac{x}{12}.$$

Since $f(x) \le 1/2$, the desired result follows.

PROBLEM 101. Let A, B, C, D, E, and F be nonnegative real numbers which satisfy

$$A + B \le E \quad \text{and} \quad C + D \le F. \tag{E.42}$$

Show that

$$\sqrt{AC} + \sqrt{BD} \le \sqrt{EF}.$$

Since one may interchange C and D in (E.42), another valid inequality is

$$\sqrt{AD} + \sqrt{BC} \le \sqrt{EF}.$$

Solution. Multiplying the inequalities in (E.42), we obtain AC + BD + (AD + BC) $\le$ EF. But $2(AD \cdot BC)^{\frac{1}{2}} \le AD + BC$ (by Problem 2), and so

$$(\sqrt{AC} + \sqrt{BD})^2 = AC + BD + 2\sqrt{AD \cdot BC} \le AC + BD + (AD + BC) \le EF.$$

PROBLEM 102. Show that, for $x > 0$ and $x \ne 1$,

$$0 \le \frac{x \log x}{x^2 - 1} \le \frac{1}{2}.$$

Solution. Consider the function

$$f(x) = 2 \log x - \frac{x^2 - 1}{x}.$$

Since

$$f'(x) = \frac{2}{x} - 1 - \frac{1}{x^2} = -\left(\frac{x - 1}{x}\right)^2 \le 0,$$

the function f is decreasing and, consequently,

$$2 \log x - \frac{x^2 - 1}{x} \begin{cases} < 0 & \text{for } x > 1, \\ > 0 & \text{for } 0 < x < 1. \end{cases}$$

Hence $(x \log x)/(x^2 - 1) \le 1/2$. The inequality $0 \le (x \log x)/(x^2 - 1)$ is obvious, because for $x > 0$, the sign of $x \log x$ is the same as the sign of $x^2 - 1$. This completes the proof of the inequalities posed. We note the following direct consequence: For $m \ge 1$,

$$\int_0^1 \frac{x^m \log x}{x^2 - 1}\, dx = \int_0^1 x^{m-1} \frac{x \log x}{x^2 - 1}\, dx \le \frac{1}{2} \int_0^1 x^{m-1}\, dx = \frac{1}{2m}.$$

PROBLEM 103. Show that, for $x > 0$,

$$x(2 + \cos x) > 3 \sin x. \tag{E.43}$$

Solution. Since $2 + \cos x > 0$ for all x, we may write (E.43) in the form

$$f(x) = x - \frac{3x}{2 + \cos x} > 0.$$

But

$$f'(x) = \left(\frac{1 - \cos x}{2 + \cos x}\right)^2 \geq 0$$

and so f(x) is seen to increase as x increases. Moreover, $f(0) = 0$. Hence the desired result.

PROBLEM 104. Show that, for $0 < x < \pi/2$,

$$2 \sin x + \tan x > 3x.$$

Solution. Since $2x^3 - 3x^2 + 1 = (x - 1)^2(2x + 1)$, we see that

$$2x^3 - 3x^2 + 1 \begin{cases} > 0 & \text{for } x > -\tfrac{1}{2} \text{ and } x \neq 1, \\ < 0 & \text{for } x < -\tfrac{1}{2}. \end{cases} \tag{E.44}$$

As $0 < \cos t < 1$ for $0 < t < \pi/2$, we get by (E.44):

$$2 \cos^3 t - 3 \cos^2 t + 1 > 0 \quad \text{for } 0 < t < \pi/2,$$

or

$$2 \cos t + \sec^2 t - 3 > 0 \quad \text{for } 0 < t < \pi/2. \tag{E.45}$$

Integrating the inequality (E.45) between the limits 0 and x we obtain

$$2 \sin x + \tan x - 3x > 0 \quad \text{for } 0 < x < \pi/2,$$

which is the desired result.

Remarks. The result in Problem 104 is due to Huygens. Another way of obtaining the result in Problem 104 is to set

$$f(t) = 2 \sin t + \tan t - 3t \quad \text{for } 0 < t < \pi/2$$

and to note that, for $0 < t < \pi/2$,

$$f'(t) = 2\cos t + \sec^2 t - 3$$

$$= (\sec^2 t)(\cos t - 1)^2(2\cos t + 1) > 0.$$

Integration of f' over the interval $(0,x)$, where $0 < x < \pi/2$, gives the desired result.

In an entirely similar manner we can show that

$$2\sinh u + \tanh u > 3u \quad \text{for } u > 0.$$

In Problem 95, Part (ii), we proved that

$$\frac{\log x}{x - 1} \le \frac{1 + x^{1/3}}{x + x^{1/3}} \quad \text{for } x > 0 \text{ and } x \neq 1.$$

Putting $x = e^{3u}$, the foregoing inequality becomes

$$\frac{3u}{e^{3u} - 1} \le \frac{1 + e^u}{e^{3u} + e^u} \tag{E.46}$$

which is invariant under the transformation $u \to -u$. Assuming that $u > 0$, (E.46) can be reworked to

$$2\sinh u + \tanh u \ge 3u.$$

PROBLEM 105. Let $x > 0$, $x \neq 1$ and suppose that n is a positive integer. Show that

$$x + \frac{1}{x^n} > 2n\,\frac{x - 1}{x^n - 1}.$$

Solution. The inequality in question transforms into

$$\frac{(x^{n+1} + 1)(x^n - 1)}{x - 1} > 2n\, x^n. \tag{E.47}$$

The identity

$$\frac{(x^{n+1} + 1)(x^n - 1)}{x - 1} = (x^{n+1} + 1)(x^{n-1} + x^{n-2} + \cdots + 1)$$

$$= x^n \sum_{k=1}^{n} (x^k + x^{-k})$$

together with

$$x^k + x^{-k} > 2$$

implies (E.47).

PROBLEM 106. Let a be a fixed real number such that $0 \le a < 1$ and k be a positive integer satisfying the condition $k > (3 + a)/(1 - a)$. Show that

$$S_k(n) = \frac{1}{n} + \frac{1}{n + 1} + \cdots + \frac{1}{nk - 1} > 1 + a$$

for any positive integer n.

Solution. Since, for $a,b > 0$ and $a \neq b$,

$$\frac{1}{a} + \frac{1}{b} \quad \frac{4}{a + b},$$

we get

$$2\, S_k(n) = \left(\frac{1}{n} + \frac{1}{nk - 1}\right) + \left(\frac{1}{n + 1} + \frac{1}{nk - 2}\right)$$

$$+ \cdots + \left(\frac{1}{nk - 1} + \frac{1}{n}\right)$$

$$> n(k - 1)\frac{4}{n(k + 1) - 1}$$

$$= \frac{4(k - 1)}{k + 1 - 1/n}$$

$$> \frac{n(k - 1)}{k + 1}.$$

The stated inequality holds for every positive integer n if

$$\frac{2(k - 1)}{k + 1} > 1 + a, \quad \text{that is,} \quad k > \frac{3 + a}{1 - a} \quad \text{for } 0 \le a < 1.$$

PROBLEM 107. Let a and b denote real numbers and r satisfy $r \ge 0$. Show that

$$|a + b|^r \le c_r(|a|^r + |b|^r), \tag{E.48}$$

where $c_r = 1$ for $r \le 1$ and $c_r = 2^{r-1}$ for $r > 1$.

Solution. Since $x \to f(x) = |x|^r$ $(r > 1)$ is a convex function, we obtain

$$\left|\frac{a + b}{2}\right|^r \le \frac{1}{2}(|a|^r + |b|^r),$$

whence

$$|a + b|^r \le 2^{r-1}(|a|^r + |b|^r).$$

For $r = 1$, the inequality (E.48) becomes $|a + b| \le |a| + |b|$.

Now, let $0 \le r < 1$. If a and b have opposite signs, the result is evidently true. Otherwise, let $t = b/a$ with $a \ne 0$. Then (E.48) becomes

$$(1 + t)^r \le 1 + t^r \quad (0 \le r < 1).$$

For $r = 0$, this result is trivially true. Consider the function

$$g(t) = (1 + t)^r - 1 - t^r \quad (0 < r < 1),$$

which vanishes at $t = 0$ and decreases as t increases. This yields (E.48) for $a \ne 0$. (E.48) also holds when $a = 0$.

PROBLEM 108. Let $0 < b \le a$. Show that

$$\frac{1}{8}\frac{(a - b)^2}{a} \le \frac{a + b}{2} - \sqrt{ab} \le \frac{1}{8}\frac{(a - b)^2}{b}.$$

Solution. The claimed inequality is trivially true in case $0 < b = a$. If $0 < b < a$, then, by Problem 2,

$$\frac{a + b}{2} > \sqrt{ab}.$$

Thus, if $0 < b < a$, then

$$0 < a + b - 2\sqrt{ab} \quad \text{and} \quad 0 > 2\sqrt{ab} - (a + b).$$

This means that

$$\frac{a - b}{2\sqrt{a}} < \sqrt{a} - \sqrt{b} < \frac{a - b}{2\sqrt{b}} \quad (0 < b < a)$$

or

$$\frac{(a - b)^2}{4a} < (\sqrt{a} - \sqrt{b})^2 = a + b - 2\sqrt{ab} < \frac{(a - b)^2}{4b} \quad (0 < b < a).$$

This completes the proof.

Remark. If $0 < b < a$, then we can easily see that

$$\frac{a + b}{2} > \frac{a + \sqrt{ab} + b}{3} > \sqrt{ab}.$$

The middle term in the foregoing inequality is sometimes referred to as the heronian mean of the positive numbers a and b.

PROBLEM 109. Consider any sequence $(a_n)_{n=1}^{\infty}$ of real numbers. Show that

$$\sum_{n=1}^{\infty} a_n \le \frac{2}{\sqrt{3}} \sum_{n=1}^{\infty} (r_n/n)^{1/2},$$

where

$$r_n = \sum_{k=n}^{\infty} a_k^2.$$

Solution. First we observe that

$$\sum_{n=k}^{\infty} \frac{1}{n^2(n + 1)^2} = \sum_{n=k}^{\infty} \left(\frac{1}{n} - \frac{1}{n + 1}\right) = \sum_{n=k}^{\infty} \left\{\int_n^{n+1} \frac{dx}{x^2}\right\}^2$$

$$\le \sum_{n=k}^{\infty} \int_n^{n+1} \frac{dx}{x^4} = \int_k^{\infty} \frac{dx}{x^4} = \frac{1}{3k^3}.$$

Using this, we have by Cauchy's Inequality

$$\sum_{n=1}^{\infty} a_n = 2\sum_{n=1}^{\infty} \frac{a_n}{n(n + 1)} \sum_{k=1}^{n} k = 2\sum_{k=1}^{\infty} k \sum_{n=k}^{\infty} \frac{a_n}{n(n + 1)}$$

$$\le 2\sum_{k=1}^{\infty} k \left\{\sum_{n=k}^{\infty} a_n^2\right\}^{1/2} \left\{\sum_{n=k}^{\infty} \frac{1}{n^2(n + 1)^2}\right\}^{1/2}$$

$$\le \frac{2}{\sqrt{3}} \sum_{k=1}^{\infty} (r_k/k)^{1/2}.$$

PROBLEM 110. Show that, for n = 1,2,3,...,

$$\left(\frac{n}{e}\right)^n < n! < e\left(\frac{n}{2}\right)^n,$$

where $\lim_{n\to\infty} \left(1 + \frac{1}{n}\right)^n = e.$

Solution. The inequality to be proved is certainly true for $n = 1,2$. Proceeding by induction, we assume that

$$\left(\frac{n}{e}\right)^n < n! < e\left(\frac{n}{2}\right)^n$$

holds for $n = 1,2,\ldots,k$ and show that the inequality is true for $n = k + 1$ as well. Thus, let

$$\left(\frac{k}{e}\right)^k < k! < e\left(\frac{k}{2}\right)^k;$$

we wish to establish that

$$\left(\frac{k + 1}{e}\right)^{k+1} < (k + 1)! < e\left(\frac{k + 1}{2}\right)^{k+1},$$

that is,

$$\frac{1}{e}\left(\frac{k + 1}{e}\right)^k < k! < \frac{e}{2}\left(\frac{k + 1}{2}\right)^k.$$

But

$$\frac{\frac{1}{e}\left(\frac{k + 1}{e}\right)^k}{\left(\frac{k}{e}\right)^k} = \frac{1}{e}\left(1 + \frac{1}{k}\right)^k < 1, \qquad \frac{\frac{e}{2}\left(\frac{k + 1}{2}\right)^k}{e\left(\frac{k}{2}\right)^k} = \frac{1}{2}\left(1 + \frac{1}{k}\right)^k > 1$$

because, for $k = 2,3,\ldots$,

$$2 < \left(1 + \frac{1}{k}\right)^k < \left(1 + \frac{1}{k + 1}\right)^{k+1} < e.$$

Remarks. Let $x_1 = 1$, $x_2 = x_3 = \cdots = x_{k+1} = 1 + 1/k$. By Problem 2,

$$x_1x_2 \cdots x_{k+1} < \left(\frac{x_1 + x_2 + \cdots + x_{k+1}}{k + 1}\right)^{k+1};$$

hence

$$\left(1 + \frac{1}{k}\right)^k < \left(\frac{1 + (k + 1)}{k + 1}\right)^{k+1} = \left(1 + \frac{1}{k + 1}\right)^{k+1}.$$

In a similar way we can show that

$$\left(1 + \frac{1}{k + 1}\right)^{k+2} < \left(1 + \frac{1}{k}\right)^{k+1}$$

by Problem 2. Indeed, let $x_1 = 1$, $x_2 = x_3 = \cdots = x_{k+2} = k/(k + 1)$. These $(k + 2)$ numbers have an arithmetic mean of $(k + 1)/(k + 2)$ and a geometric mean of $[k/(k + 1)]^{(k+1)/(k+2)}$. Hence

$$\frac{k+1}{k+2} > \left(\frac{k}{k+1}\right)^{(k+1)/(k+2)}.$$

On taking reciprocals this becomes

$$1 + \frac{1}{k+1} < \left(1 + \frac{1}{k}\right)^{(k+1)/(k+2)} \quad \text{or} \quad \left(1 + \frac{1}{k+1}\right)^{k+2} < \left(1 + \frac{1}{k}\right)^{k+1}.$$

If we consider the elementary inequalities

$$\left(1 + \frac{1}{k}\right)^{k} < e < \left(1 + \frac{1}{k}\right)^{k+1}$$

for $k = 1,2,\ldots,n - 1$, and multiply them together, we get

$$\frac{n^{n-1}}{(n-1)!} < e^{n-1} < \frac{n^{n}}{(n-1)!}.$$

This leads to the approximation

$$e\, n^{n}\, e^{-n} < n! < e\, n^{n+1}\, e^{-n}.$$

PROBLEM 111. Show that

$$1 + \frac{1}{\sqrt{2}} + \frac{1}{\sqrt{3}} + \cdots + \frac{1}{\sqrt{n}} < 2\sqrt{n} - 1.$$

Solution. We have

$$1 + \frac{1}{\sqrt{2}} + \frac{1}{\sqrt{3}} + \cdots + \frac{1}{\sqrt{n}} < 1 + \int_1^n \frac{1}{\sqrt{x}}\, dx = 2\sqrt{n} - 1.$$

PROBLEM 112. Let a, b, and x be real numbers such that $0 < a < b$ and $0 < x < 1$. Show that

$$\left(\frac{1 - x^{b}}{1 - x^{a+b}}\right)^{b} > \left(\frac{1 - x^{a}}{1 - x^{a+b}}\right)^{a}. \tag{E.49}$$

Solution. We can write (E.49) in the form

$$b \log(1 - x^{b}) - a \log(1 - x^{a}) - (b - a) \log(1 - x^{a+b}) > 0.$$

Using the series representation of $\log(1 + x)$, this inequality becomes

$$\sum_{k=1}^{\infty} \frac{(b-a)x^{(a+b)k} + ax^{ak} - bx^{bk}}{k} > 0.$$

Hence it will be enough to show that

$$(b - a)t^{b+a} + at^a - bt^b > 0 \quad (0 < t < 1),$$

that is,

$$\frac{1 - t^b}{1 - t^a} > \frac{b}{a} t^{b-a} \quad (0 < t < 1).$$

But the last inequality is a simple consequence of Cauchy's form of the Mean-Value Theorem of differential calculus:

$$\frac{1 - t^b}{1 - t^a} = \frac{b\theta^{b-1}}{a\theta^{a-1}} = \frac{b}{a} \theta^{b-a} > \frac{b}{a} t^{b-a} \quad (t < \theta < 1).$$

PROBLEM 113. Let $0 < a < 1$. Show that

$$\frac{2}{e} < a^{\frac{a}{1-a}} + a^{\frac{1}{1-a}} < 1. \tag{E.50}$$

Solution. Putting

$$f(a) = (1 + a)a^{\frac{a}{1-a}} \quad (0 < a < 1),$$

we get

$$\log f(a) = \log(1 + a) + \frac{a}{1 - a} \log a,$$

$$\frac{f'(a)}{f(a)} = \frac{1}{(1 - a)^2}\left(\log a + 2 \frac{1 - a}{1 + a}\right)$$

and, setting $g(a) = \log a + 2(1 - a)/(1 + a)$,

$$g'(a) = \frac{1}{a}\left(\frac{1 - a}{1 + a}\right)^2.$$

We conclude that

$$g(a) < g(1) = 0, \quad f'(a) < 0,$$

$$\lim_{a \to 1-} f(a) < f(a) < \lim_{a \to 0+} f(a).$$

But

$$\lim_{a \to 1-} f(a) = \frac{2}{e} \quad \text{and} \quad \lim_{a \to 0+} f(a) = 1.$$

PROBLEM 114. Let $0 < x < 2\pi$. Show that

$$-\frac{1}{2}\tan\frac{x}{4} \le \sum_{k=1}^{n} \sin kx \le \frac{1}{2}\cot\frac{x}{4}.$$

Solution. Since (see Solution of Problem 63)

$$\sum_{k=1}^{n} \sin kx = \frac{\cos\frac{x}{2} - \cos\left(n + \frac{1}{2}\right)x}{2\sin\frac{x}{2}},$$

this together with $-1 \le \cos(n + \frac{1}{2})x \le 1$ and $0 < x < 2\pi$, gives the desired result.

PROBLEM 115. Let $0 < a_k < 1$ for $k = 1,2,\ldots,n$ and $a_1 + a_2 + \cdots + a_n < 1$. Show that

$$\frac{1}{1 - \sum_{k=1}^{n} a_k} > \prod_{k=1}^{n} (1 + a_k) > 1 + \sum_{k=1}^{n} a_k \tag{E.51}$$

and

$$\frac{1}{1 + \sum_{k=1}^{n} a_k} > \prod_{k=1}^{n} (1 - a_k) > 1 - \sum_{k=1}^{n} a_k. \tag{E.52}$$

Solution. Clearly

$$(1 + a_1)(1 + a_2) = 1 + (a_1 + a_2) + a_1a_2 > 1 + (a_1 + a_2).$$

Hence

$$\begin{aligned}(1 + a_1)(1 + a_2)(1 + a_3) &> [1 + (a_1 + a_2)](1 + a_3)\\ &> 1 + (a_1 + a_2 + a_3),\end{aligned}$$

and continuing this process we see that

$$\begin{aligned}&(1 + a_1)(1 + a_2)(1 + a_3) \cdots (1 + a_n)\\ &> 1 + (a_1 + a_2 + a_3 + \cdots + a_n).\end{aligned} \tag{E.53}$$

In like manner we see that

$$(1 - a_1)(1 - a_2) = 1 - (a_1 + a_2) + a_1a_2 > 1 - (a_1 + a_2).$$

Thus, since $1 - a_3$ is positive, we have

$$(1 - a_1)(1 - a_2)(1 - a_3) > [1 - (a_1 + a_2)](1 - a_3) > 1 - (a_1 + a_2 + a_3),$$

and, generally,

$$(1 - a_1)(1 - a_2)(1 - a_3) \cdots (1 - a_n) > 1 - (a_1 + a_2 + a_3 + \cdots + a_n). \tag{E.54}$$

Next,

$$1 + a_1 = \frac{1 - a_1^2}{1 - a_1} < \frac{1}{1 - a_1},$$

so that

$$(1 + a_1)(1 + a_2) \cdots (1 + a_n) < \frac{1}{(1 - a_1)(1 - a_2) \cdots (1 - a_n)},$$

and therefore, since $a_1 + a_2 + \cdots + a_n < 1$, we have, by (E.54),

$$(1 + a_1)(1 + a_2) \cdots (1 + a_n) < \frac{1}{1 - (a_1 + a_2 + \cdots + a_n)}. \tag{E.55}$$

Similarly, we find

$$(1 - a_1)(1 - a_2) \cdots (1 - a_n) < \frac{1}{1 + (a_1 + a_2 + \cdots + a_n)}. \tag{E.56}$$

By combining the four inequalities (E.52), (E.53), (E.54), and (E.55), we obtain the inequalities (E.51) and (E.52).

PROBLEM 116. Show that

$$H_{n-1} = \frac{1}{(n - 1)!}\int_n^\infty w(t)\ e^{-t}\ dt < \frac{1}{(e - 1)^n}, \tag{E.57}$$

where t is real, n is a positive integer, and

$$w(t) = (t - 1)(t - 2) \cdots (t - n + 1).$$

Solution. We put

$$I_n = H_n e^{n+1}$$

and show that $H_{n-1} < (e - 1)^{-n}$ is the best possible estimate. Changing from t to t - n - 1 yields

$$I_n = \frac{1}{n!}\int_0^\infty (t + 1)(t + 2) \cdots (t + n)e^{-t}\, dt.$$

Now,

$$I_{n+1} - I_n = \frac{1}{(n + 1)!}\int_0^\infty t(t + 1) \cdots (t + n)e^{-t}\, dt$$

$$= K_n + e^{-1}I_{n+1},$$

where

$$K_n = \frac{1}{(n + 1)!}\int_0^1 t(t + 1) \cdots (t + n)e^{-t}\, dt. \tag{E.58}$$

Thus

$$(1 - e^{-1})I_{n+1} - I_n = K_n \tag{E.59}$$

and, in particular,

$$(1 - e^{-1})I_{n+1} > I_n$$

so that $J_n = I_n(1 - e^{-1})^{n+1}$ is a monotonic increasing sequence. Using (E.59), we have

$$J_{n+1} - J_n = K_n(1 - e^{-1})^{n+1},$$

so that

$$J_m - J_1 = \sum_{n=1}^{m-1} (J_{n+1} - J_n) = \sum_{n=1}^{m-1} K_n(1 - e^{-1})^{n+1}.$$

It follows from (E.58) that $K_n < 1$, $n = 1,2,\ldots,$ so that we may write

$$\lim_{n\to\infty} J_n = J_1 + \sum_{n=1}^{\infty} K_n(1 - e^{-1})^{n+1}. \tag{E.60}$$

Now for $t \geq 0$,

$$e^t = [1 - (1 - e^{-1})]^{-t} = 1 + t(1 - e^{-1}) + \frac{t(t + 1)(1 - e^{-1})^2}{2!} + \cdots,$$

or

$$1 = e^{-t} + te^{-t}(1 - e^{-1}) + \frac{t(t + 1)e^{-t}(1 - e^{-1})^2}{2!} + \cdots .$$

Integrating over $0 \le t \le 1$, a clearly justified termwise integration yields

$$1 = \int_0^1 e^{-t}dt + (1 - e^{-1}) \int_0^1 te^{-t}dt + \sum_{n=1}^{\infty} K_n(1 - e^{-1})^{n+1},$$

and by (E.60)

$$\lim_{n\to\infty} J_n = 1 + J_1 - \int_0^1 e^{-t}dt - (1 - e^{-1}) \int_0^1 te^{-t}dt.$$

A simple calculation shows that $J_1 = 2(1 - e^{-1})^2$ and $\lim_{n\to\infty} J_n = 1$. This is the asymptotic result $H_{n-1} \simeq (e - 1)^{-n}$ as $n \to \infty$. The inequality (E.57) follows immediately from the monotonicity of J_n.

PROBLEM 117. Let $0 < x_i < \pi$, $i = 1,2,\ldots,n$, $x = (x_1 + x_2 + \cdots + x_n)/n$. Show that

$$\prod_{i=1}^{n} \frac{\sin x_i}{x_i} \le \left(\frac{\sin x}{x}\right)^n.$$

Solution. Let $f(t) = (\sin t)/t$ and $f(x_i) = y_i$ for $i = 1,2,\ldots,n$. Then by Problem 2, $(y_1 y_2 \cdots y_n)^{1/n} \le (y_1 + y_2 + \cdots + y_n)/n$. But

$$\frac{y_1 + y_2 + \cdots + y_n}{n} = \frac{f(x_1) + f(x_2) + \cdots + f(x_n)}{n} \le f(x)$$

because f is concave on $(0,\pi)$. The latter can be seen by noting that $f(t) > 0$ for $0 < t < \pi$ and

$$\frac{f'(t)}{f(t)} = \frac{1}{\tan t} - \frac{1}{t} = \cot t - \frac{1}{t} < 0 \quad \text{for } 0 < t < \pi.$$

CHAPTER 3

SEQUENCES AND SERIES

PROBLEM 1. Let a_0 and a_1 be given and define, for $n = 2,3,\ldots,$

$$a_n = \frac{a_{n-1} + a_{n-2}}{2}.$$

Show that

$$\lim_{n\to\infty} a_n = \frac{a_0 + 2a_1}{3}.$$

Solution. We have

$$a_2 = \frac{a_0 + a_1}{2}, \quad a_3 = \frac{a_1 + a_2}{2}, \quad a_4 = \frac{a_2 + a_3}{2}, \quad \ldots;$$

hence

$$a_2 - a_1 = \frac{a_0 - a_1}{2}, \quad a_3 - a_2 = \frac{a_1 - a_2}{2}, \quad a_4 - a_3 = \frac{a_2 - a_3}{2}, \quad \ldots$$

and thus

$$a_2 - a_1 = -\frac{a_1 - a_0}{2}, \quad a_3 - a_2 = \frac{a_1 - a_2}{2} = \frac{a_1 - a_0}{2^2},$$

$$a_4 - a_3 = -\frac{a_1 - a_0}{2^3}, \quad \ldots, \quad a_n - a_{n-1} = (-1)^{n-1}\frac{a_1 - a_0}{2^{n-1}}.$$

Consequently

$$a_n - a_1 = -\frac{a_1 - a_0}{2}\left(1 - \frac{1}{2} + \frac{1}{2^2} - \cdots + (-1)^{n-2}\frac{1}{2^{n-2}}\right)$$

$$= \frac{a_1 - a_0}{3}\left\{(-1)^{n-1}\frac{1}{2^{n-1}} - 1\right\}$$

and so

$$a_n = \frac{2a_1 + a_0}{3} + (-1)^{n-1}\frac{a_1 - a_0}{3\cdot 2^{n-1}}.$$

Letting $n \to \infty$, we obtain the desired result.

PROBLEM 2. For any real number a, show that

$$\lim_{n\to\infty} \frac{a^n}{n!} = 0.$$

Solution. We assume that $a > 0$ and let the integer k be such $a < k + 1$. For $n > k$ we have

$$\frac{a^n}{n!} = \frac{a^k}{k!}\cdot\frac{a}{k + 1}\cdot\frac{a}{k + 2}\cdots\frac{a}{n}.$$

Thus

$$\frac{a^n}{n!} < \frac{a^k}{k!}\cdot\left(\frac{a}{k + 1}\right)^{n-k},$$

But $a^k/k!$ is fixed and

$$\left(\frac{a}{k + 1}\right)^{n-k} \to 0 \quad \text{as } n \to \infty.$$

PROBLEM 3. Let

$$P_n = \frac{2^3 - 1}{2^3 + 1}\cdot\frac{3^3 - 1}{3^3 + 1}\cdots\frac{n^3 - 1}{n^3 + 1}.$$

Show that $\lim_{n\to\infty} P_n = 2/3$.

Solution. Note that

$$\frac{k^3 - 1}{k^3 + 1} = \frac{k - 1}{k + 1} \cdot \frac{k^2 + k + 1}{k^2 - k + 1}.$$

But

$$\prod_{k=2}^{n} \frac{k - 1}{k + 1} = \frac{2}{n(n + 1)} \quad \text{and} \quad \prod_{k=2}^{n} \frac{k^2 + k + 1}{k^2 - k + 1} = \frac{n^2 + n + 1}{3}.$$

Hence

$$\lim_{n \to \infty} P_n = \frac{2}{3} \cdot \lim_{n \to \infty} \frac{n^2 + n + 1}{n^2 + n} = \frac{2}{3}.$$

PROBLEM 4. Let K be a positive number. Taking an arbitrary positive number x_0 and forming the sequence

$$x_n = \frac{1}{2}\left(x_{n-1} + \frac{K}{x_{n-1}}\right) \quad \text{with } n = 1,2,3,\ldots,$$

show that $\lim_{n \to \infty} x_n = \sqrt{K}$. More generally, show that if m is a positive integer and

$$x_p = \frac{m - 1}{m} x_{p-1} + \frac{K}{m\, x_{p-1}^{m-1}} \quad \text{with } p = 1,2,3,\ldots,$$

then $\lim_{p \to \infty} x_p = \sqrt[m]{K}$.

Solution. By induction we see that

$$\frac{x_n - \sqrt{K}}{x_n + \sqrt{K}} = \left\{\frac{x_0 - \sqrt{K}}{x_0 + \sqrt{K}}\right\}^{2^n}.$$

But

$$\left|\frac{x_0 - \sqrt{K}}{x_0 + \sqrt{K}}\right| < 1$$

and so

$$\lim_{n \to \infty} \left\{\frac{x_0 - \sqrt{K}}{x_0 + \sqrt{K}}\right\}^{2^n} = 0.$$

Hence

$$\lim_{n\to\infty} \frac{x_n - \sqrt{K}}{x_n + \sqrt{K}} = 0 \quad \text{and} \quad \lim_{n\to\infty} x_n = \sqrt{K}.$$

This proves the first part of the claim.

To prove the second part of the claim, we first note that $x_p^m > K$. Indeed,

$$x_p^m = x_{p-1}^m \left(1 + \frac{K - x_{p-1}^m}{m\, x_{p-1}^m}\right)^m.$$

By Bernoulli's Inequality (see Problem 35 of Chapter 2) we have that $(1 + a)^n > 1 + na$ if $n \geq 2$ and $a > 0$ or $0 > a > -1$. Thus

$$\left(1 + \frac{K - x_{p-1}^m}{m\, x_{p-1}^m}\right)^m > 1 + \frac{K - x_{p-1}^m}{x_{p-1}^m} = \frac{K}{x_{p-1}^m}.$$

We also see that

$$x_p - x_{p-1} = \frac{K - x_{p-1}^m}{m\, x_{p-1}^m} < 0.$$

Hence $(x_p)_{p=0}^{\infty}$ is a decreasing sequence of positive numbers and therefore converges to some real number λ with λ satisfying

$$\lambda = \frac{m - 1}{m}\lambda + \frac{K}{m\lambda^{m-1}}.$$

Thus $\lambda^m = K$ or $\lambda = \sqrt[m]{K}$. Moreover,

$$x_p > \sqrt[m]{K} > \frac{K}{x_p^{m-1}}.$$

Remark. It is interesting to note how the approximating sequence for $\sqrt{K}$ was formed above, namely, we took $K = x_0(K/x_0)$ and defined x_1 to be the arithmetic mean of the factors x_0 and K/x_0, then we took $K = x_1(K/x_1)$ and defined x_2 to be the arithmetic mean of the factors x_1 and K/x_1 and so forth.

PROBLEM 5. Show the convergence of the sequence

$$x_n = 1 + \frac{1}{\sqrt{2}} + \frac{1}{\sqrt{3}} + \cdots + \frac{1}{\sqrt{n}} - 2\sqrt{n}.$$

Solution. We show that the sequence is decreasing. Since

$$x_{n+1} - x_n = \frac{1}{\sqrt{n+1}} - 2(\sqrt{n+1} - \sqrt{n})$$

and

$$\sqrt{n+1} - \sqrt{n} = \frac{1}{\sqrt{n+1} + \sqrt{n}} > \frac{1}{2\sqrt{n+1}},$$

we have that $x_{n+1} < x_n$.

To see that the sequence is bounded below, we observe that

$$1 + \frac{1}{\sqrt{2}} + \frac{1}{\sqrt{3}} + \cdots + \frac{1}{\sqrt{n}} > 2\sqrt{n+1} - 2$$

(see Problem 26 of Chapter 2). Hence

$$x_n > 2(\sqrt{n+1} - \sqrt{n}) - 2 > -2.$$

But any decreasing sequence which is bounded below is convergent.

PROBLEM 6. Show that

$$\frac{3 - \sqrt{5}}{2} = \frac{1}{3} + \frac{1}{3\cdot 7} + \frac{1}{3\cdot 7\cdot 47} + \frac{1}{3\cdot 7\cdot 47\cdot 2207} + \cdots$$

(each factor in the denominator is equal to the square of the preceding factor diminished by 2).

Solution. We put

$$\frac{y_0 - \lambda}{2} = \frac{1}{y_0} + \frac{1}{y_0 y_1} + \frac{1}{y_0 y_1 y_2} + \cdots ,$$

where $y_0 = 3$, $y_1 = 7$, $y_2 = 47$, ..., and, in general,

$$y_n = y_{n-1}^2 - 2,$$

and determine λ. Noting that

$$\frac{y_1 - \lambda y_0}{2} = \frac{1}{y_1} + \frac{1}{y_1 y_2} + \cdots, \qquad \frac{y_2 - \lambda y_0 y_1}{2} = \frac{1}{y_2} + \frac{1}{y_2 y_3} + \cdots,$$

$$\ldots, \qquad \frac{y_n - \lambda y_0 y_1 \cdots y_{n-1}}{2} = \frac{1}{y_n} + \frac{1}{y_n y_{n+1}} + \cdots,$$

and, letting $n \to \infty$, we see that

$$\lambda = \lim_{n \to \infty} \frac{y_n}{y_0 y_1 \cdots y_{n-1}}.$$

From $y_n = y_{n-1}^2 - 2$ we get

$$y_n^2 - 4 = y_{n-1}^2 (y_{n-2}^2 - 4);$$

This in turn yields

$$y_n^2 - 4 = y_0^2 y_1^2 y_2^2 \cdots y_{n-1}^2 (y_0^2 - 4)$$

and so

$$\lim_{n \to \infty} \frac{y_n}{y_0 y_1 \cdots y_{n-1}} = \sqrt{y_0^2 - 4} = \sqrt{3^2 - 4} = \sqrt{5} = \lambda.$$

Remark. A more interesting way of solving the problem is to observe that y_n can be represented by the expression

$$y_n = x^{2^n} + \frac{1}{x^{2^n}},$$

where x is any of the roots $A = (3 + \sqrt{5})/2$ or $B = (3 - \sqrt{5})/2$ of $x^2 - 3x + 1 = 0$. Thus

$$y_0 = \frac{1 + x^2}{x}, \quad y_1 = \frac{1 + x^4}{x^2}, \quad y_2 = \frac{1 + x^8}{x^4}, \quad \ldots$$

and the series

$$\frac{1}{y_0} + \frac{1}{y_0 y_1} + \frac{1}{y_0 y_1 y_2} + \cdots$$

to be evaluated becomes

$$S = \frac{1}{x}\left[\frac{x^2}{1 + x^2} + \frac{x^4}{(1 + x^2)(1 + x^4)} + \frac{x^8}{(1 + x^2)(1 + x^4)(1 + x^8)} + \cdots\right].$$

But the series inside the square brackets tends to 1 or x^2, according as $x = A$ or $x = B$; this is clear from the fact that (see Problem 106 of Chapter 1)

$$\frac{a}{1+a} + \frac{b}{(1+a)(1+b)} + \frac{c}{(1+a)(1+b)(1+c)}$$

$$+ \cdots + \frac{k}{(1+a)(1+b)(1+c)\cdots(1+k)}$$

$$= 1 - \frac{1}{(1+a)(1+b)(1+c)\cdots(1+k)}$$

and $A > 1$ and $0 < B < 1$. Hence, if $x = A = (3 + \sqrt{5})/2$, then

$$1 - \frac{1}{(1+x^2)(1+x^4)\cdots(1+x^{2^n})} \to 1 \quad \text{as } n \to \infty$$

and $S = (1/x)\cdot 1 = 2/(3+\sqrt{5}) = (3 - \sqrt{5})/2$. If, on the other hand, we have $x = B = (3 - \sqrt{5})/2$, then

$$1 - \frac{1}{(1+x^2)(1+x^4)\cdots(1+x^{2^n})}$$

$$= 1 - \frac{1}{1 + x^2 + x^4 + \cdots + x^{2(2^n-1)}}$$

tends to

$$1 - \frac{1}{\frac{1}{1-x^2}} = 1 - (1 - x^2) = x^2$$

as $n \to \infty$ and $S = (1/x)\, x^2 = x = (3 - \sqrt{5})/2$.

PROBLEM 7. Show that

$$\sum_{k=1}^{\infty} \frac{1}{k^2} = \frac{\pi^2}{6}.$$

Solution. For any positive integer n, we consider

$$\int_0^{\pi/2} \cos^{2n} t\, dt.$$

Applying integration by parts twice, we get

$$\int_0^{\pi/2} \cos^{2n} t\, dt = t \cos^{2n} t \Big|_0^{\pi/2} + 2n \int_0^{\pi/2} t \cos^{2n-1} t \sin t\, dt$$

$$= n(t^2 \cos^{2n-1} t \sin t)\Big|_0^{\pi/2}$$

$$-n \int_0^{\pi/2} t^2[-(2n - 1)\cos^{2n-2} t \sin^2 t + \cos^{2n} t]dt$$

$$= -2n^2 I_{2n} + n(2n - 1)I_{2n-2},$$

where

$$I_{2n} = \int_0^{\pi/2} t^2 \cos^{2n} t\, dt.$$

On the other hand,

$$\int_0^{\pi/2} \cos^{2n} t\, dt = \int_0^{\pi/2} \cos^{2n-1} t\, d(\sin t)$$

$$= \cos^{2n-1} t \sin t\Big|_0^{\pi/2} + (2n - 1) \int_0^{\pi/2} \cos^{2n-2} t \sin^2 t\, dt,$$

that is,

$$J_{2n} = (2n - 1)J_{2n-2} - (2n - 1)J_{2n} \quad \text{or} \quad J_{2n} = \frac{2n - 1}{2n} J_{2n-2},$$

where

$$J_{2n} = \int_0^{\pi/2} \cos^{2n} t\, dt.$$

Noting that $J_0 = \pi/2$, we see that

$$J_{2n} = \frac{(2n - 1)(2n - 3)\cdots 3\cdot 1}{2n(2n - 2)\cdots 4\cdot 2}\cdot\frac{\pi}{2}.$$

We may thus conclude that

$$-2n^2 I_{2n} + n(2n - 1)I_{2n-2} = \frac{(2n - 1)!!}{(2n)!!}\,\frac{\pi}{2},$$

where, as usual,

$$(2n)!! = 2\cdot 4\cdots(2n - 2)(2n), \quad 0!! = 1;$$

$$(2n + 1)!! = 1\cdot 3\cdots(2n - 1)(2n + 1), \quad (-1)!! = 1.$$

Therefore

$$\frac{(2n)!!}{(2n-1)!!} I_{2n} - \frac{(2n-2)!!}{(2n-3)!!} I_{2n-2} = -\frac{\pi}{4}\cdot\frac{1}{n^2}$$

and so

$$\frac{(2n)!!}{(2n-1)!!} I_{2n} - \frac{0!!}{(-1)!!} I_0$$

$$= \sum_{k=1}^{n} \left(\frac{(2k)!!}{(2k-1)!!} I_{2k} - \frac{(2k-2)!!}{(2k-3)!!} I_{2k-2}\right)$$

$$= \frac{\pi}{4}\cdot \sum_{k=1}^{n} \frac{1}{k^2}.$$

This shows that

$$\frac{(2n)!!}{(2n-1)!!} I_{2n} = \frac{\pi^3}{24} - \frac{\pi}{4} \sum_{k=1}^{n} \frac{1}{k^2} = \frac{\pi}{4}\left[\frac{\pi^2}{6} - \sum_{k=1}^{n} \frac{1}{k^2}\right].$$

But

$$\lim_{n\to\infty} \frac{(2n)!!}{(2n-1)!!} I_{2n} = 0.$$

Indeed, since $(2/\pi)t \le \sin t$ for $0 \le t \le \pi/2$, we have

$$I_{2n} = \int_0^{\pi/2} t^2 \cos^{2n} t\, dt \le \left(\frac{\pi}{2}\right)^2 \int_0^{\pi/2} \sin^2 t \cos^{2n} t\, dt$$

$$= \frac{\pi^2}{4}\left[\int_0^{\pi/2} \cos^{2n} t\, dt - \int_0^{\pi/2} \cos^{2n+2} t\, dt\right]$$

$$= \frac{\pi^3}{8}\left(\frac{(2n-1)!!}{(2n)!!} - \frac{(2n+1)!!}{(2n+2)!!}\right) = \frac{\pi^3}{8}\,\frac{(2n-1)!!}{(2n+2)!!}$$

and so

$$0 < \frac{(2n)!!}{(2n-1)!!} I_{2n} \le \frac{\pi^3}{8}\cdot\frac{1}{2n+2}.$$

Remark. For related results see Problems 103 and 104.

PROBLEM 8. Show the *Formula of Wallis:*

$$\frac{\pi}{2} = \lim_{n\to\infty} \frac{2\cdot 2\cdot 4\cdot 4 \cdots 2n\cdot 2n}{1\cdot 3\cdot 3\cdot 5 \cdots (2n-1)(2n+1)}.$$

Solution. Let m be a positive integer and put

$$J_m = \int_0^{\pi/2} \sin^m x\, dx \quad \text{and} \quad J'_m = \int_0^{\pi/2} \cos^m x\, dx.$$

Integration by parts yields

$$J_m = \int_0^{\pi/2} \sin^{m-1} x\, d(-\cos x)$$

$$= -\sin^{m-1} x \cos x \Big|_0^{\pi/2} + (m - 1)\int_0^{\pi/2} \sin^{m-2} x \cos^2 x\, dx$$

$$= (m - 1)\int_0^{\pi/2} (\sin^{m-2} x)(1 - \sin^2 x)dx$$

$$= (m - 1)J_{m-2} - (m - 1)J_m.$$

Thus

$$J_m = \frac{m - 1}{m} J_{m-2}.$$

Noting that $J_0 = \pi/2$ and $J_1 = 1$, we obtain

$$J_{2n} = \frac{(2n - 1)(2n - 3) \cdots 3 \cdot 1}{2n(2n - 2) \cdots 4 \cdot 2} \cdot \frac{\pi}{2}$$

and

$$J_{2n+1} = \frac{2n(2n - 2) \cdots 4 \cdot 2}{(2n + 1)(2n - 1) \cdots 3 \cdot 1}.$$

Exactly the same results obtain for J'_m. Using the notation introduced in the Solution of Problem 7 concerning the symbol m!!, we have

$$\int_0^{\pi/2} \sin^m x\, dx = \int_0^{\pi/2} \cos^m x\, dx = \frac{(m - 1)!!}{m!!} \frac{\pi}{2} \quad \text{for m even,}$$

$$= \frac{(m - 1)!!}{m!!} \quad \text{for m odd.}$$

We now suppose that $0 < x < \pi/2$; for these x we have

$$\sin^{2n+1} x < \sin^{2n} x < \sin^{2n-1} x.$$

Integration over the interval from 0 to $\pi/2$ yields

$$J_{2n+1} < J_{2n} < J_{2n-1};$$

hence

$$\frac{(2n)!!}{(2n + 1)!!} < \frac{(2n - 1)!!}{(2n)!!}\frac{\pi}{2} < \frac{(2n - 2)!!}{(2n - 1)!!}$$

or

$$\left(\frac{(2n)!!}{(2n - 1)!!}\right)^2 \frac{1}{2n + 1} < \frac{\pi}{2} < \left(\frac{(2n)!!}{(2n - 1)!!}\right)^2 \frac{1}{2n}.$$

But the difference of the two outer expressions equals

$$\frac{1}{2n(2n + 1)}\left(\frac{(2n)!!}{(2n - 1)!!}\right)^2 < \frac{1}{2n}\frac{\pi}{2}$$

and hence tends to 0 as n becomes arbitrarily large, we see that

$$\frac{\pi}{2} = \lim_{n\to\infty}\left(\frac{(2n)!!}{(2n - 1)!!}\right)^2 \cdot \frac{1}{2n + 1}.$$

PROBLEM 9. Show the *Formula of Vieta:*

$$\frac{2}{\pi} = \sqrt{\frac{1}{2}} \cdot \sqrt{\frac{1}{2} + \frac{1}{2}\sqrt{\frac{1}{2}}} \cdot \sqrt{\frac{1}{2} + \frac{1}{2}\sqrt{\frac{1}{2} + \frac{1}{2}\sqrt{\frac{1}{2}}}} \cdots$$

Solution. Let $t \neq 0$. We commence by showing that

$$\lim_{n\to\infty}\left(\cos\frac{t}{2}\right)\left(\cos\frac{t}{2^2}\right)\cdots\left(\cos\frac{t}{2^n}\right) = \frac{\sin t}{t}. \tag{E.1}$$

Indeed,

$$\sin t = 2\left(\cos\frac{t}{2}\right)\left(\sin\frac{t}{2}\right) = 2^2\left(\cos\frac{t}{2}\right)\left(\cos\frac{t}{2^2}\right)\left(\sin\frac{t}{2^2}\right)$$

$$= \cdots = 2^n\left(\cos\frac{t}{2}\right)\left(\cos\frac{t}{2^2}\right)\cdots\left(\cos\frac{t}{2^n}\right)\left(\sin\frac{t}{2^n}\right)$$

and so

$$\left(\cos\frac{t}{2}\right)\left(\cos\frac{t}{2^2}\right)\cdots\left(\cos\frac{t}{2^n}\right) = \frac{\sin t}{2^n \sin t/2^n} = \frac{\sin t}{t}\frac{t/2^n}{\sin t/2^n}.$$

But

$$\frac{\sin t/2^n}{t/2^n} \to 1 \quad \text{as } n \to \infty.$$

From (E.1) we get, for $t = \pi/2$,

$$\frac{2}{\pi} = \lim_{n \to \infty} \left(\cos \frac{\pi}{4}\right)\left(\cos \frac{\pi}{8}\right) \cdots \left(\cos \frac{\pi}{2^{n+1}}\right).$$

Since

$$\cos \frac{\pi}{4} = \sqrt{\frac{1}{2}} \quad \text{and} \quad \cos \frac{\theta}{2} = \sqrt{\frac{1}{2} + \frac{1}{2} \cos \theta},$$

the desired result is easily obtained.

PROBLEM 10. Verify the *Formula of Machin:*

$$\frac{\pi}{4} = 4 \text{ arc tan } \frac{1}{5} - \text{arc tan } \frac{1}{239}.$$

Solution. We have

$$4 \text{ arc tan } \frac{1}{5} = 2 \text{ arc tan } \frac{1}{5} + 2 \text{ arc tan } \frac{1}{5} = 2 \text{ arc tan } \frac{\frac{2}{5}}{1 - \frac{1}{25}}$$

$$= 2 \text{ arc tan } \frac{5}{12} = \text{arc tan } \frac{5}{12} + \text{arc tan } \frac{5}{12}$$

$$= \text{arc tan } \frac{\frac{5}{12} + \frac{5}{12}}{1 - \frac{25}{144}} = \text{arc tan } \frac{120}{119}.$$

Further

$$\text{arc tan } \frac{120}{119} + \text{arc tan } \left(-\frac{1}{239}\right)$$

$$= \text{arc tan } \frac{\frac{120}{119} - \frac{1}{239}}{1 + \frac{120}{119}\frac{1}{239}} = \text{arc tan } 1 = \frac{\pi}{4}.$$

Remark. From the expansion

$$\text{arc tan } x = x - \frac{x^3}{3} + \frac{x^5}{5} - \cdots + (-1)^{k-1} \frac{x^{2k-1}}{2k - 1} + \cdots,$$

where $-1 \le x \le 1$, we see that

$$\frac{\pi}{4} = 1 - \frac{1}{3} + \frac{1}{5} - \cdots + (-1)^{k-1} \frac{1}{2k - 1} + \cdots$$

To calculate π accurate to seven decimal places, we would have to consider the first ten million terms of the last series. However, by Machin's Formula,

$$\pi = 16\left(\frac{1}{5} - \frac{1}{3}\cdot\frac{1}{5^2} + \frac{1}{5}\cdot\frac{1}{5^5} - \frac{1}{7}\cdot\frac{1}{5^7} + \frac{1}{9}\cdot\frac{1}{5^9} - \frac{1}{11}\cdot\frac{1}{5^{11}} + \cdots\right)$$

$$- 4\left(\frac{1}{239} - \frac{1}{3}\cdot\frac{1}{239^3} + \cdots\right)$$

with the displayed number of terms being sufficient to obtain $\pi = 3.1415926\ldots$

PROBLEM 11. Let q be a positive integer. Show that

$$\sum_{n=1}^{\infty} \frac{1}{n(n + 1)(n + 2)\cdots(n + q)} = \frac{1}{q\cdot q!}.$$

Solution. The following identity is easily established by induction:

$$\sum_{n=1}^{k} \frac{1}{n(n + 1)\cdots(n + q)} = \frac{1}{q}\left\{\frac{1}{q!} - \frac{1}{(k + 1)(k + 2)\cdots(k + q)}\right\}.$$

Therefore the claim follows immediately.

Remarks. The identity

$$\sum_{n=1}^{k} n(n + 1)\cdots(n + q) = \frac{1}{q + 2}\, k(k + 1)\cdots(k + q + 1)$$

can readily be established by induction as well.

It is also of interest to note that, if p is an arbitrary real number different from -1, -2, -3, ..., then

$$\sum_{n=1}^{\infty} \frac{1}{(p + n)(p + n + 1)} = \sum_{n=1}^{\infty} \left\{\frac{1}{p + n} - \frac{1}{p + n + 1}\right\} = \frac{1}{p + 1},$$

$$\sum_{n=1}^{\infty} \frac{1}{(p + n)(p + n + 1)(p + n + 2)}$$

$$= \sum_{n=1}^{\infty} \frac{1}{2}\left\{\frac{1}{(p + n)(p + n + 1)} - \frac{1}{(p + n + 1)(p + n + 2)}\right\}$$

$$= \frac{1}{2(p + 1)(p + 2)},$$

and in general, for $q \geq 1$ and q an integer,

$$\sum_{n=1}^{\infty} \frac{1}{(p+n)(p+n+1)\cdots(p+n+q)} = \frac{1}{q(p+1)(p+2)\cdots(p+q)}.$$

Letting $p = 0$, we obtain the result in Problem 11.

To see that

$$\sum_{n=1}^{\infty} \frac{1}{(p+n)(p+n+1)\cdots(p+n+q)} = \frac{1}{q(p+1)(p+2)\cdots(p+q)}$$

is true under the stated conditions, we only need to observe that

$$\frac{1}{(p+n)(p+n+1)\cdots(p+n+q)} - \frac{1}{(p+n+1)(p+n+2)\cdots(p+n+q+1)} = \frac{q+1}{(p+n)(p+n+1)\cdots(p+n+q+1)}$$

and proceed by induction on q.

Finally, we note that the result in Problem 11 is equivalent to

$$\sum_{n=1}^{\infty} \frac{1}{\binom{q+n}{q+1}} = \frac{q+1}{q}.$$

PROBLEM 12. Let

$$S_n = \sum_{k=1}^{n}\left(\sqrt{1+\frac{k}{n^2}} - 1\right).$$

Show that $\lim_{n\to\infty} S_n = 1/4$.

Solution. Since

$$\frac{x}{2+x} < \sqrt{1+x} - 1 < \frac{x}{2} \quad \text{for } 1 + x > 0,$$

we see that, putting $x = k/n^2$,

$$\frac{k}{2n^2+k} < \sqrt{1+\frac{k}{n^2}} - 1 < \frac{k}{2n^2}.$$

Hence

$$\sum_{k=1}^{n} \frac{k}{2n^2 + k} < S_n < \frac{1}{2n^2} \sum_{k=1}^{n} k.$$

But

$$\frac{1}{2n^2} \sum_{k=1}^{n} k = \frac{n(n + 1)}{4n^2} \to \frac{1}{4} \quad \text{as } n \to \infty.$$

On the other hand,

$$\lim_{n\to\infty} \left\{ \frac{1}{2n^2} \sum_{k=1}^{n} k - \sum_{k=1}^{n} \frac{k}{2n^2 + k} \right\} = \lim_{n\to\infty} \sum_{k=1}^{n} \frac{k^2}{2n^2(2n^2 + k)}.$$

But

$$\sum_{k=1}^{n} \frac{k^2}{2n^2(2n^2 + k)} < \sum_{k=1}^{n} \frac{k^2}{4n^4} = \frac{n(n + 1)(2n + 1)}{24n^4}$$

(see Problem 15 of Chapter 1) and so

$$\lim_{n\to\infty} \left\{ \frac{1}{2n^2} \sum_{k=1}^{n} k - \sum_{k=1}^{n} \frac{k}{2n^2 + k} \right\} = 0$$

and

$$\lim_{n\to\infty} \sum_{k=1}^{n} \frac{k}{2n^2 + k} = \frac{1}{4}.$$

Hence the desired conclusion.

PROBLEM 13. Show that

$$\lim_{n\to\infty} \left(1 + \frac{1}{n}\right)^n = 1 + \frac{1}{1!} + \frac{1}{2!} + \frac{1}{3!} + \cdots + \frac{1}{n!} + \cdots$$

Solution. By Problems 2 and 3 of Chapter 1, the sequence

$$x_n = \left(1 + \frac{1}{n}\right)^n \qquad n = 1,2,3,\ldots$$

is increasing and bounded, hence converges to a number e, where $2 < e < 3$. The number e is the base of the natural logarithms. It is also clear that

$$x_n = \left(1 + \frac{1}{n}\right)^n = 1 + \binom{n}{1}\frac{1}{n} + \cdots + \binom{n}{k}\frac{1}{n^k} + \cdots + \binom{n}{n}\frac{1}{n^n}$$

$$= 1 + 1 + \frac{1}{2!}\left(1 - \frac{1}{n}\right) + \frac{1}{3!}\left(1 - \frac{1}{n}\right)\left(1 - \frac{2}{n}\right)$$

$$+ \cdots + \frac{1}{k!}\left(1 - \frac{1}{n}\right)\cdots\left(1 - \frac{k - 1}{n}\right)$$

$$+ \cdots + \frac{1}{n!}\left(1 - \frac{1}{n}\right)\cdots\left(1 - \frac{n - 1}{n}\right).$$

Since $k < n$,

$$x_n > 2 + \frac{1}{2!}\left(1 - \frac{1}{n}\right) + \frac{1}{3!}\left(1 - \frac{1}{n}\right)\left(1 - \frac{2}{n}\right)$$

$$+ \cdots + \frac{1}{k!}\left(1 - \frac{1}{n}\right)\cdots\left(1 - \frac{k - 1}{n}\right).$$

Letting $n \to \infty$, we have

$$e \geq 2 + \frac{1}{2!} + \frac{1}{3!} + \cdots + \frac{1}{k!} = y_k.$$

Clearly

$$x_n < y_n \leq e$$

and so $\lim_{n \to \infty} y_n$ exists and is equal to e.

We also note that

$$y_{n+m} - y_n = \frac{1}{(n + 1)!} + \frac{1}{(n + 2)!} + \cdots + \frac{1}{(n + m)!}$$

$$< \frac{1}{(n + 1)!}\left\{1 + \frac{1}{n + 2} + \frac{1}{(n + 2)^2} + \cdots + \frac{1}{(n + 2)^{m-1}}\right\}$$

$$< \frac{1}{(n + 1)!}\frac{n + 2}{n + 1}.$$

Keeping n fixed and letting $m \to \infty$, we get

$$e - y_n \leq \frac{1}{(n + 1)!}\frac{n + 2}{n + 1}.$$

Thus

$$0 < e - y_n < \frac{1}{n!n}$$

because $(n + 2)/(n + 1) < 1/n$.

Remarks. The foregoing estimate can be used to calculate e. It can also be used to show that e is an irrational number. Indeed, suppose that e is rational and $e = p/q$, where p and q are positive integers. Then $0 < q!(e - y_q) < 1/q$. By our assumption $q!e$ is an integer. Since

$$q!y_q = q!\left(1 + 1 + \frac{1}{2!} + \cdots + \frac{1}{q!}\right)$$

is an integer, we see that $q!(e - y_q)$ is an integer. Since $q \geq 1$,

$$0 < q!(e - y_q) < 1/q$$

would imply the existence of an integer between 0 and 1 and we have reached a contradiction.

The number $e = 2.7182818\ldots$

PROBLEM 14. Show that the sequence

$$c_n = 1 + \frac{1}{2} + \cdots + \frac{1}{n} - \log n$$

converges.

Solution. Let $d_n = c_n - 1/n$. Then

$$c_{n+1} - c_n = \frac{1}{n+1} - \log n \quad \text{and} \quad d_{n+1} - d_n = \frac{1}{n} - \log\left(1 + \frac{1}{n}\right).$$

Since, for $n < x < n + 1$, we have $1/(n + 1) < 1/x < 1/n$, and so

$$\int_n^{n+1} \frac{dx}{n+1} < \int_n^{n+1} \frac{dx}{x} < \int_n^{n+1} \frac{dx}{n},$$

it follows that

$$\frac{1}{n+1} < \log\left(1 + \frac{1}{n}\right) < \frac{1}{n}.$$

This inequality shows that the sequence c_n decreases monotonically, whereas the sequence d_n increases monotonically. Furthermore, $d_n < c_n$; hence $d_1 = 0$ is a lower bound for the c_n. In other words, the sequence c_n converges to a limit c; c is known as *Euler's Constant* and $c = 0.5772156649\ldots$

Remarks. The foregoing shows that the harmonic series

$$1 + \frac{1}{2} + \frac{1}{3} + \cdots + \frac{1}{n} + \cdots$$

diverges. Moreover it is easy to see that the sequence c_n increases very slowly; for example, $c_{1000000} = 14.39\ldots$

PROBLEM 15. Consider the alternating series

$$1 - \frac{1}{2} + \frac{1}{3} - \frac{1}{4} + \frac{1}{5} - \frac{1}{6} + \cdots + (-1)^{n+1}\frac{1}{n} + \cdots$$

Examine the following rearrangement of this series: Let the first p positive terms be followed by the first q negative terms, then the next p positive terms be followed by the next q negative terms, etc. Show that the resulting series converges to $\log(2\sqrt{p/q})$.

Solution. Let

$$H_n = 1 + \frac{1}{2} + \cdots + \frac{1}{n}.$$

By Problem 14,

$$H_n = 1 + \frac{1}{2} + \cdots + \frac{1}{n} = \log n + c + x_n,$$

where c is Euler's Constant and $x_n \to 0$ as $n \to \infty$. Thus

$$\frac{1}{2} + \frac{1}{4} + \cdots + \frac{1}{2m} = \frac{1}{2}H_m = \frac{1}{2}\log m + \frac{1}{2}c + \frac{1}{2}x_n$$

and

$$1 + \frac{1}{3} + \cdots + \frac{1}{2k-1} = H_{2k} - \frac{1}{2}H_k$$

$$= \log 2 + \frac{1}{2}\log k + \frac{1}{2}c + x_{2k} - \frac{1}{2}x_k.$$

The foregoing shows that the partial sums of the series in question are of the form $\log(2\sqrt{p/q}) + z_n$, where $z_n \to 0$ as $n \to \infty$.

PROBLEM 16. Let a and d be positive numbers and consider the arithmetic progression

$$a,\quad a + d,\quad a + 2d,\quad \ldots,\quad a + (n - 1)d.$$

Let A_n denote the arithmetic mean and G_n the geometric mean of the foregoing progression. Show that

$$\lim_{n\to\infty} \frac{G_n}{A_n} = \frac{2}{e},$$

where e is the base of the natural logarithms.

Solution. Setting c = a/d, we get

$$\frac{G_n}{A_n} = \frac{\sqrt[n]{\frac{c}{n}\cdot\frac{c+1}{n}\cdot\frac{c+2}{n}\cdots\frac{c+n-1}{n}}}{\frac{c}{n}+\frac{n-1}{2n}}.$$

It is clear that

$$\frac{c}{n} + \frac{n-1}{2n} \to \frac{1}{2} \quad \text{as } n \to \infty.$$

To see that

$$\sqrt[n]{\frac{c}{n}\cdot\frac{c+1}{n}\cdot\frac{c+2}{n}\cdots\frac{c+n-1}{n}} \to \frac{1}{e} \quad \text{as } n \to \infty$$

we observe that

$$\frac{1}{n}\left\{\log\frac{c}{n} + \log\frac{c+1}{n} + \log\frac{c+2}{n} + \cdots + \log\frac{c+n-1}{n}\right\}$$

tends to

$$\int_0^1 \log x \, dx = x \log x - x \Big|_0^1 = -1$$

as $n \to \infty$.

Remark. Note the special case

$$\lim_{n\to\infty} \frac{\sqrt[n]{n!}}{n} = \frac{1}{e}.$$

PROBLEM 17. Let [t] denote the greatest integer less than or equal to t. Show that

$$w_n = \frac{1}{n}\sum_{k=1}^{n}\left(\left[\frac{2n}{k}\right] - 2\left[\frac{n}{k}\right]\right) \to 2\log 2 - 1 \quad \text{as } n \to \infty.$$

Solution. We see that

$$\int_0^1 \left(\left[\frac{2}{x}\right] - 2\left[\frac{1}{x}\right]\right) dx = \lim_{n\to\infty}\sum_{k=1}^{n-1}\int_{1/k+1}^{1/k}\left(\left[\frac{2}{x}\right] - 2\left[\frac{1}{x}\right]\right) dx$$

$$= \lim_{n\to\infty}\sum_{k=1}^{n-1}\left(\frac{1}{k+\frac{1}{2}} - \frac{1}{k+1}\right)$$

$$= 2\left(\frac{1}{3} - \frac{1}{4} + \frac{1}{5} - \frac{1}{6} + \cdots\right) = 2 \log 2 - 1$$

by Problem 15, or noting that, for positive integers p and q,

$$\int_0^1 \frac{x^{p-1}}{1 + x^q}\,dx = \frac{1}{p} - \frac{1}{p + q} + \frac{1}{p + 2q} - \frac{1}{p + 3q} + \cdots.$$

PROBLEM 18. Let $p_1, p_2, \ldots, p_k$ be positive and $a_1 > a_2 > \cdots > a_k > 0$. Show that

$$\lim_{n\to\infty} \sqrt[n]{p_1a_1^n + p_2a_2^n + \cdots + p_ka_k^n} = a_1.$$

Solution. We have

$$p_1a_1^n + p_2a_2^n + \cdots + p_ka_k^n$$

$$= p_1a_1^n\left[1 + \frac{p_2}{p_1}\left(\frac{a_2}{a_1}\right)^n + \frac{p_3}{p_1}\left(\frac{a_3}{a_1}\right)^n + \cdots + \frac{p_k}{p_1}\left(\frac{a_k}{a_1}\right)^n\right].$$

But the expression in square brackets tends to 1 as $n \to \infty$. Moreover, $\sqrt[n]{p_1}$ tends to 1 as $n \to \infty$. Indeed, let $x_n = \sqrt[n]{p_1} - 1$. Then $x_n > 0$ if $p_1 > 1$ and $-1 < x_n < 0$ if $0 < p_1 < 1$. In either case

$$1 + nx_n < (1 + x_n)^n = p_1$$

by Bernoulli's Inequality (see Problem 35 of Chapter 2) and so $x_n \to 0$ as $n \to \infty$. If $p_1 = 1$, then $x_n = 0$ for all n and there is nothing to prove.

PROBLEM 19. Let c for the moment be an arbitrary real number and put

$$x_1 = \frac{c}{2} \quad\text{and}\quad x_{n+1} = \frac{c}{2} + \frac{x_n^2}{2} \quad\text{for}\quad n = 1,2,3,\ldots$$

Investigate the convergence of the sequence $(x_n)_{n=1}^{\infty}$.

Solution. We observe that if the limit of the given sequence exists, say $\lim_{n\to\infty} x_n = a$, then

$$a = \frac{c}{2} + \frac{a^2}{2}, \quad \text{or} \quad a^2 - 2a + c = 0.$$

Thus

$$a = 1 \pm \sqrt{1 - c},$$

and so the sequence $(x_n)_{n=1}^{\infty}$ can not converge for $c > 1$.

(a) We assume now that $0 < c \le 1$. Then $x_n > 0$. It is easy to verify that $x_n < 1$ and $x_n < x_{n+1}$ for $n = 1,2,3,\ldots$ Thus $\lim_{n\to\infty} x_n$ exists and cannot be larger than 1; hence the limit is equal to $1 - \sqrt{1 - c}$.

(b) Suppose that $-3 \le c < 0$. Then $x_n \ge c/2$ and $x_n < 0$ for $n = 1,2,3,\ldots$ Indeed, the first claim is obvious and we need only to have a closer look at the inequality $x_n < 0$ for $n = 1,2,3,\ldots$ But the inequality is true for $n = 1$. Assuming that $x_n < 0$ for some positive integer n, we see that $x_n \le |c|/2$,

$$x_n^2 \le \frac{|c|^2}{4} < c \qquad (\text{because } |c|/4 < 1)$$

and x_{n+1} is seen to have the same sign as $c/2$, that is, x_{n+1} is negative.

At any rate the sequence $(x_n)_{n=1}^{\infty}$ is bounded. However, the sequence is not monotonic. Noting that

$$x_{2k+1} - x_{2k-1} = \frac{1}{2}(x_{2k}^2 - x_{2k-2}^2)$$

and

$$x_{2k+2} - x_{2k} = \frac{1}{2}(x_{2k+1}^2 - x_{2k-1}^2),$$

it can easily be established by induction that

$$x_{2k+1} > x_{2k-1} \quad \text{and} \quad x_{2k+2} < x_{2k} \quad \text{for } k = 1,2,3,\ldots$$

In other words,

$$a' = \lim_{n\to\infty} x_{2k-1} \quad \text{and} \quad a'' = \lim_{n\to\infty} x_{2k}$$

exist. We now show that $a' = a''$. In

$$x_{n+1} = c/2 + x_n^2/2$$

we let $n \to \infty$; we first let $n \to \infty$ with n even, and then with n odd, and we obtain

$$a' = c/2 + (a'')^2/2, \qquad a'' = c/2 + (a')^2/2.$$

Thus

$$a' - a'' = \frac{1}{2}\{(a'')^2 - (a')^2\}$$

and so $(a' - a'')(a' + a'' + 2) = 0$. As we shall see at once, the second factor cannot vanish for $c > -3$, and so $a' = a''$ must hold. Indeed, if we set $a'' = a' - 2$ in

$$a'' = \frac{c}{2} + \frac{(a')^2}{2},$$

we would get

$$(a')^2 + 2a' + (4 + c) = 0$$

which has no real root for $c > -3$.

For $c = -3$ both factors $(a' - a'')$ and $(a' + a'' + 2)$ vanish because then $a' = a'' = -1$.

Thus in all cases $a' = a''$. If we denote this common value by a we see that $a = 1 - \sqrt{1 - c}$ because the limit of the negative-valued sequence $(x_n)_{n=1}^{\infty}$ cannot be (strictly) positive.

Finally, for $c < -3$ the seqquence $(x_n)_{n=1}^{\infty}$ does not converge; for example, for $c = -4$ we obtain the sequence

$$-2,\ 0,\ -2,\ 0,\ -2,\ 0,\ \ldots$$

and this sequence has no limit.

PROBLEM 20. Let x be real and consider the equation $x = a^x$ with $a > 0$. Setting

$$x_1 = a^x,\quad x_2 = a^{x_1},\quad x_3 = a^{x_2},\quad \ldots,\quad x_n = a^{x_{n-1}},\quad \ldots,$$

investigate the limit $L(x) = \lim_{n\to\infty} x_n$.

Solution. We observe that it is sufficient to suppose x to be positive; if $x < 0$, we may take $x_1 = a^x$ in place of x.

First we show that $\lim_{n\to\infty} x_n = \infty$ if $a > e^{1/e} = 1.44466\ldots$ Indeed, since $\log x \le x/e$, where e is the base of the natural logarithms, we have for the considered case $\log x \le x \log a$ or $x < a^x$; that is, $x < x_1$. Thus we see that

$$x < x_1 < x_2 < \cdots < x_{n-1} < x_n.$$

We now show that the difference

$$x_{k+1} - x_k = a^{x_k} - x_k$$

is larger than a certain positive number. We consider the function

$$g(x) = a^x - x$$

and form its derivative

$$g'(x) = a^x \log a - 1.$$

We see that, if $a > e$, the derivative is positive and the function g has its least value at $x = 0$; but we have $g(0) = 1$, and it follows from this that $g(x) > 1$.

If $e^{1/e} < a < e$, the derivative vanishes for

$$x = -\frac{\log \log a}{\log a}$$

and we obtain

$$a^x - x > \frac{1 + \log \log a}{\log a}.$$

Denoting by A the positive number $(1 + \log \log a)/(\log a)$, we have that $g(x) > A$, that is, all differences $x_{k+1} - x_k$ are larger than A. We thus see that x_n becomes arbitrarily large as $n \to \infty$.

We now take up the case $1 < a < e^{1/e}$. The equation $a^x = x$ has two real roots in this case, one being between 1 and e and the other between e and ∞. We denote the first root by α and the second by β.

The function $x_1 - x = a^x - x = g(x)$ is positive for $x = 0$ and $x = \infty$; it only remains to study its values for all $x > 0$. We form the derivative $g'(x) = a^x \log a - 1$ and denote its positive root by x_0; we evidently get that

$$a^{x_0} = \frac{1}{\log a}.$$

As $1 < a < e^{1/e}$, we have $0 < \log a < 1/e$, and so $1/(\log a) > e$; we have

$$a^{x_0} > e, \quad \text{or} \quad x_0 \log a > 1.$$

Thus the function g decreases in the interval from 0 to x_0 attaining its minimal value at x_0 and then increases for all $x > x_0$. By substituting x_0 for x in g(x), we get

$$g(x_0) = a^{x_0} - x_0 = \frac{1}{\log a} - x_0 = \frac{1 - x_0 \log a}{\log a} < 0;$$

thus the equation $g(x) = 0$ has two simple roots, one between 0 and x_0 and the other between x_0 and ∞.

It is easy to see that the number e is between the real roots of the equation $g(x) = 0$. Indeed,

$$g(e) = a^e - e < (e^{1/e}) - e = 0.$$

We consider the three intervals from 0 to α, from α to β, and from β to ∞.

The function g is positive in the first and third interval and negative in the second. Consequently, if we take the initial value x in the first or third interval, the consecutive values of x_k increase with k. On the other hand, if the initial value x comes from the second interval, the numbers x_k form a decreasing sequence.

By putting $x = \beta + \delta$, with $\delta > 0$, we have

$$a^{\beta+\delta} - \beta - \delta = (a^{\delta} - 1)\beta - \delta < \delta(\log \beta - 1)$$

because $a^{\beta} - \beta = 0$ and $(a^{\delta} - a^{0})/\delta = a^{t}(\log a)$ for some t such $0 < t < \delta$ by the Mean Value Theorem of differential calculus; but $\log a = (\log \beta)/\beta$ and $a^{t} > 1$ in the considered case. Thus we see that

$$x_{k+1} - x_k > \delta(\log \beta - 1),$$

and so, however small δ is, we have $\lim_{n\to\infty} x_n = \infty$ for $x > \beta$.

As $a > 1$, it follows that

$$a^x < e^{\gamma}$$

for $x < \gamma$. By taking, instead of γ, the root α, we have, for $x < \alpha$

$$a^x < a^{\alpha},$$

that is to say, $x_1 < \alpha$. We also note that $x_n < \alpha$; but since, for $x < \alpha$, we have

$$x_n > x_{n-1} > \cdots > x_1 > x,$$

we see that x_n tends to a limit which does not exceed α; but as this limit is equal to one of the roots of the equation $a^x = x$ we have that this limit is equal to α, that is, we can write $\lim_{n\to\infty} x_n = \alpha$.

In the same way we can show that if x is between α and β, we have

$$x > x_1 > x_2 > \cdots > x_n > \alpha,$$

and so $\lim_{n\to\infty} x_n = \alpha$. If $x = \beta$, we have

$$x = x_1 = x_2 = \cdots = x_n = \beta$$

and so $\lim_{n\to\infty} x_n = \beta$. The limiting case $a = e^{1/e}$ gives

$$\lim_{n\to\infty} x_n = e \quad \text{for } x \le e$$

and

$$\lim_{n\to\infty} x_n = \infty \quad \text{for } x > e.$$

We come to the case $a < 1$.

When considering the function g, we noted that its derivative is constantly negative for $a < 1$; thus the function is decreasing; it is positive for $x = 0$ and negative for $x = 1$, hence has one real root between 0 and 1. We denote this root by α.

We consider for the moment the equation

$$a^{a^x} = x.$$

This equation evidently does not have real roots for $a > e^{1/e}$; for $1 < a < e^{1/e}$ the two roots α and β of the equation $a^x = x$ are the only roots of the equation

$$a^{a^x} = x.$$

Always staying with the case $a < 1$ and putting $a = 1/b$, we obtain $b > 1$. We take up the function

$$h(x) = a^{a^x} - x.$$

By putting $y = a^x$, we can write the derivative h' in the form

$$h'(x) = a^y\, y \log^2 a - 1 = b^{-y}\, y \log^2 b - 1.$$

The second derivative is

$$h''(x) = \log^3 a\; a^y\, y\, (1 + \log \log ay) = -\log^3 b\; b^{-y}\, y\, (1 - \log \log by).$$

If we suppose $b < e$, we have

$$1 - \log b\, b^{-x} > 0,$$

and we obtain

$$h''(x) < 0;$$

consequently, the first derivative h' is a decreasing function. But we have

$$h'(0) = \frac{\log^2 b}{b} - 1 < 0,$$

and so $h'(x) < 0$.

The function h is decreasing and since

$$h(0) = a > 0, \quad h(1) = a^a - 1 < 0,$$

it becomes zero for the root α of the equation $a^x = x$.

For the case $b > e$, the equation $h''(x) = 0$ admits a real root between 0 and 1; this root is equal to

$$x_0 = \frac{\log \log b}{\log b}.$$

For $0 \le x < x_0$ we have $h''(x) > 0$; but for $x_0 < x \le 1$ we have $h''(x) < 0$. In the first interval the function h' increases when we start from the value

$$\frac{\log^2 b}{b} - 1$$

which is negative for all values of b and attains its maximum for $x = x_0$. After this value the function begins to decrease.

If $h'(x_0) < 0$, $h'(x)$ will be negative for all values of x and so $h(x)$ has the only root α which will be at the same time root of the equation $a^x = x$.

If one has $h'(x_0) > 0$, it is not difficult to convince oneself that the equation $h(x) = 0$, in addition to the root of the equation $a^x = x$, admits two others. But, since

$$h'(x_0) = \frac{1}{e} \log b - 1,$$

we prove that if $\log b > e$, the equation $h(x) = 0$ has two real roots included between 0 and 1, distinct from α. In this case

$$b > e^e, \quad a < \frac{1}{e^e} = 0.065948\ldots$$

The function h clearly has one root α which belongs to the equation $a^x = x$. We substitute this root into the expression of the first derivative; we have the relation

$$h'(\alpha) = \alpha^2 \log^2 b - 1 = \log^2 \alpha - 1 \geq 0,$$

where $h'(\alpha)$ becomes zero for $b = e^e$, $\alpha = 1/e$, and $h'(\alpha)$ is positive for $b > e^e$.

Thus it is seen that for ε positive and very small

$$h(\alpha - \varepsilon) < 0, \quad h(\alpha + \varepsilon) > 0.$$

But, because of the inequalities $h(0) > 0$, $h(1) < 0$, we arrive at the result that the equation $h(x) = 0$ has two other real roots, one between 0 and α and the other between α and 1. We denote the first root by α_1 and the second by α_2, and have

$$\alpha_2 > \alpha_1, \quad \alpha_2 = a^{\alpha_1}, \quad \alpha_1 = a^{\alpha_2}.$$

Thus we see that, for the case $1/(e^e) \leq a < 1$, we have

$$a^{x_k} \gtrless a^\alpha$$

if $x_k \gtrless \alpha$; but $a^{x_k} = x_{k+1}$ and $a^\alpha = \alpha$ from which

$$x_{k+1} \lessgtr \alpha.$$

When $x < \alpha$, we clearly have

$$x < \alpha, \quad x_2 < \alpha, \quad x_4 < \alpha, \quad \ldots, \quad x_{2n} < \alpha,$$

$$x_1 > \alpha, \quad x_3 > \alpha, \quad \ldots, \quad x_{2n+1} > \alpha.$$

Since the $h(x)$ and $g(x)$ have the same sign, we have the inequalities

$$x_{2k+2} - x_{2k} = h(x_{2k}) > 0,$$

$$x_{2k+1} - x_{2k-1} = h(x_{2k-1}) < 0,$$

and we note that the expressions with even indices increase, remaining less than α; consequently they tend to a limit. In the same way, the expressions with odd indices decrease, remaining larger than α and tending to a limit. It is clear that these expressions cannot posess a limit different from the root of the function h and so

$$\lim_{n\to\infty} x_{2n} = \lim_{n\to\infty} x_{2n+1} = \alpha.$$

For $x > \alpha$ the reasoning is the same. The expressions with even index approach the limit α decreasingly.

Finally, we take up the case $a < 1/(e^e)$.

We consider the four intervals

$$I(0,\alpha_1),\quad II(\alpha_1,\alpha),\quad III(\alpha,\alpha_2),\quad IV(\alpha_2,\infty).$$

The signs of the functions g and h for these intervals are tabulated next:

	I	II	III	IV
g(x)	+	+	-	-
h(x)	+	-	+	-

The table of signs of the functions g and h shows that the root α is to be found between two consecutive terms of our sequence

$$x,\quad x_1,\quad x_2,\quad x_3,\quad \ldots,\quad x_n,\quad \ldots$$

We will show that the terms with even index are to be found in the same interval. Indeed, if x is in the first interval, x_1 will be in the fourth and conversely.

This can be shown in the following manner: If $x < \alpha_1$, we have

$$a^x > a^{\alpha_1},$$

or, in other words,

$$x_1 > \alpha_2,$$

and conversely; if $x > \alpha_2$, we have

$$x_1 < \alpha_1.$$

We note that all terms x_{2k} with even index are situated in the same interval and all other terms x_{2k+1} in the other.

If x is in the second interval, all terms x_{2k} are in the same interval, but all terms with odd index are in the third interval and vice versa. Indeed, if $\alpha_1 < x < \alpha$, we have

$$a^{\alpha_1} > a^x > a^\alpha \quad \text{or} \quad \alpha_2 > x_1 > \alpha.$$

In the same way, if $\alpha < x < \alpha_2$, we have

$$a^\alpha > a^x > a^{\alpha_2} \quad \text{or} \quad \alpha > x_1 > \alpha_1.$$

From the preceding discussion we conclude that if x is in one of the two first intervals I, II, we have $x_{2n} \to \alpha_1$, $x_{2n+1} \to \alpha_2$ as $n \to \infty$. For the intervals III, IV, or $x > \alpha$, we have $x_{2n} \to \alpha_2$, $x_{2n+1} \to \alpha_1$ as $n \to \infty$.

We summarize the results in the following fashion:

Theorem. The limit L(x) is determined as follows:

1. If $a > e^{1/e}$, then
   $L(x) = \infty$ for all real values of x.
2. If $1 < a \le e^{1/e}$, then
   $L(x) = \alpha$ for $-\infty < x < \alpha$,
   $L(\beta) = \beta$ for $x = \beta$,
   $L(x) = \infty$ for $\beta < x < +\infty$.
3. If $1/(e^e) \le a \le 1$, then $L(x) = \alpha$.
4. If $0 < a < 1/(e^e)$ and putting $L_1(x) = \lim_{n\to\infty} x_{2n}$ and $L_2(x) = \lim_{n\to\infty} x_{2n+1}$, then
   $L_1(x) = \alpha_1$, $L_2(x) = \alpha_2$ for $-\infty < x < \alpha$,
   $L_1(\alpha) = L_2(\alpha) = \alpha$ for $x = \alpha$,
   $L_1(x) = \alpha_2$, $L_2(x) = \alpha_1$ for $\alpha < x < +\infty$.

Here α and β are roots of $a^x = x$ and $\alpha < \beta$; moreover, $\alpha_1 < \alpha < \alpha_2$ and $a^{\alpha_1} = \alpha_2$, $a^{\alpha_2} = \alpha_1$.

Remark. The foregoing Theorem is actually due to Euler (De formulis exponentialibus replicatis, Acta Academia Scientiarum Imperialis Petropolitanae, 1777.)

PROBLEM 21. Show that for any irrational number α, there exist infinitely many rational numbers p/q such that

$$\left|\alpha - \frac{p}{q}\right| < q^{-2}.$$

Solution. Let n be a positive integer. Consider the n + 1 real numbers

$$0, \quad \alpha - [\alpha], \quad 2\alpha - [2\alpha], \quad \ldots, \quad n\alpha - [n\alpha] \tag{E.2}$$

(here [r] denotes the greatest integer contained in r) and their distribution in the half-open intervals

$$\left[\frac{j}{n}, \frac{j+1}{n}\right) \quad \text{with } j = 0,1,2,\ldots,n-1.$$

The union of these n intervals is the half-open interval [0,1) and hence contain the n + 1 numbers (E.2). It is clear that two of these numbers (E.2) lie in the same interval; this is the "pigeonhole principle": if there are n + 1 objects in n boxes, there must be at least one box containing more than one object. Call these numbers

$$n_1\alpha - [n_1\alpha] \quad \text{and} \quad n_2\alpha - [n_2\alpha] \quad \text{with} \quad 0 \le n_1 < n_2 \le n.$$

Since the intervals are of length 1/n and are not closed at both ends, we must have that

$$|n_2\alpha - [n_2\alpha] - n_1\alpha + [n_1\alpha]| < \frac{1}{n}.$$

Write q for the positive integer $n_2 - n_1$ and p for $[n_2\alpha] - [n_1\alpha]$; we have that $|q\alpha - p| < 1/n$ with $q \le n$. Hence, given any positive integer n, there exist integers q and p with $n \ge q > 0$ such that

$$|q\alpha - p| < \frac{1}{n} \quad \text{or} \quad \left|\alpha - \frac{p}{q}\right| < \frac{1}{nq}. \tag{E.3}$$

The latter relation implies the inequality stated in the problem, because $n \ge q$ implies that

$$\frac{1}{nq} \le \frac{1}{q^2}.$$

To show that there are infinitely many pairs (p,q), suppose, on the contrary, that there are only finitely many, say $(p_1,q_1),\ldots,(p_m,q_m)$. We prove that this supposition is false by finding another pair (p,q) satisfying (E.3). We define ε as the minimum of

$$\left|\alpha - \frac{p_1}{q_1}\right|, \quad \ldots, \quad \left|\alpha - \frac{p_m}{q_m}\right|.$$

Since α is irrational, ε is positive. Choose n so that $1/n < \varepsilon$, and then by the first part of the proof which led to (E.3) we can find a rational number p/q so that

$$\left|\alpha - \frac{p}{q}\right| < \frac{1}{nq} < \frac{1}{n} < \varepsilon.$$

By the definition of ε it follows that p/q is different from p_i/q_i for $i = 1,2,\ldots,m$.

Remark. The "pigeonhole principle" has already been used in Problems 35 and 36 in Chapter 1.

PROBLEM 22. Let α be an irrational number. Show that the set of all numbers of the form $\alpha x + y$, with x and y integers, is dense in the real number system, that is, given any real number r and any $\varepsilon > 0$, there are integers x and y such that $|\alpha x + y - r| < \varepsilon$ holds.

Solution. From Problem 21 we see that the inequality

$$|\alpha x + y| < \varepsilon < 1$$

can be satisfied by nonzero integers x and y. Hence let

$$\alpha x_1 + y_1 = \beta, \quad |\beta| < \varepsilon.$$

Let g be the largest integer contained in r/β; then

$$g \leq \frac{r}{\beta} < g + 1.$$

Thus, putting $x = gx_1$ and $y = gy_1$, we get

$$|\alpha x + y - r| = |\alpha gx_1 + gy_1 - r| = |g\alpha - r| < |\beta| < \varepsilon.$$

PROBLEM 23. The decimal fraction

$$\alpha = 0.1234567891011121314\ldots$$

(the positive integers written consecutively) represents an irrational number. Show that, for $n = 0,1,2,3,\ldots$,

$$10^n\alpha - [10^n\alpha]$$

is dense in the interval [0,1].

Solution. We obtain $10^n\alpha - [10^n\alpha]$ by shifting the decimal point in α

by n places to the right and deleting all digits to the left of the decimal point. Let

$$\beta = 0.\beta_1\beta_2\ldots\beta_k$$

be a finite decimal fraction. We choose n such that $10^n\alpha - [10^n\alpha]$ begins with the digits $\beta_1, \beta_2, \ldots, \beta_k$ and is followed by r zeros; then we have

$$|10^n\alpha - [10^n\alpha] - \beta| < \frac{1}{10^{k+r}}.$$

PROBLEM 24. Let be an irrational number and consider the sequence

$$n\alpha - [n\alpha] \quad \text{for } n = 1,2,3,\ldots$$

Show that the set of accumulation points (i.e., the set of all subsequential limits) is [0,1].

Solution. By Problem 21, there are integers q_1 and p such that, for any $\varepsilon > 0$,

$$|q_1\alpha - p| < \varepsilon.$$

The point $q_1\alpha - [q_1\alpha]$ is therefore within a distance of ε of either 0 or 1. The sequence of points

$$q_1\alpha - [q_1\alpha], \quad 2q_1\alpha - [2q_1\alpha], \quad 3q_1\alpha - [3q_1\alpha], \quad \ldots$$

continued so long as may be necessary, mark a chain (in one direction or the other) across the interval [0,1] such that the distance between consecutive points on the chain is less than ε. Hence for any point r of (0,1) there is a point $kq_1\alpha - [kq_1\alpha]$ or $n\alpha - [n\alpha]$ within a distance ε and so the set of points $n\alpha - [n\alpha]$ for n = 1,2,3,... is dense in the interval [0,1].

Remarks. Let e be the base of the natural logarithms. Since e is an irrational number, the set of accumulation points of the sequence

$$ne - [ne] \quad \text{with } n = 1,2,3,\ldots$$

is the interval (0,1). However, the sequence

$$n!e - [n!e] \quad \text{with } n = 1,2,3,\ldots$$

has 0 as its only point of accumulation.

Indeed, by Taylor's Theorem,

$$e = 1 + \frac{1}{1!} + \frac{1}{2!} + \cdots + \frac{1}{n!} + \frac{e^{\theta_n}}{(n + 1)!} \quad \text{with } 0 < \theta_n < 1,$$

hence

$$n!e = n! + \frac{n!}{1!} + \frac{n!}{2!} + \cdots + 1 + \frac{e^{\theta_n}}{n + 1}.$$

For $n \geq 2$, we have

$$\frac{e^{\theta_n}}{n + 1} < \frac{1}{n + 1} < 1,$$

hence

$$n!e - [n!e] < \frac{e^{\theta_n}}{n + 1} < \frac{e}{n + 1}.$$

Another interesting example of a sequence with an unusual set of accumulation points is provided by the following situation. For every integer $p \geq 2$ there are exactly $p - 1$ numbers of the form $1/k + 1/m$ for which the sum of the positive integers k and m equals p. For $p = 2,3,4,\ldots$, consider these numbers enumerated. Then we obtain the sequence

$$2, \frac{3}{2}, \frac{3}{2}, \frac{4}{3}, 1, \frac{4}{3}, \frac{5}{4}, \frac{5}{6}, \frac{5}{6}, \ldots$$

It is easily checked that this sequence has the accumulation points

$$0, 1, \frac{1}{2}, \frac{1}{3}, \frac{1}{4}, \ldots$$

and no other accumulation points.

PROBLEM 25. Consider the following enumeration of the proper fractions

$$\frac{0}{1}, \frac{1}{1}, \frac{0}{2}, \frac{1}{2}, \frac{2}{2}, \frac{0}{3}, \frac{1}{3}, \frac{2}{3}, \ldots$$

Here we have ordered the proper fractions into groups according to increasing denominators and, within each group, according to increasing numerators and we have also listed fractions in their unreduced form. It is clear that in this enumeration the fraction p/q has index

$$k = k(p,q) = + \tfrac{1}{2}q(q + 1).$$

For each k, we now cover the k-th term in this enumeration of rational numbers between 0 and 1 by an interval of length 2^{-k} in such a way that the k-th term is at the center of its covering interval. The sum of the lengths of these covering intervals is

$$\sum_{k=1}^{\infty} \frac{1}{2^k} = 1.$$

Show that the point $\sqrt{2}/2$ does not belong to any covering interval however.

Solution. Suppose that $t = \sqrt{2}/2$ belongs to one of the covering intervals. Then there would be integers p and q ($q \geq p \geq 0$, $q \geq 1$) such that

$$\left|\frac{p}{q} - t\right| < \frac{1}{2}\,\frac{1}{2^{p+(1/2)q(q+1)}} \leq \frac{1}{2^{(1/2)q(q+1)+1}}.$$

But for every p and q with $q \geq 1$, we have, since $\sqrt{2}$ is an irrational number,

$$|2p^2 - q^2| \geq 1$$

and therefore

$$\left|\frac{p}{q} - \frac{\sqrt{2}}{2}\right| = \frac{|(p/q)^2 - 1/2|}{(p/q) + (\sqrt{2}/2)} = \frac{|2p^2 - q^2|}{2pq + q^2\sqrt{2}} > \frac{1}{4q^2}$$

$$\geq \frac{1}{2^{(1/2)q(q+1)}}.$$

PROBLEM 26. The harmonic series $\Sigma_{n=1}^{\infty}$ 1/n diverges, but if we omit the terms corresponding to the integers whose decimal representations contain a specified digit at least once (the digit 0, for example), the resulting series converges. Verify these claims.

Solution. By Problem 14, the harmonic series $\Sigma_{n=1}^{\infty}$ 1/n is seen to be divergent. In the Remarks to Problem 14 we also noted that the divergence was very slow.

Another way of seeing that the harmonic series is divergent is to note that the sequence of partial sums of this series does not form a Cauchy sequence; indeed, setting

$$s_n = 1 + \frac{1}{2} + \cdots + \frac{1}{n},$$

we have that

$$s_{2n} - s_n = \frac{1}{n+1} + \frac{1}{n+2} + \cdots + \frac{1}{2n}.$$

Denoting the sum on the right-hand side of the last equation by t_n, it is clear that $1/2 < t_n < 1$. In fact,

$$\log\left(2 - \frac{1}{n+1}\right) < t_n < \log 2$$

because, for any positive integers m and j with $m < j$,

$$\log \frac{j+1}{m} < \frac{1}{m} + \frac{1}{m+1} + \cdots + \frac{1}{j} < \log \frac{j}{m-1}.$$

The last inequality is a consequence of

$$\frac{1}{k+1} < \int_k^{k+1} \frac{dx}{x} < \frac{1}{k} \quad \text{for } k = 1,2,3,\ldots$$

Thus t_n tends to log 2 as $n \to \infty$.

The rest of the claim seems surprising at first sight; but for very large n the integers that do not have a particular digit in their decimal representations are quite scarce. Indeed, there are $9 \cdot 10^{n-1}$ integers with n-digit decimal representations, but only 9^n of them omit 0, and $9^n/9 \cdot 10^{n-1} \to 0$ as $n \to \infty$.

The n-digit integers which do not have any digit 0 can be written as $a_1 \ldots a_n$ (understood to mean, $10^{n-1}a_1 + 10^{n-2}a_2 + \cdots + a_n$), where $0 < a_k \le 9$. Since all the terms are positive, the convergence of $\Sigma\, 1/k$ over all k of this form is equivalent to the convergence of

$$S = \sum_{n=1}^{\infty} \Sigma \frac{1}{a_1 \ldots a_n}.$$

For a given n there are 9^n terms in the inner sum, and each exceeds 10^{n-1}, so

$$S \le \sum_{n=1}^{\infty} \frac{9^n}{10^{n-1}} = 90.$$

To go further, we notice that of the n-digit integers, one-ninth have leading digit 1, one-ninth have leading digit 2, and so on; so of the 9^n n-digit integers, 9^{n-1} are between 10^{n-1} and $2 \cdot 10^{n-1}$, and so on. These integers then contribute at most

$$10^{-n+1}\, 9^{n-1}(1 + 1/2 + \cdots + 1/9) = (0.9)^{n-1}(1 + 1/2 + \cdots + 1/9)$$

to S, and consequently

$$S < \left(1 + \frac{1}{2} + \cdots + \frac{1}{9}\right) \sum_{n=1}^{\infty} (0.9)^{n-1} = 10\left(1 + \frac{1}{2} + \cdots + \frac{1}{9}\right) < 28.3.$$

On the other hand, the integers between 10^{n-1} and $2\ 10^{n-1}$ are less than $2 \cdot 10^{n-1}$, and so on, so that for $n \geq 2$ the n-digit integers contribute at least

$$(0.9)^{n-1}\left(\frac{1}{2} + \cdots + \frac{1}{10}\right)$$

to S. Hence

$$S > 1 + \frac{1}{2} + \cdots + \frac{1}{9} + \sum_{n=1}^{\infty} (0.9)^{n-1}\left(\frac{1}{2} + \cdots + \frac{1}{10}\right) > 20.189.$$

PROBLEM 27. Prove that every positive rational number is the sum of a finite number of distinct terms of the harmonic series

$$1 + \frac{1}{2} + \frac{1}{3} + \cdots + \frac{1}{n} + \cdots$$

Solution. Let A and B be positive integers. Then by the divergence of the series under consideration there is a unique nonnegative integer n_0 such that (taking $1/0 = 0$)

$$\sum_{j=0}^{n_0} \frac{1}{j} < \frac{A}{B} \leq \sum_{j=0}^{n_0+1} \frac{1}{j}.$$

If equality holds the desired representation is at hand and so we assume that

$$\frac{A}{B} < \sum_{j=0}^{n_0+1} \frac{1}{j}.$$

Then $A/B - \Sigma_{j=0}^{n_0} 1/j = C/D < 1/(n_0 + 1)$. Take n_1 as the unique positive integer such that $1/(n_1 + 1) \leq C/D < 1/n_1$. We again suppose inequality as the problem is otherwise solved, and put

$$\frac{C}{D} - \frac{1}{n_1 + 1} + \frac{E}{F} > 0.$$

But $E/F = \{C(n_1 + 1) - D\}/D(n_1 + 1)$ and $C(n_1 + 1) - D < C$ so that with E/F in lowest terms we must have $E < C$. Then

$$\frac{E}{F} < \frac{1}{n_1(n_1 + 1)}$$

and so the unique integer n_2 such that $1/(n_2 + 1) \le E/F < 1/n_2$ satisfies $n_2 > n_1$. In a finite number of steps we must obtain the desired representation, since if the equality does not occur before, it must occur when the numerator of the reduced fraction has become 1.

PROBLEM 28. Consider an infinite series whose n-th term is $\pm(1/n)$, the $\pm$ signs being determined according to a pattern that repeats periodically in blocks of eight. There are 2^8 possible patterns. Show that a necessary and sufficient condition for the series to be conditionally convergent is that there be four "+" signs and four "-" signs in the block of eight.

Solution. Let u_n be the n-th term $\pm 1/n$ and $S_n = u_1 + \cdots + u_n$. Since $u_n \to 0$ as $n \to \infty$, $(S_n)_{n=1}^{\infty}$ will converge if and only if $(S_{8n})_{n=1}^{\infty}$ does. Using the fact that

$$\frac{1}{n} - \frac{1}{n + k} = \frac{k}{n(n + k)},$$

that $\Sigma_{n=1}^{\infty} (1/n^2)$ converges (see Problem 7), and that $\Sigma_{n=1}^{\infty} (1/n)$ diverges (see Problem 26), one shows that with four "+" signs and four "-" signs in each block, $(S_{8n})_{n=1}^{\infty}$ converges as the term-by-term sum of four convergent sequences while an imbalance of signs makes $(S_{8n})_{n=1}^{\infty}$ divergent as the sum of a convergent and a divergent sequence.

PROBLEM 29. Let $p_n > 0$, $p_1 \ge p_2 \ge p_3 \ge \cdots$, the series

$$p_1 + p_2 + p_3 + \cdots + p_n + \cdots$$

be divergent and the series

$$s_1p_1 + s_2p_2 + s_3p_3 + \cdots + s_np_n + \cdots,$$

where s_n assume only the values -1, 1, be convergent. Show that

$$\liminf_{n\to\infty} \frac{s_1 + s_2 + \cdots + s_n}{n} \le 0 \le \limsup_{n\to\infty} \frac{s_1 + s_2 + \cdots + s_n}{n}.$$

Solution. Put $S_n = s_1 + s_2 + \quad + s_n$, $S_0 = 0$. Then

$$s_1p_1 + s_2p_2 + \cdots + s_np_n = \sum_{k=1}^{n} (S_k - S_{k-1})p_k$$

$$= \sum_{k=1}^{n-1} S_k(p_k - p_{k+1}) + S_np_n.$$

If we had $S_n > cn$ for $n > m$, $c > 0$, then

$$s_1p_1 + s_2p_2 + \cdots + s_np_n$$

$$> \sum_{k=1}^{m} S_k(p_k - p_{k+1}) + c\sum_{k=m+1}^{n-1} k(p_k - p_{k+1}) + cnp_n$$

$$= K + c\sum_{k=m+1}^{n} p_k$$

would follow; here K is independent of n and so the right-hand side tends to $+\infty$.

PROBLEM 30. Let $p_n > 0$, $p_1 \ge p_2 \ge p_3 \ge \cdots$ and the series

$$s_1p_1 + s_2p_2 + s_3p_3 + \cdots + s_np_n + \cdots$$

in which the factors $s_1, s_2, s_3, \ldots$ assume only the values -1, 1, be convergent. Show that

$$\lim_{n\to\infty} (s_1 + s_2 + \cdots + s_n)p_n = 0.$$

Note the two extreme cases $s_1 = s_2 = s_3 = \cdots$ and $s_1 = -s_2 = s_3 = -s_4 = \cdots$.

Solution. Let $S_n = s_1 + s_2 + \cdots + s_n$ as in Problem 29. Then the sequence

$$S_1, S_2, S_3, \ldots, S_n, \ldots \tag{E.4}$$

has the property that between two terms having opposite signs there is a vanishing term. We distinguish two cases: 1. in the sequence (E.4) infinitely many terms vanish; 2. disregarding a finite set of terms, all terms of (E.4) have the same sign. Let them be positive for example. In case 1, let the index M be so chosen that $S_M = 0$ and for $M \le m < n$ such that

$$\left|\sum_{k=m+1}^{n} s_k p_k\right| = \left|\sum_{k=m+1}^{n} [(S_k - S_m) - (S_{k-1} - S_m)]p_k\right|$$
$$= \left|\sum_{k=m+1}^{n-1} (S_k - S_m)(p_k - p_{k+1}) + (S_n - S_m)p_n\right| < \varepsilon \tag{E.5}$$

holds. Let S_m be the closest term on the left of S_n in (E.4) which vanishes such that $S_{m+1}, S_{m+2}, \ldots, S_n$ have the same sign. Then it follows from inequality (E.5) that $|(S_n - S_m)p_n| = |S_n p_n| < \varepsilon$. In case 2, let M be so chosen that inequality (E.5) holds for $M \le m < n$ and that moreover $S_M, S_{M+1}, S_{M+2}, \ldots$ are positive. Let S_m be their minimum. Since in this case $S_k - S_m \ge 0$ for $k > m$, it follows from the estimate (E.5) that $(S_n - S_m)p_n < \varepsilon$, that is, $S_n p_n < \varepsilon + S_m p_n$. But m is fixed and p_n tends to 0; hence for sufficiently large n we have $S_n p_n < \varepsilon$.

PROBLEM 31. Let the terms of the convergent series

$$p_1 + p_2 + p_3 + \cdots + p_n + \cdots = s$$

satisfy the inequalities

$$p_1 \ge p_2 \ge p_3 \ge \cdots,$$

$$0 < p_n \le p_{n+1} + p_{n+2} + p_{n+3} + \cdots.$$

Show that any point σ in the half-closed interval $0 < \sigma \le s$ can be represented by an infinite subseries

$$p_{t_1} + p_{t_2} + p_{t_3} + \cdots + p_{t_n} + \cdots = \sigma.$$

Solution. Put, for $n = 1,2,3,\ldots$, $k = 0,1,2,\ldots$,

$$p_n + p_{n+1} + \cdots + p_{n+k} = P_{n,k} \quad \text{and} \quad \lim_{k\to\infty} P_{n,k} = P_n.$$

Let p_{n_1} be the first term such that $p_{n_1} < \sigma$; then either there exists some

k_1 such that $P_{n_1,k_1} < \sigma$, $P_{n_1,k_1+1} \geq \sigma$, $k_1 \geq 0$, or we have $P_{n_1} \leq \sigma$. In the second case we have, since $P_{n_1} \geq P_{n_1-1} \geq \sigma$ (for $n_1 = 1$ these inequalities are to read: $P_1 = s \geq \sigma$), $P_{n_1} = \sigma$, that is, σ can be represented by an infinite subseries. In the first case we further determine the first term p_{n_2} such that $n_2 > n_1 + k_1$, $P_{n_1,k_1} + p_{n_2} < \sigma$; then either there is some k_2 satisfying $P_{n_1} + P_{n_2,k_2} < \sigma$, $P_{n_1,k_1} + P_{n_2,k_2+1} \geq \sigma$, $k_2 \geq 0$, or we have $P_{n_1,k_1} + P_{n_2} \leq \sigma$. In the second case, since

$$P_{n_1,k_k} + P_{n_2} \geq P_{n_1,k_1} + P_{n_2-1} \geq \sigma$$

($n_2 > n_1 + k_1 + 1$, because $P_{n_1,k_1} + p_{n_1+k_1+1} = P_{n_1,k_1+1} \geq 0$), $P_{n_1,k_1} + P_{n_2} = \sigma$, that is, σ is again representable by an infinite subseries. If this process never terminates (that is, if the first case always occurs), then

$$\sigma = P_{n_1,k_1} + P_{n_2,k_2} + P_{n_3,k_3} + \cdots.$$

PROBLEM 32. Find the series $p_1 + p_2 + \cdots + p_n + \cdots$ such that

$$p_1 = \frac{1}{2}, \quad p_n = p_{n+1} + p_{n+2} + p_{n+3} + \cdots, \quad n = 1,2,3,\ldots$$

and note that in this case each σ mentioned in Problem 31 can only be represented by infinite subseries.

Solution. From $p_n = p_{n+1} + p_{n+2} + p_{n+3} + \cdots$ and

$$p_{n+1} = p_{n+2} + p_{n+3} + \cdots$$

we get $p_n = 2p_{n+1}$. Hence

$$p_1 = \frac{1}{2}, \quad p_2 = \frac{1}{4}, \quad \ldots, \quad p_n = \frac{1}{2^n}.$$

The representation by non-terminating dyadic fractions is unique.

PROBLEM 33. If the series $a_1 + a_2 + a_3 + \cdots$ is absolutely convergent and every subseries

$$a_k + a_{2k} + a_{3k} + \cdots, \quad k = 1,2,3,\ldots$$

has sum 0, then $a_1 = a_2 = a_3 = \cdots = 0$.

Solution. Since the series $s_k = a_k + a_{2k} + a_{3k} + \cdots$ is of the same type as $s_1 = a_1 + a_2 + a_3 + \cdots$, it is sufficient to show that $a_1 = 0$.

Let $p_1 + p_2, \ldots, p_m$ be the first m prime numbers. Then

$$\begin{aligned} s_1 &- (s_{p_1} + s_{p_2} + \cdots + s_{p_m}) \\ &+ (s_{p_1p_2} + s_{p_1p_3} + \cdots) \\ &- (s_{p_1p_2p_3} + s_{p_1p_2p_4} + \cdots) \\ &+ \cdots \\ &+ (-1)^m s_{p_1p_2\cdots p_m} \end{aligned}$$

contains only a_1 and not those terms a_n whose index n is not divisible by the primes $p_1, p_2, \ldots, p_m$, in fact, it contains every such a_n only once (see Problem 17 of Chapter 1). This means

$$|a_1| \le \sum_{n=p_m+1}^{\infty} |a_n|,$$

and so $a_1 = 0$.

PROBLEM 34. Verify *Stirling's Formula:* If n is a positive integer, then

$$1\cdot 2\cdot 3\cdots n = \sqrt{2\pi n}\, n^n e^{-n + \frac{\theta}{12\, n}},$$

where θ is between 0 and 1.

Solution. Let

$$t_n = \sqrt{\left(\frac{2}{1}\right)^3\left(\frac{3}{2}\right)^5\left(\frac{4}{3}\right)^7 \cdots \left(\frac{n}{n-1}\right)^{2n-1}}.$$

Then

$$t_n = \frac{n^{n+\frac{1}{2}}}{1\cdot 2\cdot 3 \cdots n}$$

or

$$1\cdot 2\cdot 3 \cdots n = \frac{1}{t_n} n^{n+\frac{1}{2}}. \tag{E.6}$$

We calculate t_n. We get

$$\log t_n = \frac{3}{2}\log\frac{2}{1} + \frac{5}{2}\log\frac{3}{2} + \frac{7}{2}\log\frac{4}{3} + \cdots + \frac{2n-1}{2}\log\frac{n}{n-1}.$$

But

$$\log\frac{n}{n-1} = \frac{2}{1}\frac{1}{2n-1} + \frac{2}{3}\frac{1}{(2n-1)^3} + \frac{2}{5}\frac{1}{(2n-1)^5} + \cdots$$

and so

$$\frac{2n-1}{2}\log\frac{n}{n-1} = 1 + u_{n-1},$$

where

$$u_{n-1} = \frac{1}{3}\left\{\frac{1}{(2n-1)^2} + \frac{1}{5}\frac{1}{(2n-1)^4} + \cdots\right\}.$$

Thus

$$\log t_n = n - 1 + (u_1 + u_2 + \cdots + u_{n-1}) = n - 1 + S_{n-1}. \tag{E.7}$$

In order to calculate S_{n-1}, we study the series

$$u_1 + u_2 + u_3 + \cdots$$

We have

$$u_{n-1} < \frac{1}{3}\left\{\frac{1}{(2n-1)^2} + \frac{1}{(2n-1)^3} + \cdots\right\} = \frac{1}{12}\left\{\frac{1}{n-1} - \frac{1}{n}\right\}.$$

We obtain, by addition,

$$S_{n-1} < \frac{1}{12}\left(1 - \frac{1}{n}\right).$$

Thus S_{n-1} tends to a limit S, smaller than 1/12. The terms being positive, we have

$$S_{n-1} < S. \tag{E.8}$$

We also obtain

$$S - S_{n-1} = u_n + u_{n+1} + \cdots < \frac{1}{12\ n}$$

or

$$S_{n-1} > S - \frac{1}{12\ n}. \tag{E.9}$$

Using the inequalities (E.8) and (E.9), we can put

$$S_{n-1} = S - \frac{\theta}{12\ n},$$

where is between 0 and 1. Inequality (E.7) gives

$$\log t_n = n - 1 + S - \frac{\theta}{12\ n}.$$

Therefore

$$\frac{1}{t_n} = e^{1-S}\ e^{-n + \frac{\theta}{12\ n}}$$

holds. Substitution in (E.6) gives

$$1 \cdot 2 \cdot 3 \cdots n = C\ n^{n + \frac{1}{2}}\ e^{-n + \frac{\theta}{12\ n}}, \tag{E.10}$$

where C is the constant e^{1-S} to be determined. Let

$$f(n) = \frac{2}{1} \cdot \frac{2}{3} \cdot \frac{4}{3} \cdot \frac{4}{5} \cdot \frac{6}{5} \cdot \frac{6}{7} \cdot \frac{8}{7} \cdots \frac{2n}{2n - 1}.$$

It is easy to see that

$$f(n) = \frac{2^{4n-1}}{n} \left(\frac{(1 \cdot 2 \cdot 3 \cdots n)^2}{1 \cdot 2 \cdot 3 \cdots 2n} \right)^2.$$

Now, by (E.10),

$$(1 \cdot 2 \cdot 3 \cdots n)^2 = C^2\ n^{2n+1}\ e^{-2n + \frac{\theta}{6n}},$$

$$1 \cdot 2 \cdot 3 \cdots 2n = C\ (2n)^{2n + \frac{1}{2}}\ e^{-2n + \frac{\theta}{24\ n}}.$$

where θ', the same as θ, is between 0 and 1. Thus

$$f(n) = \frac{C^2}{4}\ e^{\frac{4\theta - \theta'}{12\ n}},$$

where θ and θ' are between 0 and 1. Hence

$$\lim_{n\to\infty} f(n) = \frac{C^2}{4}.$$

But, by the Formula of Wallis (see Problem 8),

$$\lim_{n\to\infty} f(n) = \frac{\pi}{2}.$$

Thus $C = \sqrt{2\pi}$. Hence, finally

$$n! = \sqrt{2\pi n}\, n^n e^{-n + \frac{\theta}{12n}} \quad \text{with } 0 < \theta < 1.$$

Remark. To obtain

$$\log \frac{n}{n-1} = \frac{2}{1}\cdot\frac{1}{2n-1} + \frac{2}{3}\cdot\frac{1}{(2n-1)^3} + \frac{2}{5}\cdot\frac{1}{(2n-1)^5} + \cdots$$

we have used the familiar expansion

$$\log \frac{1+x}{1-x} = 2x\left(1 + \frac{1}{3}x^2 + \frac{1}{5}x^4 + \cdots + \frac{1}{2m+1}x^{2m} + \cdots\right), \qquad |x| < 1,$$

setting $x = 1/(2n-1)$; the latter is a simple consequence of

$$\log(1+x) = x - \frac{x^2}{2} + \frac{x^3}{3} - \cdots + (-1)^{m-1}\frac{x^m}{m} + \cdots$$

for $-1 < x \le 1$.

PROBLEM 35. Show that

$$\sum_{k=1}^{\infty} \left(k \log \frac{2k+1}{2k-1} - 1\right) = \frac{1}{2}(1 - \log 2).$$

Solution. Let

$$s_n = \sum_{k=1}^{n} \left(k \log \frac{2k+1}{2k-1} - 1\right) = \log \frac{1}{1}\cdot\frac{1}{3}\cdot\frac{1}{5}\cdot\frac{1}{7} \cdots \frac{1}{2n-1}(2n+1)^n - n.$$

It is easy to see that

$$s_n = \log \frac{2^{n-1}(n-1)!\,(2n+1)^n}{(2n-1)!} - n.$$

Using Stirling's Formula (see Problem 34), we see that the sequence $(s_n)_{n=1}^{\infty}$ tends to the same limit as the sequence $(t_n)_{n=1}^{\infty}$, where

$$t_n = \log \frac{2^{n-1} \sqrt{2\pi(n-1)}\ (n-1)^{n-1}\ e^{1-n}\ (2n+1)^n}{\sqrt{2\pi(2n-1)}\ (2n-1)^{2n-1}\ e^{1-2n}},$$

as $n \to \infty$. But, as $n \to \infty$,

$$t_n = \frac{1}{2} \log \frac{n-1}{2n-1} + \log \left(\frac{2n-2}{2n-1}\right)^{n-1} \left(\frac{2n+1}{2n-1}\right)^n \to -\frac{1}{2} \log 2 + \frac{1}{2}.$$

Indeed, as $n \to \infty$,

$$\left(\frac{2n-1}{2n-2}\right)^{n-1} = \left(1 + \frac{1}{2n-2}\right)^{n-1} \to \sqrt{e}$$

and thus

$$\lim_{n\to\infty} \left(\frac{2n-2}{2n-1}\right)^{n-1} = \frac{1}{\sqrt{e}}.$$

Moreover, as $n \to \infty$,

$$\left(\frac{2n+1}{2n-1}\right)^n = \left(1 + \frac{1}{n-\frac{1}{2}}\right)^n \to e$$

because

$$\left(1 + \frac{1}{n}\right)^n < \left(1 + \frac{1}{n-\frac{1}{2}}\right)^n < \left(1 + \frac{1}{n-1}\right)^{n-1}\left(1 + \frac{1}{n-1}\right)$$

and passage to the limit as $n \to \infty$ gives that

$$\left(1 + \frac{1}{n}\right)^n \quad \text{and} \quad \left(1 + \frac{1}{n-1}\right)^{n-1}\left(1 + \frac{1}{n-1}\right)$$

have the common limit e.

PROBLEM 36. Let

$$Q_n = \left(1 + \frac{1}{n}\right)^{n^2} \frac{n!}{n^n \sqrt{n}}.$$

Show that $(Q_n)_{n=1}^{\infty}$ is monotonely decreasing and find its limit.

Solution. We have

$$Q_{n+1}/Q_n = \left(1 + \frac{1}{n+1}\right)^{(n+1)^2} \left(1 + \frac{1}{n}\right)^{-(n^2+n+\frac{1}{2})}$$

$$= \left(1 - \frac{1}{(n+1)^2}\right)^{(n+1)^2} \left(1 + \frac{1}{n}\right)^{n+\frac{1}{2}} = e^{S_1+S_2},$$

where

$$S_1 = -\sum_{k=1}^{\infty} \frac{1}{(k+1)(n+1)^{2k}}$$

and

$$S_2 = \sum_{k=3}^{\infty} (-1)^k \left(\frac{1}{k} - \frac{1}{2(k-1)}\right)\frac{1}{n^{k-1}}$$

are convergent series for all positive integers n. Note that S_1 is a series of negative terms and S_2 is an alternating series whose terms in absolute value decrease monotonely. Thus the sums of both series are less than their respective first terms. Thus it follows that

$$S_1 + S_2 < -\frac{1}{2(n+1)^2} + \frac{1}{12\,n^2} < 0 \quad \text{for } n = 1,2,3,\ldots$$

This proves that $Q_{n+1}/Q_n < 1$; that is, $(Q_n)_{n=1}^{\infty}$ is monotonely decreasing.

Since $Q_n > 0$, it follows that $(Q_n)_{n=1}^{\infty}$ converges. The limit of the sequence $(Q_n)_{n=1}^{\infty}$ is easily calculated if n! is replaced by $\sqrt{2\pi n}(n/e)^n$ (see Problem 34). Thus

$$\lim_{n\to\infty} Q_n = \lim_{n\to\infty} \left(1 + \frac{1}{n}\right)^{n^2} e^{-n} \sqrt{2\pi}$$

$$= \sqrt{2\pi} \lim_{n\to\infty} \exp\left\{n^2 \log\left(1 + \frac{1}{n}\right) - n\right\}$$

$$= \sqrt{2\pi} \exp\left\{\lim_{n\to\infty} \left(-\frac{1}{2} + \frac{1}{3n} - \frac{1}{4n^2} + \cdots\right)\right\}$$

$$= \sqrt{2\pi/e}.$$

PROBLEM 37. Let n be a positive integer larger than 1. Show that

$$(n!)! > \{(n-1)!\}^{n!}\, e \left(\frac{n}{e}\right)^{n!}.$$

Solution. The inequality

$$\sum_{k=2}^{m} \log k > \int_1^m \log x \, dx, \quad m > 2$$

gives

$$m! > m^m e^{1-m},$$

so that

$$n^m m! > m^m e \left(\frac{n}{e}\right)^m.$$

Replacing m by n! we get the desired result.

PROBLEM 38. Let $a_n \to 0$ and $b_n \to 0$ as $n \to \infty$. Suppose, moreover, that $(b_n)_{n=1}^{\infty}$ is strictly decreasing for all sufficiently large n. Show that

$$\lim_{n\to\infty} \frac{a_n}{b_n} = \lim_{n\to\infty} \frac{a_n - a_{n+1}}{b_n - b_{n+1}}$$

provided that the second quotient is convergent or properly divergent.

Solution. Suppose first that the limit is finite and equal to s; then if $\varepsilon > 0$ is given, m can be found so that

$$s - \varepsilon < \frac{a_n - a_{n+1}}{b_n - b_{n+1}} < s + \varepsilon \quad \text{and} \quad b_n - b_{n+1} > 0$$

if $n \geq m$; thus, for $n \geq m$,

$$(s - \varepsilon)(b_n - b_{n+1}) < a_n - a_{n+1} < (s + \varepsilon)(b_n - b_{n+1}).$$

Changing n to n + 1, n + 2, ..., n + p - 1 and adding the results, we find

$$(s - \varepsilon)(b_n - b_{n+p}) < a_n - a_{n+p} < (s + \varepsilon)(b_n - b_{n+p}).$$

Taking the limit as $p \to \infty$, we obtain

$$(s - \varepsilon)b_n \leq a_n \leq (s + \varepsilon)b_n$$

because by assumption $a_{n+p} \to 0$ and $b_{n+p} \to 0$ as $p \to \infty$. Since b_n is positive, we have, for $n \geq m$

$$\left|\frac{a_n}{b_n} - s\right| \leq \varepsilon \quad \text{or} \quad \lim_{n\to\infty} \frac{a_n}{b_n} = s.$$

On the other hand, if $(a_n - a_{n+1})/(b_n - b_{n+1})$ properly diverges to ∞, we can find m so that $(a_n - a_{n+1})/(b_n - b_{n+1}) > k$, for $n \geq m$, however large k may be. By the same argument as before, we get

$$a_n - a_{n+p} > k(b_n - b_{n+p}),$$

which leads to $a_n \geq kb_n$, or $a_n/b_n \geq k$ for $n \geq m$. Thus

$$\lim_{n\to\infty} \frac{a_n}{b_n} = \infty.$$

PROBLEM 39. Let $(x_n)_{n=1}^{\infty}$ and $(y_n)_{n=1}^{\infty}$ be two sequences of real numbers. Suppose that the second sequence properly diverges to $+\infty$ and that it is strictly increasing for all sufficiently large n. Show that

$$\lim_{n\to\infty} \frac{x_n}{y_n} = \lim_{n\to\infty} \frac{x_n - x_{n-1}}{y_n - y_{n-1}}$$

provided that the limit on the right-hand side exists (be it finite or infinite).

Solution. We assume first that the limit is finite, that is,

$$\lim_{n\to\infty} \frac{x_n - x_{n-1}}{y_n - y_{n-1}} = r,$$

where r is a finite real number. Then, for any $\varepsilon > 0$, there exists a natural number k such that, for $n \geq k$, we have

$$\left| \frac{x_n - x_{n-1}}{y_n - y_{n-1}} - r \right| < \frac{\varepsilon}{2} \quad \text{and} \quad y_n - y_{n-1} > 0.$$

We see therefore that all fractions

$$\frac{x_{k+1} - x_k}{y_{k+1} - y_k}, \frac{x_{k+2} - x_{k+1}}{y_{k+1} - y_{k+1}}, \ldots, \frac{x_{n-1} - x_{n-2}}{y_{n-1} - y_{n-2}}, \frac{x_n - x_{n-1}}{y_n - y_{n-1}}$$

are situated between $r - \varepsilon/2$ and $r + \varepsilon/2$. By Problem 28 in Chapter 2, the fraction

$$\frac{x_n - x_k}{y_n - y_k}$$

must also be between $r - \varepsilon/2$ and $r + \varepsilon/2$ and hence

$$\left|\frac{x_n - x_k}{y_n - y_k} - r\right| < \frac{\varepsilon}{2}.$$

We now use the identity

$$\frac{x_n}{y_n} - r = \frac{x_k - ry_k}{y_n} + \left\{1 - \frac{y_k}{y_n}\right\}\left\{\frac{x_n - x_k}{y_n - y_k} - r\right\}$$

and we get that

$$\left|\frac{x_n}{y_n} - r\right| \le \left|\frac{x_k - ry_k}{y_n}\right| + \left|\frac{x_n - x_k}{y_n - y_k} - r\right|$$

(noting that $y_k/y_n \to 0$ as $n \to \infty$). We know already that the second summand on the right-hand side of the foregoing inequality is smaller than $\varepsilon/2$ for $n \ge k$; the first summand (whose numerator is a fixed quantity) also becomes smaller than $\varepsilon/2$ for $n \ge n'$ because $y_n \to \infty$. Choosing $n' \ge k$, we see that, for $n \ge n'$,

$$\left|\frac{x_n}{y_n} - r\right| < \varepsilon.$$

This proves the claim for the considered special case.

To finish the proof, we observe that the case of an infinite limit easily reduces to the case of a finite limit. Take, for example, the case where $(x_n - x_{n-1})/(y_n - y_{n-1}) \to \infty$ as $n \to \infty$, that is, the sequence properly diverges to $+\infty$. Then for all sufficiently large n, $x_n - x_{n-1} > y_n - y_{n-1}$; hence $x_n \to \infty$ with $y_n \to \infty$ and the sequence $(x_n)_{n=1}^{\infty}$ must be strictly increasing for all sufficiently large n. Therefore we can use what we have proved in the beginning of this proof and apply that result to the reciprocal expression, namely y_n/x_n:

$$\lim_{n\to\infty} \frac{y_n}{x_n} = \lim_{n\to\infty} \frac{y_n - y_{n-1}}{x_n - x_{n-1}} = 0$$

and we conclude that $\lim_{n\to\infty} x_n/y_n = \infty$.

PROBLEM 40. Let $a_n \to a$ as $n \to \infty$. Show that $b_n \to a$ as $n \to \infty$, where $b_n = (a_1 + a_2 + \cdots + a_n)/n$.

Solution. We set $x_n = a_1 + a_2 + \cdots + a_n$ and $y_n = n$ and apply the result in Problem 39.

PROBLEM 41. Let $(p_n)_{n=1}^{\infty}$ be a sequence of positive numbers tending to the limit p as $n \to \infty$ and $p > 0$. Show that the sequence of geometric means

$$\sqrt[n]{p_1 p_2 \cdots p_n} \to p \quad \text{as } n \to \infty.$$

Solution. Let

$$\sqrt[n]{p_1 p_2 \cdots p_n} = T_n.$$

Then

$$\lim_{n\to\infty} \log T_n = \lim_{n\to\infty} \frac{\log p_1 + \log p_2 + \cdots + \log p_n}{n} = \lim_{n\to\infty} \log p_n$$

(by Problem 40). Thus

$$\lim_{n\to\infty} \log \frac{T_n}{p_n} = 0 \quad \text{or} \quad \lim_{n\to\infty} T_n = \lim_{n\to\infty} p_n = p.$$

Remark. The results in Problems 40 and 41 have many applications; for example,

(i) $\dfrac{1 + \frac{1}{2} + \cdots + \frac{1}{n}}{n} \to 0$ as $n \to \infty$ because $1/n \to 0$ as $n \to \infty$;

(ii) $\sqrt[n]{1 \cdot \frac{2}{1} \cdot \frac{3}{2} \cdots \frac{n}{n-1}} \to 1$ as $n \to \infty$ because $n/(n-1) \to 1$ as $n \to \infty$;

(iii) $\dfrac{1 + \sqrt{2} + \sqrt[3]{3} + \cdots + \sqrt[n]{n}}{n} \to 1$ as $n \to \infty$ because $\sqrt[n]{n} \to 1$ as $n \to \infty$;

(iv) $\left(1 + \frac{1}{n}\right)^n \to e$ as $n \to \infty$ implies

$$\sqrt[n]{\left(\frac{2}{1}\right)^1 \left(\frac{3}{2}\right)^2 \left(\frac{4}{3}\right)^3 \cdots \left(\frac{n+1}{n}\right)^n} = \sqrt[n]{\frac{(n+1)^n}{n!}} = \frac{n+1}{\sqrt[n]{n!}} \to e$$

or

$$\frac{\sqrt[n]{n!}}{n} \to \frac{1}{e} \quad \text{as } n \to \infty. \quad \text{(See Remark to Problem 16.)}$$

PROBLEM 42. Let $(t_n)_{n=1}$ be a sequence of positive numbers and suppose that t_{n+1}/t_n converges to a positive number t as $n \to \infty$. Show that

$$\lim_{n\to\infty} \sqrt[n]{t_n} = t.$$

Solution. Let $p_1 = t_1$, $p_2 = t_2/t_1$, ..., $p_n = t_n/t_{n-1}$, ... By Problem 41,

$$\lim_{n\to\infty} \sqrt[n]{p_1 p_2 \cdots p_n} = \lim_{n\to\infty} p_n$$

or

$$\lim_{n\to\infty} \sqrt[n]{t_n} = \lim_{n\to\infty} \frac{t_n}{t_{n-1}} = \lim_{n\to\infty} \frac{t_{n+1}}{t_n} = t.$$

PROBLEM 43. Let $a > 0$. Show that

$$\lim_{n\to\infty} \frac{1^{a-1} + 2^{a-1} + 3^{a-1} + \cdots + n^{a-1}}{n^a} = \frac{1}{a}.$$

Solution. We use the result in Problem 39 and let

$$x_n = 1^{a-1} + 2^{a-1} + \cdots + n^{a-1} + (n+1)^{a-1} \quad \text{and} \quad y_n = (n+1)^a.$$

Then

$$\frac{x_n - x_{n-1}}{y_n - y_{n-1}} = \frac{(n+1)^{a-1}}{(n+1)^a - n^a}.$$

But, as $n \to \infty$,

$$\frac{(n+1)^a - n^a}{(n+1)^{a-1}} = \frac{\left(1 + \frac{1}{n}\right)^a - 1^a}{\left(1 + \frac{1}{n}\right)^{a-1} \frac{1}{n}} = \frac{\left(1 + \frac{1}{n}\right)^a - 1^a}{\frac{1}{n}} \left(1 + \frac{1}{n}\right)^{1-a} \to a$$

because

$$\lim_{n\to\infty} \frac{\left(1 + \frac{1}{n}\right)^a - 1^a}{\frac{1}{n}} = \frac{d}{dx} x^a \Big|_{x=1} = a, \quad \lim_{n\to\infty} \left(1 + \frac{1}{n}\right)^{1-a} = 1.$$

Remark. Note that

$$\frac{1}{n}\cdot\frac{1^{a-1} + 2^{a-1} + 3^{a-1} + \cdots + n^{a-1}}{n^{a-1}} \to \int_0^1 x^{a-1}\, dx = \frac{1}{a} \quad \text{as } n \to \infty.$$

PROBLEM 44. Let $(u_n)_{n=1}^{\infty}$ be a sequence with positive terms and

$$\lim_{n\to\infty} \frac{u_{n+1}}{u_n} = h,$$

where $h > 0$. Show that

$$\lim_{n\to\infty} \left(\frac{u_n}{\sqrt[n]{u_1 u_2 \cdots u_n}}\right)^{1/n} = \sqrt{h}.$$

Solution. Put

$$\frac{u_2}{u_1} = p_1,\ \frac{u_3}{u_2} = p_2,\ \ldots,\ \frac{u_n}{u_{n-1}} = p_{n-1},\ \ldots;$$

we get

$$p_1 p_2 \cdots p_{n-1} = \frac{u_n}{u_1},$$

$$p_2 p_3 \cdots p_{n-1} = \frac{u_n}{u_2},$$

$$\cdots$$

$$p_{n-1} = \frac{u_n}{u_{n-1}};$$

thus

$$\left(\frac{u_n}{\sqrt[n]{u_1 u_2 \cdots u_n}}\right)^{1/n} = \left(p_1^1 p_2^2 p_3^3 \cdots p_{n-1}^{n-1}\right)^{1/n^2}.$$

Now, for the sequence

p_1,

p_2, p_2,

$p_3, \; p_3, \; p_3,$

$\cdots$

$p_{n-1}, \; p_{n-1}, \; \ldots, \; p_{n-1},$

$\cdots$

we have $\lim_{n\to\infty} p_{n-1} = h$; thus, by Problem 41,

$$\lim_{n\to\infty} \left(p_1^1 p_2^2 p_3^3 \cdots p_{n-1}^{n-1}\right)^{\frac{1}{2}n(n-1)} = h.$$

Consequently

$$\lim_{n\to\infty} \left(p_1^1 p_2^2 p_3^3 \cdots p_{n-1}^{n-1}\right)^{1/n^2} = \sqrt{h} = \lim_{n\to\infty} \left(\frac{u_n}{\sqrt[n]{u_1 u_2 \cdots u_n}}\right)^{1/n}.$$

PROBLEM 45. Let

$$\lim_{n\to\infty} n^x a_n = a.$$

Show that

$$\lim_{n\to\infty} n^x \sqrt[n]{a_1 a_2 \cdots a_n} = a\, e^x.$$

Solution. Put $\sqrt[x]{a_n} = b_n$. Then

$$\lim_{n\to\infty} nb_n = \sqrt[x]{a}.$$

Thus, by Problem 41,

$$\lim_{n\to\infty} \sqrt[n]{b_1(2b_2)\cdots(nb_n)} = \lim_{n\to\infty} \sqrt[n]{n!\,b_1 b_2 \cdots b_n} = \sqrt[x]{a}$$

and since

$$\frac{n}{\sqrt[n]{n!}} \to e \quad \text{as } n\to\infty$$

(see Part (iv) of Remark to Problem 41), we get

$$\lim_{n \to \infty} n \sqrt[n]{b_1 b_2 \cdots b_n} = \sqrt[x]{a}\, e$$

or

$$\lim_{n \to \infty} n^x \sqrt[n]{a_1 a_2 \cdots a_n} = a\, e^x.$$

PROBLEM 46. Let $p > 1$. Show that

$$\frac{1}{n}\left[\left(\frac{n}{n+1}\right)^p + \left(\frac{n}{n+2}\right)^p + \cdots\right]$$

tends to $1/(p-1)$ as $n \to \infty$.

Solution. Noting that $(n+2)^{-p}$ is the smallest value of $y = x^{-p}$ on the interval $[n+1,n+2]$, $(n+2)^{-p}$ the smallest value of $y = x^{-p}$ on $[n+2,n+3]$, and so forth, we see that

$$\frac{1}{(n+2)^p} + \frac{1}{(n+3)^p} + \cdots < \int_{n+1}^{\infty} x^{-p}\, dx = \frac{1}{(p-1)(n+1)^{p-1}}$$

or

$$\frac{1}{(n+1)^p} + \frac{1}{(n+2)^p} + \cdots < \frac{1}{(n+1)^p} + \frac{1}{(p-1)(n+1)^{p-1}}.$$

Similarly, since $(n+1)^{-p}$ is the largest value of $y = x^{-p}$ on the interval $[n+1,n+2]$, $(n+2)^{-p}$ the largest value of $y = x^{-p}$ on $[n+2,n+3]$, and so forth, we obtain

$$\frac{1}{(n+1)^p} + \frac{1}{(n+2)^p} + \cdots > \frac{1}{(p-1)(n+1)^{p-1}}.$$

Thus

$$\frac{1}{(n+1)^{p-1}} < (p-1)\left[\frac{1}{(n+1)^p} + \frac{1}{(n+2)^p} + \cdots\right]$$

$$< \frac{1}{(n+1)^{p-1}} + \frac{p-1}{(n+1)^p}.$$

Multiplying by n^{p-1}, we get

$$\left(\frac{n}{n+1}\right)^{p-1} < (p-1)\frac{1}{n}\left[\left(\frac{n}{n+1}\right)^p + \left(\frac{n}{n+2}\right)^p + \cdots\right]$$

$$< \left(\frac{n}{n+1}\right)^{p-1} + \left(\frac{n}{n+1}\right)^{p-1} \frac{p-1}{n+1}.$$

Setting

$$S_n = \frac{1}{n}\left[\left(\frac{n}{n+1}\right)^p + \left(\frac{n}{n+2}\right)^p + \cdots\right] \quad \text{and} \quad \lim_{n\to\infty} S_n = S,$$

we see that $1 \le (p - 1)S \le 1$ and so $S = 1/(p - 1)$.

Remarks. Note that

$$S_n = \frac{1}{n}\left[\left(\frac{n}{n+1}\right)^p + \left(\frac{n}{n+2}\right)^p + \cdots\right]$$

represents the sum of the areas of approximating rectangles under the curve $y = x^{-p}$ over the interval $(1,\infty)$; indeed,

$$\frac{1}{n}\left(\frac{n}{n+1}\right)^p = \int_1^{1+1/n} \frac{dx}{\left(1 + \frac{1}{n}\right)^p} < \int_1^{1+1/n} \frac{dx}{x^p},$$

$$\frac{1}{n}\left(\frac{n}{n+2}\right)^p = \int_{1+1/n}^{1+2/n} \frac{dx}{\left(1 + \frac{2}{n}\right)^p} < \int_{1+1/n}^{1+2/n} \frac{dx}{x^p},$$

$$\cdots$$

and so

$$S_n < \int_1^\infty \frac{dx}{x^p} = \frac{1}{p-1}, \quad S_n < S_{n+1}, \quad \lim_{n\to\infty} S_n = \int_1^\infty \frac{dx}{x^p}.$$

The same method of proof can be used to evaluate many other limits. For example, dividing the interval $[0,1]$ into n subintervals of equal length and considering the sum of areas of approximating rectangles under the curve $y = (1 + x)^{-1}$ leads to the result that

$$\frac{1}{n+1} + \frac{1}{n+2} + \cdots + \frac{1}{2n} \to \int_0^1 \frac{dx}{1+x} = \log 2 \quad \text{as } n \to \infty.$$

Again, dividing the interval $[0,1]$ into n subintervals of equal length and considering the sum of areas of approximating rectangles under the curve $y = x^{-1/2}$ shows that

$$\frac{1}{\sqrt{n}}\left\{\frac{1}{\sqrt{1}} + \frac{1}{\sqrt{2}} + \cdots + \frac{1}{\sqrt{n}}\right\} \to \int_0^1 \frac{dx}{\sqrt{x}} = 2 \quad \text{as } n \to \infty.$$

This latter limit can, of course, also be derived with ease from the result in Problem 5.

To compute the limit $\lim_{n\to\infty} S_n$, where

$$S_n = \frac{1}{\sqrt{4n^2 - 1}} + \frac{1}{\sqrt{4n^2 - 2^2}} + \cdots + \frac{1}{\sqrt{4n^2 - n^2}},$$

we note first that

$$S_n = \frac{1}{n}\left\{\frac{1}{\sqrt{4 - \left(\frac{1}{n}\right)^2}} + \frac{1}{\sqrt{4 - \left(\frac{2}{n}\right)^2}} + \cdots + \frac{1}{\sqrt{4 - \left(\frac{n}{n}\right)^2}}\right\};$$

hence

$$\lim_{n\to\infty} S_n = \int_0^1 \frac{dx}{\sqrt{4 - x^2}} = \arcsin \frac{x}{2}\Bigg|_0^1 = \frac{\pi}{6}.$$

In a similar way we can show that $\lim_{n\to\infty} A_n = 1/e$, where

$$A_n = \frac{\sqrt[n]{n!}}{n}.$$

Indeed, taking logarithms, we get

$$\lim_{n\to\infty} \log A_n = \lim_{n\to\infty} \frac{1}{n}\left(\log\frac{1}{n} + \log\frac{2}{n} + \cdots + \log\frac{n}{n}\right)$$

$$= \int_0^1 \log x\, dx = (x \log x - x)\Bigg|_0^1 = -1$$

and the desired result follows.

PROBLEM 47. Verify the following claims:

(a) If $0 < y < 1$, then $\displaystyle\lim_{n\to\infty} \sum_{j=1}^{n-1} \frac{jy^j}{n - j} = 0.$

(b) If $0 < y < 1$, then $\displaystyle\lim_{n\to\infty} ny^n \sum_{i=1}^{n} \frac{1}{iy^i} = \frac{1}{1 - y}.$

(c) If $x > 1$, then $\displaystyle\lim_{n\to\infty} \frac{n}{x^n} \sum_{i=1}^{n} \frac{x^{i-1}}{i} = \frac{1}{x - 1}.$

Solution. Let

$$S_n = \sum_{j=1}^{n} \frac{jy^j}{n+1-j} = \frac{y}{n} + \frac{2y^2}{n-1} + \frac{3y^3}{n-2} + \cdots + \frac{(n-1)y^{n-1}}{2} + \frac{ny^n}{1}$$

and

$$T_n = \frac{y + 2^2y^2 + 3^2y^3 + \cdots + n^2y^n}{n}$$

It is clear that $0 \le S_n \le T_n$. Moreover

$$\frac{(n+1)^2y^{n+1}}{n^2y^n} = \left(1 + \frac{1}{n}\right)^2 y \to y \quad \text{as } n \to \infty$$

and $0 < y < 1$; it follows that n^2y^n tends to 0 as $n \to \infty$ because n^2y^n and $(n+1)^2y^{n+1}$ tend to the same limit when $n \to \infty$. We see, therefore, by Problem 40, that T_n tends to 0 as $n \to \infty$; thus S_n tends to 0 as $n \to \infty$. This completes the proof of Part (a).

In order to verify Part (b), we observe that

$$\sum_{i=1}^{n} \frac{n}{i} y^{n-i} = \sum_{j=0}^{n-1} \frac{n}{n-j} y^j = \sum_{j=0}^{n-1} y^j + \sum_{j=0}^{n-1} \frac{jy^j}{n-j}. \tag{E.11}$$

But $\Sigma_{j=0}^{\infty} y^j = 1/(1-y)$ for $0 < y < 1$. Passing to the limit as $n \to \infty$ in (E.11) and using the result in Part (a), we obtain the desired result.

The claim in Part (c) follows at once from the result in Part (b); we only need to set $y = 1/x$.

PROBLEM 48. Show that

$$\lim_{n \to \infty} \frac{1}{2} \cdot \frac{3}{4} \cdot \frac{5}{6} \cdots \frac{2n-1}{2n} = 0.$$

Solution. The claim follows at once from the result in Problem 37 of Chapter 2.

PROBLEM 49. For each positive integer n, let $f(n) = (n!)^{1/n}$. Show that $f(n+1)/f(n)$, for $n = 1,2,\ldots$ is monotonely decreasing.

Solution. We show equivalently that

$$F_n = \frac{f(n+1)}{f(n)} \bigg/ \frac{f(n)}{f(n-1)} < 1 \quad \text{for } n = 2,3,\ldots$$

Consider

$$(F_n)^{n(n+1)/2} = [(n-1)!]^{1/(n-1)} \frac{1}{n}\left(\frac{n+1}{n}\right)^{n/2}.$$

Since the geometric mean is less than the arithmetic mean (see Problem 2 of Chapter 2),

$$[(n-1)!]^{1/(n-1)} < \frac{(n-1) + (n-2) + \cdots + 3 + 2 + 1}{n}$$

$$= \frac{n(n-1)}{2} < \frac{n}{2}.$$

Therefore

$$(F_n)^{n(n+1)/2} < \frac{1}{2}\left(1 + \frac{1}{n}\right)^{n/2} < \frac{1}{2} e^{\frac{1}{2}} < 1.$$

Thus $F_n < 1$ and the desired result follows.

Remark. It is not difficult to see that

$$1 < \frac{f(n+1)}{f(n)} < \frac{n+1}{n};$$

consult Problem 34.

PROBLEM 50. Let

$$a_n = b_n e^{-\frac{1}{12n}} \quad \text{and} \quad b_n = (n!e^n)n^{-(n+\frac{1}{2})}.$$

Show that each interval (a_n, b_n), $n = 1,2,3,\ldots$, contains the interval (a_{n+1}, b_{n+1}) as a subinterval.

Solution. Since, for $-1 < x < 1$,

$$\log \frac{1+x}{1-x} = 2x\left(1 + \frac{1}{3}x^2 + \frac{1}{5}x^4 + \cdots\right),$$

setting $x = (2n+1)^{-1}$, we get

$$\log \frac{n+1}{n} = \frac{2}{2n+1}\left[1 + \frac{1}{3(2n+1)^2} + \frac{1}{5(2n+1)^4} + \cdots\right].$$

Thus

$$\left(n + \frac{1}{2}\right) \log \left(1 + \frac{1}{n}\right) = 1 + \frac{1}{3(2n+1)^2} + \frac{1}{5(2n+1)^4} + \cdots,$$

which is larger than 1, but less than

$$1 + \frac{1}{3}\left[\frac{1}{(2n+1)^2} + \frac{1}{(2n+1)^4} + \cdots\right] = 1 + \frac{1}{12\,n(n+1)}$$

and so

$$e < \left(1 + \frac{1}{n}\right)^{n+\frac{1}{2}} < e^{1 + \frac{1}{12\,n(n+1)}}.$$

Now

$$\frac{b_n}{b_{n+1}} = \frac{\left(1 + \frac{1}{n}\right)^{n+\frac{1}{2}}}{e}$$

and so

$$1 < \frac{b_n}{b_{n+1}} < e^{\frac{1}{12\,n(n+1)}} = \frac{e^{\frac{1}{12\,n}}}{e^{\frac{1}{12(n+1)}}}.$$

Thus $b_n > b_{n+1}$ and

$$b_n\, e^{-\frac{1}{12\,n}} < b_{n+1}\, e^{-\frac{1}{12(n+1)}}$$

or $a_n < a_{n+1}$.

PROBLEM 51. Show that the sequence

$$a_n = \left(1 + \frac{1}{n}\right)^{n+p}, \quad n = 1,2,3,\ldots,$$

is monotonely decreasing if and only if $p \geq 1/2$.

Solution. The fact that a_n is decreasing for $p \geq 1/2$ is clear from the expansion

$$\log a_n = \frac{2(n + p)}{2n + 1}\left[1 + \frac{1}{3(2n + 1)^2} + \frac{1}{5(2n + 1)^4} + \cdots\right]$$

$$= \left(1 + \frac{p - \frac{1}{2}}{n + \frac{1}{2}}\right)\left\{1 + \frac{1}{3(2n + 1)^2} + \frac{1}{5(2n + 1)^4} + \cdots\right\}$$

(see the Solution of Problem 50). This leads to

$$\log a_n = \frac{\frac{1}{2} - p}{n + \frac{1}{2}} + \frac{1}{12\, n^2} + 0(n^{-3}),$$

where $0(x_n)$, with $x_n > 0$, denotes a quantity that divided by x_n remains bounded; thus

$$\log a_{n+1} - \log a_n = \frac{\frac{1}{2} - p}{\left(n + \frac{1}{2}\right)\left(n + \frac{3}{2}\right)} + 0(n^{-3}),$$

hence a_n increases for n larger than a certain subscript n_0 if $p < 1/2$. If $p \le 0$ this is true already for $n \ge 1$ as can be seen by expanding $\left(1 + \frac{1}{n}\right)^n$ with the help of the binomial formula.

PROBLEM 52. Show that the sequence

$$a_n = \left(1 + \frac{1}{n}\right)^n\left(1 + \frac{x}{n}\right), \qquad n = 1,2,3,\ldots,$$

is monotonely decreasing if and only if $x \ge 1/2$.

Solution. Since

$$a_n = \left(1 + \frac{1}{n}\right)^{n+\frac{1}{2}} \frac{1 + \frac{x}{n}}{\left(1 + \frac{1}{n}\right)^{\frac{1}{2}}},$$

the first factor is seen to be decreasing by Problem 51; the square of the second factor is

$$1 + \frac{2x - 1}{n + 1} + \frac{x^2}{n(n + 1)}.$$

The condition $x \ge 1/2$ is therefore sufficient. But

$$\log a_n = 2n\left[\frac{1}{2n + 1} + \frac{1}{3(2n + 1)^3} + \frac{1}{5(2n + 1)^5} + \cdots\right]$$

$$+ 2\left\{\frac{x}{2n + x} + \frac{1}{3}\left(\frac{x}{2n + x}\right)^3 + \frac{1}{5}\left(\frac{x}{2n + x}\right)^5 + \cdots\right\}$$

$$= \frac{2n}{2n + 1} + \frac{2x}{2n + x} + \frac{1}{12 n^2} + O(n^{-3}).$$

Since

$$\log a_n - \log a_{n+1} = \frac{4x - 2}{4x^2} + O(n^{-3}),$$

the condition is also seen to be necessary.

Remark. Note that

$$\log \left(1 + \frac{x}{n}\right) = \log \frac{1 + \dfrac{x}{2n + x}}{1 - \dfrac{x}{2n + x}}.$$

PROBLEM 53. Show that for any positive integer n,

$$\frac{e}{2n + 2} < e - \left(1 + \frac{1}{n}\right)^n < \frac{e}{2n + 1}.$$

Solution. The first inequality means

$$\left(1 + \frac{1}{n}\right)^{n+1} < e\left(1 + \frac{1}{2n}\right)$$

and this is a consequence of the following inequality

$$f(x) = x + x \log\left(1 + \frac{x}{2}\right) - (1 + x) \log (1 + x) > 0, \qquad 0 < x \le \frac{1}{n}.$$

(Note that

$$f'(x) = \frac{x}{x + 2} - \log \frac{1 + x}{1 + \frac{x}{2}} > \frac{x}{x + 2} - \frac{1 + x}{1 + \frac{x}{2}} + 1 = 0, \qquad f(0) = 0.)$$

The second inequality is equivalent to

$$e < \left(1 + \frac{1}{n}\right)^n \left(1 + \frac{1}{2n}\right)$$

and is a consequence of the result in Problem 52.

PROBLEM 54. The number

$$e = \lim_{n \to \infty} \left(1 + \frac{1}{n}\right)^n$$

is contained in the interval

$$\left(1 + \frac{1}{n}\right)^n < e < \left(1 + \frac{1}{n}\right)^{n+1}$$

(see Problem 13 of this Chapter and Problem 2 in Chapter 1). In which quarter of the interval is it contained?

Solution. The number e is situated in the second quarter of the interval because

$$\left(1 + \frac{1}{n}\right)^n\left(1 + \frac{1}{4n}\right) < e < \left(1 + \frac{1}{n}\right)^n\left(1 + \frac{1}{2n}\right), \qquad n = 1,2,3,\ldots;$$

The first inequality follows from Problem 53 because

$$1 + \frac{1}{4n} < \left(1 + \frac{1}{n}\right)\left(1 + \frac{1}{2n}\right)^{-1},$$

the second inequality is contained in Problem 52.

PROBLEM 55. Show that the sequence

$$a_n = \left(1 + \frac{x}{n}\right)^{n+1}, \qquad n = 1,2,3,\ldots,$$

is monotonely decreasing if and only if $0 < x \le 2$.

Solution. We see that a_n is decreasing for $0 < x \le 2$ because

$$\log a_n = (n + 1) \log \frac{1 + \dfrac{x}{2n + x}}{1 - \dfrac{x}{2n + x}} = (2n + 2) \sum_{k=1}^{\infty} \frac{1}{2k - 1}\left(\frac{x}{2n + x}\right)^{2k-1}$$

$$= \sum_{k=1}^{\infty} \frac{x^{2k - 1}}{2k - 1} \frac{1}{(2n + x)^{2k-2}} + (2 - x) \sum_{k=1}^{\infty} \frac{x^{2k-1}}{2k - 1} \frac{1}{(2n + x)^{2k-1}}.$$

Furthermore

$$\log a_n = x + \frac{x^3}{3} \frac{1}{(2n + x)^2} + \frac{x(2 - x)}{2n + x} + O(n^{-3}),$$

$$\log a_n - \log a_{n+1} = \frac{2x(2 - x)}{(2n + x)(2n + x + 2)} + O(n^{-3}),$$

that is, $\log a_n - \log a_{n+1} < 0$ for n sufficiently large, if $x < 0$ or $x > 2$. For $x = 0$ we have $a_n = 1$, $n = 1,2,3,\ldots$.

PROBLEM 56. What is the smallest amount that may be invested at interest rate i, compounded anually, in order that one may withdraw 1 dollar at the end of the first year, ..., n^2 dollars at the end of the n-th year, in perpetuity?

Solution. The present value of n^2 dollars n years from now at rate i per year is $n^2(1 + i)^{-n}$. Thus the required sum is

$$\sum_{n=1}^{\infty} \frac{n^2}{(1 + i)^n}.$$

Since $(1 - x)^{-1} = \Sigma_{n=0}^{\infty} x^n$, it follows by differentiation that

$$x(1 - x)^{-2} = \sum_{n=1}^{\infty} n x^n, \quad (x + x^2)(1 - x)^{-3} = \sum_{n=1}^{\infty} n^2 x^n,$$

all series being convergent for $-1 < x < 1$. Taking $x = (1 + i)^{-1}$, the required sum is found to be

$$\frac{(1 + i)(2 + i)}{i^3}.$$

PROBLEM 57. Show that

$$\lim_{n\to\infty} \sqrt{1 + 2\sqrt{1 + 3\sqrt{1 + \cdots \sqrt{1 + (n - 1)\sqrt{1 + n}}}}} = 3.$$

Solution. We have that

$$3 = \sqrt{1 + 2\cdot 4} = \sqrt{1 + 2\sqrt{16}} = \sqrt{1 + 2\sqrt{1 + 3\sqrt{25}}}$$

$$= \sqrt{1 + 2\sqrt{1 + 3\sqrt{1 + 4\sqrt{36}}}}.$$

This leads to conjecturing the relation

$$3 = \sqrt{1 + 2\sqrt{1 + 3\sqrt{1 + \cdots \sqrt{1 + n\sqrt{(n + 2)^2}}}}}$$

for all $n \geq 1$. Proceeding by induction we verify that

$$(n + 2)^2 = 1 + (n + 1)(n + 3) = 1 + (n + 1)\sqrt{(n + 3)^2}.$$

This given, we must have

$$3 \geq \sqrt{1 + 2\sqrt{1 + 3\sqrt{1 + \cdots \sqrt{1 + (n - 1)\sqrt{1 + n}}}}}.$$

To get an inequality in the other direction going, observe that for any $\alpha > 1$

$$\sqrt{1 + n\alpha} \leq \sqrt{\alpha}\sqrt{1 + n}.$$

A repetition of this argument gives then

$$3 \leq (n + 2)^a \sqrt{1 + 2\sqrt{1 + 3\sqrt{1 + \cdots \sqrt{1 + (n - 1)\sqrt{1 + n}}}}},$$

where $a = 2^{1-n}$.

Remarks. In Problem 57 we showed that

$$\sqrt{1 + 2\sqrt{1 + 3\sqrt{1 + \cdots}}} = 3.$$

This formula can easily be conjectured along the following lines: since

$$n(n + 2) = n\sqrt{1 + (n + 1)(n + 3)}$$

and letting $n(n + 2) = f(n)$, we see that

$$f(n) = n\sqrt{1 + f(n + 1)} = n\sqrt{1 + (n + 1)\sqrt{1 + f(n + 2)}}$$

$$= n\sqrt{1 + (n + 1)\sqrt{1 + (n + 2)\sqrt{1 + f(n + 3)}}}$$

$$= \cdots,$$

that is,

$$n(n + 2) = n\sqrt{1 + (n + 1)\sqrt{1 + (n + 2)\sqrt{1 + (n + 3)\sqrt{1 + \cdots}}}}.$$

Putting $n = 1$, we have

$$\sqrt{1 + 2\sqrt{1 + 3\sqrt{1 + \cdots}}} = 3.$$

In a similar manner, since $n(n + 3) = n\sqrt{n + 5 + (n + 1)(n + 4)}$ and supposing that $g(n) = n(n + 3)$, we have

$$g(n) = n\sqrt{n + 5 + g(n + 1)} = n\sqrt{n + 5 + (n + 1)\sqrt{n + 6 + g(n + 2)}}$$

and so forth; we may conjecture that (taking $n = 1$):

$$\sqrt{6 + 2\sqrt{7 + 3\sqrt{8 + 4\sqrt{9 + \cdots}}}} = 4.$$

PROBLEM 58. Let $a_1 > b_1 > 0$ be given. We form the numbers

$$a_2 = \frac{a_1 + b_1}{2} \quad \text{and} \quad b_2 = \sqrt{a_1 b_1},$$

$$a_3 = \frac{a_2 + b_2}{2} \quad \text{and} \quad b_3 = \sqrt{a_2 b_2},$$

...

$$a_{n+1} = \frac{a_n + b_n}{2} \quad \text{and} \quad b_{n+1} = \sqrt{a_n b_n},$$

...

Show that the sequences $(a_n)_{n=1}^{\infty}$ and $(b_n)_{n=1}^{\infty}$ tend to a common limit $L(a_1,b_1)$ and prove that

$$L(a_1,b_1) = \frac{\pi}{2G},$$

where

$$G = \int_0^{\pi/2} \frac{dx}{\sqrt{a_1^2 \cos^2 x + b_1^2 \sin^2 x}}.$$

Solution. We observe that $a_1 > a_2 > b_2 > b_1$ and that in general

$$a_1 > a_2 > \cdots > a_n > a_{n+1} > b_{n+1} > b_n > \cdots > b_2 > b_1$$

and hence $(a_n)_{n=1}^{\infty}$ is monotonely decreasing and bounded and $(b_n)_{n=1}^{\infty}$ is monotonely increasing and bounded because the a_n's and the b_n's are, in fact, the consecutive arithmetic and geometric means of the initially given numbers a_1 and b_1 with $a_1 > b_1 > 0$. Indeed, it is evident that $a_1 > a_2$ and $b_2 > b_1$ (because $a_1 > b_1 > 0$). To see that $a_2 > b_2$ we might refer to Problem 2 of Chapter 2 or note directly that

$$\frac{a_1 + b_1}{2} - \sqrt{a_1 b_1} = \frac{(\sqrt{a_1} - \sqrt{b_1})^2}{2} > 0 \quad \text{for } a_1 \neq b_1.$$

In the same way we can show that

$$a_n > a_{n+1} > b_{n+1} > b_n.$$

Moreover, it is easy to see that

$$a_1 > a_n > b_n > b_1.$$

The sequences $(a_n)_{n=1}^{\infty}$ and $(b_n)_{n=1}^{\infty}$ are therefore convergent; let

$$\alpha = \lim_{n\to\infty} a_n \quad \text{and} \quad \beta = \lim_{n\to\infty} b_n.$$

But

$$a_{n+1} = \frac{a_n + b_n}{2}$$

and so, for $n \to \infty$, we get

$$\alpha = \frac{\alpha + \beta}{2};$$

hence $\alpha = \beta$. We denote this common limit by $L = L(a_1,b_1)$.

Let

$$G = \int_0^{\pi/2} \frac{dx}{\sqrt{a_1^2 \cos^2 x + b_1^2 \sin^2 x}} \quad (a_1 > b_1 > 0).$$

We put

$$\sin x = \frac{2a_1 \sin t}{(a_1 + b_1) + (a_1 - b_1) \sin^2 t};$$

as t changes from 0 to $\pi/2$, so grows x from 0 to $\pi/2$. Differentiation gives

$$\cos x \, dx = 2a_1 \frac{(a_1 + b_1) - (a_1 - b_1) \sin^2 t}{[(a_1 + b_1) + (a_1 - b_1) \sin^2 t]^2} \cos t \, dt.$$

But

$$\cos x = \frac{\sqrt{(a_1 + b_1)^2 - (a_1 - b_1)^2 \sin^2 t}}{(a_1 + b_1) + (a_1 - b_1) \sin^2 t} \cos t,$$

and thus

$$dx = 2a_1 \frac{(a_1 + b_1) - (a_1 - b_1) \sin^2 t}{(a_1 + b_1) + (a_1 - b_1) \sin^2 t} T,$$

where

$$T = \frac{dt}{\sqrt{(a_1 + b_1)^2 - (a_1 - b_1)^2 \sin^2 t}}.$$

On the other hand

$$\sqrt{a_1^2 \cos^2 x + b_1^2 \sin^2 x} = a_1 \frac{(a_1 + b_1) - (a_1 - b_1) \sin^2 t}{(a_1 + b_1) + (a_1 - b_1) \sin^2 t}$$

and thus

$$\frac{dx}{\sqrt{a_1^2 \cos^2 x + b_1^2 \sin^2 x}} = \frac{dt}{\sqrt{[(a_1 + b_1)/2]^2 \cos^2 t + a_1 b_1 \sin^2 t}}.$$

Putting

$$a_2 = \frac{a_1 + b_1}{2} \quad \text{and} \quad b_2 = \sqrt{a_1 b_1}$$

we get

$$G = \int_0^{\pi/2} \frac{dx}{\sqrt{a_1^2 \cos^2 x + b_1^2 \sin^2 x}} = \int_0^{\pi/2} \frac{dt}{\sqrt{a_2^2 \cos^2 t + b_2^2 \sin^2 t}}.$$

By repeated application of this transformation, we get

$$G = \int_0^{\pi/2} \frac{dx}{\sqrt{a_n^2 \cos^2 x + b_n^2 \sin^2 x}} \qquad (n = 1,2,3,\ldots),$$

where a_n and b_n are defined by the recursive formula

$$a_n = \frac{a_{n-1} + b_{n-1}}{2}, \qquad b_n = \sqrt{a_{n-1}b_{n-1}}.$$

As we know already, these two sequences converge to the common value $L(a_1,b_1) = L$. It is easy to see that

$$\frac{\pi}{2a_n} < G < \frac{\pi}{2b_n};$$

passage to the limit as $n \to \infty$ gives

$$G = \frac{\pi}{2L(a_1,b_1)} \quad \text{or} \quad L(a_1,b_1) = \frac{\pi}{2G}.$$

Remarks. The foregoing formula $L(a_1,b_1) = \pi/2G$ and its derivation is due to Gauss. We consider an application of this formula and compute the integral

$$G = \int_0^{\pi/2} \frac{dx}{\sqrt{1 + \cos^2 x}} = \int_0^{\pi/2} \frac{dx}{\sqrt{2\cos^2 x + \sin^2 x}}.$$

Here $a_1 = \sqrt{2}$ and $b_1 = 1$; the numerical sequences $(a_n)_{n=1}^{\infty}$ and $(b_n)_{n=1}^{\infty}$ converge rapidly to L in this case and a_5 and b_5 are approximately equal to 1.198154. Hence we may put $L \simeq 1.198154$ and obtain the approximate value

$$G = \frac{\pi}{2L} \simeq 1.3110138\ldots$$

The integral

$$G = \int_0^{\pi/2} \frac{dx}{\sqrt{a^2\cos^2 x + b^2\sin^2 x}} \qquad (a > b > 0)$$

may be changed into a complete elliptic integral of the first kind by setting

$$G = \frac{1}{a}\int_0^{\pi/2} \frac{dx}{\sqrt{1 - \frac{a^2 - b^2}{a^2}\sin^2 x}} = \frac{1}{a} K\left(\frac{\sqrt{a^2 - b^2}}{a}\right)$$

and can be computed with the help of tables (see Jahnke, E., Emde, F., and Lösch, F. (1960), Tables of Higher Functions, McGraw-Hill Book Co., Inc., New York, N.Y.).

We now consider the complete elliptic integral of the first kind

$$K(k) = \int_0^{\pi/2} \frac{dx}{\sqrt{1 - k^2 \sin^2 x}};$$

for every value of the module k it may be obtained from G if we set

$$a = 1 \quad \text{and} \quad b = \sqrt{1 - k^2} = k'.$$

In order to apply to this integral the method of Gauss, we first compute

$$a_1 = \frac{1 + \sqrt{1 - k^2}}{2} = \frac{1 + k'}{2}, \qquad b_1 = \sqrt{k'},$$

$$k_1 = \frac{\sqrt{a_1^2 - b_1^2}}{a_1} = \frac{1 - k'}{1 + k'}, \qquad \frac{1}{a_1} = 1 + k_1;$$

thus

$$\int_0^{\pi/2} \frac{dx}{\sqrt{1 - k^2 \sin^2 x}} = (1 + k_1) \int_0^{\pi/2} \frac{dt}{\sqrt{1 - k_1^2 \sin^2 t}}$$

or

$$K(k) = (1 + k_1)K(k_1).$$

By repeated application of this formula we obtain

$$K(k) = (1 + k_1)(1 + k_2) \cdots (1 + k_n)K(k_n),$$

where the sequence of numbers (k_n), as can be verified by induction is given by

$$k_n = \frac{1 - \sqrt{1 - k_{n-1}^2}}{1 + \sqrt{1 - k_{n-1}^2}} \qquad (k_0 = k)$$

so that

$$0 < k_n < 1 \quad \text{and} \quad k_n < k_{n-1}^2$$

holds; these inequalities are responsible for the fact that k_n tends to 0 rapidly as $n \to \infty$. We also have that

$$0 < K(k_n) - \frac{\pi}{2} = \int_0^{\pi/2} \frac{dx}{\sqrt{1 - k_n^2 \sin^2 x}} - \frac{\pi}{2}$$

$$= \int_0^{\pi/2} \frac{1 - \sqrt{1 - k_n^2 \sin^2 x}}{\sqrt{1 - k_n^2 \sin^2 x}} dx < \frac{\pi}{2} \frac{1 - \sqrt{1 - k_n^2}}{\sqrt{1 - k_n^2}}.$$

From this we get that

$$K(k_n) \to \frac{\pi}{2} \quad \text{as } n \to \infty;$$

hence

$$K(k) = \lim_{n\to\infty} (1 + k_1)(1 + k_2) \cdots (1 + k_n).$$

On this can be based a method of approximate calculation of the integral $K(k)$; for sufficiently large n,

$$K(k) \approx \frac{\pi}{2}(1 + k_1)(1 + k_2) \cdots (1 + k_n).$$

PROBLEM 59. Find the sum of the series

$$\cos^3 x - \frac{1}{3} \cos^3 3x + \frac{1}{3^2} \cos^3 3^2x - \frac{1}{3^3} \cos^3 3^3x + \cdots$$

Solution. Consider the following succession of identities

$$4 \cos^3 x = \cos 3x + 3 \cos x,$$

$$4 \cos^3 3x = \cos 3^2x + 3 \cos 3x,$$

$$4 \cos^3 3^2x = \cos 3^3x + 3 \cos 3^2x,$$

$$\cdots$$

$$4 \cos^3 3^nx = \cos 3^{n+1}x + 3 \cos 3^nx.$$

If we multiply the first of these identities by 1, the second by $(-3)^{-1}$, the third by 3^{-2}, ..., the n-th by $(-3)^{-n}$, and denote by S_n the sum of the first n terms of the proposed series, we see that

$$4S_n = 3 \cos x + (-3)^n \cos 3^{n+1}x.$$

Letting $n \to \infty$, we get that $S_n \to (3/4) \cos x$.

Remarks. In the following, let f and g be functions such that

$$f(x) = af(bx) + cg(x),$$

where a, b, c are given nonzero constants. Then

$$af(bx) = a^2 f(b^2 x) + acg(bx),$$

$$a^2 f(b^2 x) = a^3 f(b^3 x) + a^2 cg(b^2 x),$$

$$\cdots$$

$$a^{n-1} f(b^{n-1} x) = a^n f(b^n x) + a^{n-1} cg(b^{n-1} x);$$

thus

$$f(x) = a^n f(b^n x) + c[g(x) + ag(bx) + \cdots + a^{n-1} g(b^{n-1} x)].$$

If the product $a^n f(b^n x)$ converges to a limit L as $n \to \infty$, we get

$$f(x) = L + c \lim_{n \to \infty} [g(x) + ag(bx) + \cdots + a^{n-1} g(b^{n-1} x)]$$

and the series

$$g(x) + ag(bx) + a^2 g(b^2 x) + \cdots$$

is convergent to

$$\frac{f(x) - L}{c}.$$

If the product $a^n f(b^n x)$ does not converge as $n \to \infty$, the series is divergent or indeterminate.

In an entirely similar manner we obtain the set of equalities

$$f(x) = af(bx) + cg(x),$$

$$a^{-1} f(b^{-1} x) = f(x) + a^{-1} cg(b^{-1} x),$$

$$a^{-2} f(b^{-2} x) = a^{-1} f(b^{-1} x) + a^{-2} cg(b^{-2} x),$$

$$\cdots$$

$$a^{-n} f(b^{-n} x) = a^{1-n} f(b^{1-n} x) + a^{-n} cg(b^{-n} x);$$

thus

$$af(bx) = a^{-n}f(b^{-n}x) - c[g(x) + a^{-1}g(b^{-1}x) + \cdots + a^{-n}g(b^{-n}x)].$$

Therefore, if $a^{-n}f(b^{-n}x)$ converges to M as $n \to \infty$, we obtain that

$$g(x) + a^{-1}g(b^{-1}x) + a^{-2}g(b^{-2}x) + \cdots = \frac{M - af(bx)}{c}.$$

We now consider a number of applications of these results; among these applications will be the solution to Problem 59. Recall the functional equation $f(x) = af(bx) + cg(x)$, where a, b, and c are given nonzero constants.

Case 1: Consider the identity

$$\sin x = 3\sin(x/3) - 4\sin^3(x/3);$$

here $f(x) = \sin x$, $g(x) = \sin^3(x/3)$, $a = 3$, $b = 1/3$, $c = -4$. We have

$$L = \lim_{n\to\infty} [3^n \sin(3^{-n}x)] = x, \qquad M = \lim_{n\to\infty} [3^{-n}\sin(3^n x)] = 0;$$

thus

$$\sin^3\frac{x}{3} + 3\sin^3\frac{x}{3^2} + 3^2\sin^3\frac{x}{3^3} + \cdots = \frac{x - \sin x}{4},$$

$$\sin^3\frac{x}{3} + \frac{1}{3}\sin^3 x + \frac{1}{3^2}\sin^3 3x + \cdots = \frac{3}{4}\sin\frac{x}{3}$$

or, more simply,

$$\sin^3 x + \frac{1}{3}\sin^3 3x + \frac{1}{3^2}\sin^3 3^2x + \cdots = \frac{3}{4}\sin x.$$

Case 2: Consider the identity

$$\cos x = -3\cos(x/3) + 4\cos^3(x/3);$$

here $f(x) = \cos x$, $g(x) = \cos^3(x/3)$, $a = -3$, $b = 1/3$, $c = 4$. The product $(-3)^n \cos(3^{-n}x)$ increases indefinitely in absolute value as $n \to \infty$, hence the series

$$\cos^3\frac{x}{3} - 3\cos^3\frac{x}{3^2} + 3^2\cos^3\frac{x}{3^3} - \cdots$$

is divergent. But $\lim_{n\to\infty} [(-3)^{-n}\cos^3(3^n x)] = M = 0$; hence, on changing x to 3x,

$$\cos^3 x - \frac{1}{3}\cos^3 3x + \frac{1}{3^2}\cos^3 3^2x - \frac{1}{3^3}\cos^3 3^3x + \cdots = \frac{3}{4}\cos x.$$

This, however, is the solution to Problem 59.

Case 3: Consider the identity

$$\cot x = 2 \cot 2x + \tan x;$$

here $f(x) = \cot x$, $g(x) = \tan x$, $a = 2$, $b = 2$, $c = 1$. The product $2^n \cot(2^n x)$ does not converge as $n \to \infty$; but

$$\lim_{n\to\infty} [2^{-n} \cot(2^{-n}x)] = 1/x$$

and so we get

$$\tan x + \frac{1}{2}\tan\frac{x}{2} + \frac{1}{2^2}\tan\frac{x}{2^2} + \cdots = \frac{1}{x} - \frac{2}{\tan 2x}$$

and, in particular,

$$\tan\frac{\pi}{8} + \frac{1}{2}\tan\frac{\pi}{16} + \frac{1}{4}\tan\frac{\pi}{32} + \cdots = 2\left(\frac{4}{\pi} - 1\right).$$

Case 4: Consider the identity

$$\arctan x = \arctan(bx) + \arctan\frac{(1 - b)x}{1 + bx^2}.$$

If $0 < b < 1$, we may conclude that arc tan x equals

$$\arctan\frac{(1 - b)x}{1 + bx^2} + \arctan\frac{(1 - b)bx}{1 + b^3x^2} + \arctan\frac{(1 - b)b^2x}{1 + b^5x^2} + \cdots$$

and so, in particular,

$$\arctan\frac{1}{3} + \arctan\frac{2}{9} + \arctan\frac{4}{33}$$

$$+ \cdots + \arctan\frac{2^{2n-1}}{2^{2n-1} + 1} + \cdots = \frac{\pi}{4}.$$

If $b > 1$, we find

$$\arctan\frac{(b - 1)x}{1 + bx^2} + \arctan\frac{(b - 1)bx}{1 + b^3x^2}$$

$$+ \arctan\frac{(b - 1)b^2x}{1 + b^5x^2} + \cdots = \text{arc cot } x.$$

PROBLEM 60. For n = 1,2,3,..., let

$$S_n = \left\{ \sqrt{2} \cdot \sqrt[4]{2} \cdot \sqrt[8]{2} \cdots \sqrt[2^n]{2} \right\}.$$

Find

$$\lim_{n \to \infty} S_n.$$

Solution. Since

$$\log S_n = \left(\frac{1}{2} + \frac{1}{4} + \frac{1}{8} + \cdots + \frac{1}{2^n} \right) \log 2,$$

we see that

$$\lim_{n \to \infty} \log S_n = \left(\frac{1}{1 - \frac{1}{2}} - 1 \right) \log 2 = \log 2$$

and so S_n tends to 2 as $n \to \infty$.

PROBLEM 61. Show that $\sqrt{5}$ is equal to the ratio S/S', where

$$S = \frac{1}{1^2} + \frac{1}{2^2} + \frac{1}{5^2} + \frac{1}{13^2} + \frac{1}{34^2} + \frac{1}{89^2} + \cdots$$

and

$$S' = \frac{1}{1} - \frac{2}{3} + \frac{3}{8} - \frac{4}{21} + \frac{5}{55} - \frac{6}{144} + \cdots$$

Solution. Denoting by Y_n the n-th term of the sequence

1, 2, 5, 13, 34, 89, ...,

appearing in the first series, we see that any three consecutive terms satisfy the relation

$$Y_{n+2} - 3\,Y_{n+1} + Y_n = 0;$$

denoting by U_n the n-th term of the sequence

1, 3, 8, 21, 55, 144, ...,

appearing in the second series, we see that any three consecutive terms satisfy the relation

$$U_{n+2} - 3\,U_{n+1} + U_n = 0.$$

Hence, using the theory of difference equations, we get

$$Y_n = C_1\,a^n + \frac{C_2}{a^n}$$

and

$$U_n = C_1'\,a^n + \frac{C_2'}{a^n}$$

with a and 1/a being the roots of the equation

$$x^2 - 3x + 1 = 0. \tag{E.12}$$

The constants are determined by the conditions

$$C_1\,a + \frac{C_2}{a} = 1, \qquad C_1\,a^2 + \frac{C_2}{a^2} = 2,$$

$$C_1'\,a + \frac{C_2'}{a} = 1, \qquad C_1'\,a^2 + \frac{C_2'}{a^2} = 3;$$

hence, noting that

$$a + \frac{1}{a} = 3,$$

we have

$$C_1 = \frac{C_2}{a} = \frac{1}{\sqrt{5}a} \quad \text{and} \quad C_1' = -C_2' = \frac{1}{\sqrt{5}}.$$

Denoting by T_n the n-th term of the first series, we see that

$$T_n = \frac{5\,a^{2n-1}}{(a^{2n-1} + 1)^2};$$

denoting by T_n' the n-th term of the second series, we have

$$T_n' = \pm\,\frac{n\,a^n\sqrt{5}}{a^{2n} - 1}$$

with the + sign taken when n is odd and the - sign when n is even. We therefore have

$$S = 5a\left\{\frac{1}{(a+1)^2} + \frac{a^2}{(a^3+1)^2} + \frac{a^4}{(a^5+1)^2} + \cdots\right\}$$

and

$$S' = a\sqrt{5}\left\{\frac{1}{a^2-1} - \frac{2a}{a^4-1} + \frac{3a^2}{a^6-1} - \cdots\right\}.$$

By taking for a that root of the equation (E.12) which is larger than unity, we can develop all terms of S' in convergent series and we find

$$\begin{aligned}\frac{S'}{a\sqrt{5}} = {} & a^{-2} + a^{-4} + a^{-6} + a^{-8} + \cdots \\ & - 2(a^{-3} + a^{-7} + a^{-11} + a^{-15} + \cdots) \\ & + 3(a^{-4} + a^{-10} + a^{-16} + a^{-22} + \cdots) \\ & - 4(a^{-5} + a^{-13} + a^{-21} + a^{-29} + \cdots) \\ & + \cdots\end{aligned}$$

But

$$\frac{1}{a^2} - \frac{2}{a^3} + \frac{3}{a^4} - \frac{4}{a^5} + \cdots = \frac{1}{(a+1)^2},$$

$$\frac{1}{a^4} - \frac{2}{a^7} + \frac{3}{a^{10}} - \frac{4}{a^{13}} + \cdots = \frac{a^2}{(a^3+1)^2},$$

and so forth. Hence, under the stated condition for a,

$$\frac{1}{(a+1)^2} + \frac{1}{(a^3+1)^2}a^2 + \frac{1}{(a^5+1)^2}a^4 + \cdots$$

$$= \frac{1}{a^2-1} - \frac{2a}{a^4-1} + \frac{3a^2}{a^6-1} - \cdots$$

and so $S/S' = \sqrt{5}$.

Remarks. A similar representation of $\sqrt{5}$ is given by

$$\frac{\frac{1}{1} + \frac{1}{2} + \frac{1}{5} + \frac{1}{13} + \cdots}{\frac{1}{1} - \frac{1}{4} + \frac{1}{11} - \frac{1}{29} + \cdots}$$

in which the denominators satisfy the relation $D_{n+2} - 3D_{n+1} + D_n = 0$.

Here the problem reduces to showing that if

$$T = (1 + q)f(q) = (1 + q)\left\{\frac{1}{1+q} + \frac{q}{1+q^3} + \frac{q^2}{1+q^5} + \cdots\right\}$$

and

$$T' = (1 - q)g(q) = (1 - q)\left\{\frac{1}{1-q} - \frac{q}{1-q^3} + \frac{q^2}{1-q^5} - \cdots\right\},$$

where $q = (3 - \sqrt{5})/2$, that is q is the smaller of the two roots of $x^2 - 3x + 1 = 0$, then $T/T' = (1 + q)/(1 - q) = \sqrt{5}$. To do this, it suffices to verify that $f(q) = g(q)$. But

$$\begin{aligned} f(q) = 1 &- q + q^2 - q^3 + q^4 - \cdots \\ &+ q - q^4 + q^7 - q^{10} + q^{13} - \cdots \\ &+ q^2 - q^7 + q^{12} - q^{17} + q^{22} - \cdots \\ &+ q^3 - q^{10} + q^{17} - q^{24} + q^{31} - \cdots \\ &+ q^4 - q^{13} + q^{22} - q^{31} + q^{40} - \cdots \\ &+ \cdots \end{aligned}$$

and

$$\begin{aligned} g(q) = 1 &+ q + q^2 + q^3 + q^4 + \cdots \\ &- (q + q^4 + q^7 + q^{10} + q^{13} + \cdots) \\ &+ (q^2 + q^7 + q^{12} + q^{17} + q^{22} + \cdots) \\ &- (q^3 + q^{10} + q^{17} + q^{24} + q^{31} + \cdots) \\ &+ (q^4 + q^{13} + q^{22} + q^{31} + q^{40} + \cdots) \\ &- \cdots \end{aligned}$$

The method of proof in Problem 61 can, of course, be modified to cover other suitable situations.

PROBLEM 62. In the Solution of Problem 26 and in the Remarks to Problem 46 we noted that, for $n = 1,2,3,\ldots,$

$$\frac{1}{n+1} + \frac{1}{n+2} + \frac{1}{n+3} + \cdots + \frac{1}{2n}$$

tends to log 2 as $n \to \infty$. Use this result to show that

$$1 - \frac{1}{2} + \frac{1}{3} - \frac{1}{4} + \frac{1}{5} - \frac{1}{6} + \cdots = \log 2.$$

Solution. The desired result follows immediately from the Identity of Catalan (see Problem 42 of Chapter 1).

Remark. Note that the result in Problem 62 is contained as a special case in Problem 15.

PROBLEM 63. Let a, b, m, p, and q be positive integers. Show that, if $m = p^2 - q$, then the sequence of fractions

$$\frac{a}{b}, \quad \frac{a_1}{b_1} = \frac{pa + mb}{a + pb}, \quad \frac{a_2}{b_2} = \frac{pa_1 + mb_1}{a_1 + pb_1}, \quad \ldots, \quad \frac{a_n}{b_n} = \frac{pa_{n-1} + mb_{n-1}}{a_{n-1} + pb_{n-1}}, \ \ldots$$

converges to $\sqrt{m}$ as $n \to \infty$.

Solution. By definition,

$$\frac{a_{n+1}}{b_{n+1}} = \frac{pa_n + mb_n}{a_n + pb_n} = p - \cfrac{q}{\cfrac{a_n}{b_n} + p}$$

and

$$\frac{a_n}{b_n} = p - \cfrac{q}{\cfrac{a_{n-1}}{b_{n-1}} + p}.$$

Thus

$$\frac{a_{n+1}}{b_{n+1}} = p - \cfrac{q}{2p - \cfrac{q}{2p - \cfrac{q}{2p} - \ddots - \cfrac{q}{\cfrac{a}{b} + p}}}.$$

Hence the sought for limit will be

$$x = p - \cfrac{q}{2p - \cfrac{q}{2p - \cfrac{q}{2p} \ddots}}$$

But, if we develop $\sqrt{m}$ into a continued fraction, we have

$$\sqrt{m} = \sqrt{p^2 - q} = p - \frac{q}{p + \sqrt{p^2 - q}}$$

$$= p - \cfrac{q}{2p - \cfrac{q}{2p - \cfrac{q}{2p} \ddots}}$$

Thus we see that $x = \sqrt{m}$.

PROBLEM 64. Let

$$b_n = \sum_{k=0}^{n} \binom{n}{k}^{-1}$$

for $n \geq 1$. Show that $\lim_{n \to \infty} b_n = 2$.

Solution. Note that

$$b_n = \frac{n+1}{2n} b_{n-1} + 1 \qquad \text{for } n \geq 2.$$

PROBLEM 65. Let $(p_n)_{n=1}^{\infty}$ be a sequence of positive real numbers. Show that if the series

$$\sum_{n=1}^{\infty} \frac{1}{p_n}$$

converges, then so does the series

$$\sum_{n=1}^{\infty} \frac{n^2}{(p_1 + p_2 + \cdots + p_n)^2} p_n.$$

Solution. Put $q_n = p_1 + \cdots + p_n$ and $q_0 = 0$. We get the estimate

$$S_m = \sum_{n=1}^{m} \left(\frac{n}{q_n}\right)^2 (q_n - q_{n-1})$$

in terms of $T = \Sigma_{n=1}^{\infty} 1/p_n$. We observe that

$$S_m \le \frac{1}{p_1} + \sum_{n=2}^{m} \frac{n^2}{q_n q_{n-1}} (q_n - q_{n-1}) = \frac{1}{p_1} + \sum_{n=2}^{m} \frac{n^2}{q_{n-1}} - \sum_{n=2}^{m} \frac{n^2}{q_n}$$

$$= \frac{1}{p_1} + \sum_{n=1}^{m-1} \frac{(n+1)^2}{q_n} - \sum_{n=2}^{m} \frac{n^2}{q_n} \le \frac{5}{p_1} + 2\sum_{n=2}^{m} \frac{n}{q_n} + \sum_{n=2}^{m} \frac{1}{q_n}.$$

By the Cauchy-Schwarz Inequality (see Problem 14 in Chapter 2),

$$\left\{\sum_{n=1}^{m} \frac{n}{q_n}\right\}^2 \le \left\{\sum_{n=2}^{m} \frac{n^2}{q_n^2} p_n\right\}\left\{\sum_{n=2}^{m} \frac{1}{p_n}\right\}$$

and thus $S_m \le 5/p_1 + 2\sqrt{S_m T} + T$. But this inequality implies that

$$\sqrt{S_m} \le T + \sqrt{2T + 5/p_1}.$$

PROBLEM 66. Suppose that $u_0, u_1, u_2, \ldots$ is a sequence of real numbers such that

$$u_n = \sum_{k=1}^{\infty} u_{n+k}^2 \qquad \text{for } n = 0,1,2,\ldots$$

Show that if $\Sigma_{n=0}^{\infty} u_n$ converges then $u_k = 0$ for all k.

Solution. Clearly $0 \le u_{n+1} \le u_n$ for $n = 1,2,\ldots$ If $\Sigma_{j=0}^{\infty} u_j$ converges, we take $k \ge 1$ so that $\Sigma_{j=k+1}^{\infty} u_j < 1$. Then

$$u_{k+1} \le u_k = \sum_{j=k+1}^{\infty} u_j^2 \le u_{k+1}\left(\sum_{j=k+1}^{\infty} u_j\right) \le u_{k+1}.$$

Hence $u_k = u_{k+1}$ and so $u_{k+1} = 0$, implying $u_j = 0$ for $j > k$ since

$$u_{k+1} = \sum_{j=k+1}^{\infty} u_j^2.$$

By induction $u_j = 0$ for $j < k + 1$.

PROBLEM 67. Given a sequence $(x_n)_{n=1}^{\infty}$ such that $x_n - x_{n-2} \to 0$ as $n \to \infty$. Show that

$$\lim_{n\to\infty} \frac{x_n - x_{n-1}}{n} = 0.$$

Solution. For $\varepsilon > 0$, let n_0 be sufficiently large so that $|x_n - x_{n-2}| < \varepsilon$ for all $n \geq n_0$. Observe that for $n \geq n_0$,

$$x_n - x_{n-1} = (x_n - x_{n-2}) - (x_{n-1} - x_{n-3}) + (x_{n-2} - x_{n-4})$$
$$- \cdots + \{(x_{n_0+2} - x_{n_0}) - (x_{n_0+1} - x_{n_0-1})\}.$$

Thus

$$|x_n - x_{n-1}| \leq (n - n_0)\varepsilon + |x_{n_0+1} - x_{n_0-1}|$$

and so $(x_n - x_{n-1})/n$ tends to zero as $n \to \infty$.

PROBLEM 68. Let $(x_n)_{n=1}^{\infty}$ be a given sequence and let $y_n = x_{n-1} + 2x_n$ for $n = 2,3,\ldots$ Suppose that the sequence $(y_n)_{n=2}^{\infty}$ converges. Show that the sequence $(x_n)_{n=1}^{\infty}$ converges.

Solution. Let $\bar{y} = \lim_{n\to\infty} y_n$ and set $\bar{x} = \bar{y}/3$. We wish to show that $\bar{x} = \lim_{n\to\infty} x_n$. For $\varepsilon > 0$ there is an n_0 such that for all n larger than n_0, $|y_n - \bar{y}| < \varepsilon/2$. Now

$$\varepsilon/2 > |y_n - \bar{y}| = |x_{n-1} + 2x_n - 3\bar{x}| = |2(x_n - \bar{x}) + (x_{n-1} - \bar{x})|$$
$$\geq 2|x_n - \bar{x}| - |x_{n-1} - \bar{x}|.$$

This may be rewritten as $|x_n - \bar{x}| < \varepsilon/4 + \frac{1}{2}|x_{n-1} - \bar{x}|$, which can be iterated to give

$$|x_{n+m} - \bar{x}| < \frac{\varepsilon}{4}\left(\sum_{i=1}^{m} 2^{-i}\right) + 2^{-(m+1)}|x_{n-1} - \bar{x}| < \frac{\varepsilon}{2} + 2^{-(m+1)}|x_{n-1} - \bar{x}|.$$

By taking m large enough,

$$2^{-(m+1)}|x_{n-1} - \bar{x}| < \frac{\varepsilon}{2}.$$

Thus for all sufficiently large k, $|x_k - \bar{x}| < \varepsilon$.

PROBLEM 69. Let $0 < x_1 < 1$ and $x_{n+1} = x_n(1 - x_n)$ for $n = 1,2,3,\ldots$ Show that $nx_n \to 1$ as $n \to \infty$.

Solution. Observe that

$$\begin{aligned}(n + 1)x_{n+1} &= nx_n + x_n - (n + 1)x_n^2 \\ &= nx_n + x_n\{1 - (n + 1)x_n\}.\end{aligned} \tag{E.13}$$

To see that nx_n is increasing, we need to show that $1 - (n + 1)x_n \geq 0$. From the graph of $y = x(1 - x)$ we note that $x_2 \leq 1/4$ and that $x_n \leq a \leq 1/2$ implies $x_{n+1} \leq a(1 - a)$. So by induction

$$(n + 1)x_n \leq (n + 1)\frac{1}{n}\left(1 - \frac{1}{n}\right) = 1 - \frac{1}{n^2} \leq 1.$$

Furthermore, $nx_n < (n + 1)x_n \leq 1$ and so nx_n is bounded above by 1. Thus nx_n converges to a limit L with $0 < nx_n < L \leq 1$. Now summing (E.13) from 2 to n we obtain

$$\begin{aligned}1 &\geq (n + 1)x_{n+1} \\ &= 2x_2 + x_2(1 - 3x_2) + x_3(1 - 4x_3) + \cdots + x_n\{1 - (n + 1)x_n\}.\end{aligned} \tag{E.14}$$

If $L \neq 1$ then $\{1 - (n + 1)x_n\} \geq (1 - L)/2$ for all large n and thus (E.14) would show that $\Sigma_{n=1}^{\infty} x_n$ is convergent. However, $nx_n \geq x_1$ and so $\Sigma_{n=1}^{\infty} x_n \geq x_1 \Sigma_{n=1}^{\infty} 1/n$. But $\Sigma_{n=1}^{\infty} 1/n$ is divergent.

PROBLEM 70. We say that a sequence of points in an interval is dense when every subinterval contains at least one point of the sequence. With this agreement in mind, let $P_1, P_2, P_3, \ldots$ be a sequence of distinct points which is dense in the interval (0,1). The points

$$P_1, P_2, P_3, \ldots, P_{n-1}$$

decompose the interval into n parts, and P_n decomposes one of these into two parts. Let a_n and b_n be the lengths of these two intervals. Show that

$$\sum_{n=1}^{\infty} a_n b_n (a_n + b_n) = \frac{1}{3}.$$

Solution. The identity to be proved can be written in the form

$$-(a_1 + b_1)^3 + \sum_{n=1}^{\infty} (a_n + b_n)^3 - a_n^3 - b_n^3 = 0.$$

Consider the partial sum

$$-a_1^3 - b_1^3 + (a_2 + b_2)^3 - a_2^3 - b_2^3 + (a_3 + b_3)^3 - a_3^3 - b_3^3$$

$$+ \cdots + (a_n + b_n)^3 - a_n^3 - b_n^3.$$

Each of the positive terms cancels one of the preceding negative terms. Because the sequence $P_1, P_2, P_3, \ldots$ is dense, every negative term eventually gets cancelled out. If we choose n sufficiently large, the only terms that are not cancelled out in the above sum are the cubes of the lengths of disjoint intervals, each of which is of length ε at most. The sum of the terms that are not cancelled out is therefore majorized by ε. It follows that the above sum is at most ε in absolute value for n large. Now let $n \to \infty$.

PROBLEM 71. Let N denote the set of all positive integers. What is the set of limit points of the set $\{\sqrt{a} - \sqrt{b} : a, b \in N\}$?

Solution. The set contains all integral multiples of each of its members. Since it has arbitrarily small members $\sqrt{n+1} - \sqrt{n}$, noting that

$$(\sqrt{n+1} - \sqrt{n})(\sqrt{n+1} + \sqrt{n}) = 1$$

and so

$$0 < \sqrt{n+1} - \sqrt{n} < \frac{1}{2\sqrt{n}},$$

every real number is a limit point of it.

Remark. Generalization of the result in Problem 71: If $a_1, a_2, \ldots$ and $b_1, b_2, \ldots$ are unbounded monotone increasing sequences of real numbers and $\lim_{n\to\infty} (a_{n+1} - a_n) = 0$, then $\{a_n - b_n : n, m \in N\}$ is dense in the set of real numbers.

Indeed, let r be any real number and $\varepsilon > 0$. Then there exists k so that $a_{n+1} - a_n < \varepsilon$ for all $n \geq k$. Pick m large enough so that $b_m \geq a_k - r$ and let $n \geq k$ be determined (uniquely) by the condition $a_n \leq b_m + r < a_{n+1}$. This gives $|(a_n - b_m) - r| < \varepsilon$.

In Problem 71, $a_n = b_n = \sqrt{n}$ satisfies the assumptions of the foregoing general proposition.

PROBLEM 72. For $k \geq 0$, let S be the set of all numbers of the form

$$s = \sqrt{k \pm \sqrt{k \pm \cdots \pm \sqrt{k}}}$$

with arbitrary finite sequence of signs. If $k \geq 2$, then all s in S are real. Show: (i) if $k = 2$, then S is dense in the interval $(0,2)$; (ii) if $k > 2$, then S is dense in no interval.

Solution. The set T of expressions of the form

$$t = \pm\sqrt{k \pm \sqrt{k \pm \cdots \pm \sqrt{k}}}$$

is precisely the set of zeros of the polynomial $P_n(t)$ for some $n = 1,2,3,\ldots,$ where

$$P_1(t) = P(t) = t^2 - k, \quad P_n(t) = P_1(P_{n-1}(t)), \quad n = 2,3,\ldots$$

In the case $k = 2$ we may put $t = 2\cos x$, whence $P_n(t) = 2\cos(2^n x)$ and the zeros of P_n are $2\cos\{(2k + 1)2^{-n}\pi\}$, $k = 0,1,2,\ldots,$ so that T is a dense subset of $(-2,2)$ and hence S a dense subset of $(0,2)$.

Let $k > 2$. It will suffice to show that S contains none of its own limit points. Evidently $M = \sup S$ satisfies $M = (k + M)^{1/2}$, whence $M = \frac{1}{2}\{1 + (1 + 4k)^{1/2}\}$. Now $k > 2$ implies $M < k$, whence $m = \inf S = (k - M)^{1/2} > 0$. Therefore the greatest s in S less than $\sqrt{k}$ is less than $(k - m)^{1/2}$, and the least s in S greater than $\sqrt{k}$ is greater than $(k + m)^{1/2}$. It can be seen that if s has a deleted neighborhood disjoint from S, then each of $(k + s)^{1/2}$ and $(k - s)^{1/2}$ has such a neighborhood. It follows by induction that each s in S has a neighborhood containing no other point of S.

PROBLEM 73. Let $x_0 = 1$, $x_{n+1} = 1/(1 + x_n)$ for $n \geq 0$. Show that $(x_n)_{n=0}^{\infty}$ converges.

Solution. It is clear that $1/2 \le x_n \le 1$ for all n. Moreover, for arbitrary n, $k \ge 0$,

$$x_{n+1+k} - x_{n+1} = -\frac{x_{n+k} - x_n}{(1 + x_{n+k})(1 + x_n)}$$

and so

$$|x_{n+1+k} - x_{n+1}| = \frac{|x_{n+k} - x_n|}{(1 + x_{n+k})(1 + x_n)} \le \frac{1}{\left(1 + \frac{1}{2}\right)^2}|x_{n+k} - x_n|$$

$$= \frac{4}{9}|x_{n+k} - x_n|.$$

By induction on n we get

$$|x_{n+k} - x_n| \le \left(\frac{4}{9}\right)^n |x_k - x_0| \le 2\cdot\left(\frac{4}{9}\right)^n.$$

But $(4/9)^n \to 0$ as $n \to \infty$ and the sequence $(a_n)_{n=0}$ is a Cauchy sequence and hence converges; its limit a clearly satisfies the equation $a^2 + a = 1$ and a cannot be negative; thus $a = (\sqrt{5} - 1)/2$.

Remark. Note that the sequence

$$x_0 = 1, \quad x_{n+1} = \frac{2 + x_n}{1 + x_n} \quad (n \ge 0)$$

can be treated similarly; it converges to $\sqrt{2}$.

PROBLEM 74. Let $u_1, u_2, u_3, \ldots, v_1, v_2, v_3, \ldots$ be sequences of positive terms, the first sequence being bounded above, and in the second sequence $\lim_{n\to\infty} v_n = 1$. Show that the limit superiors satisfy

$$\overline{\lim_{n\to\infty}}\, u_n v_n = \overline{\lim_{n\to\infty}}\, u_n.$$

Solution. It is clear that as $u_1, u_2, u_3, \ldots$ are positive and bounded above and the sequence $v_1, v_2, v_3, \ldots$ has positive terms only and is convergent (hence bounded), the sequence of positive terms $u_1v_1, u_2v_2, u_3v_3, \ldots$ is bounded. If possible, let

$$\overline{\lim_{n\to\infty}}\, u_n v_n = L' > L = \overline{\lim_{n\to\infty}}\, u_n.$$

Take $2\varepsilon = L' - L$, and let the least upper bound of $u_1, u_2, u_3, \ldots$ be K. Since $\lim_{n \to \infty} v_n = 1$, there is a positive integer k_1, such that

$$|v_n - 1| < \frac{\varepsilon}{2K}, \quad \text{when } n \geq k_1.$$

Thus $|u_n v_n - u_n| = |u_n||v_n - 1| < K \frac{\varepsilon}{2K} < \frac{1}{2}$, when $n \geq k_1$. Therefore $u_n v_n < u_n + \frac{1}{2}\varepsilon$, when $n \geq k_1$. But since

$$\overline{\lim_{n \to \infty}}\, u_n = L,$$

there is a positive integer k_2 such that $u_n < L + \frac{1}{2}\varepsilon$, when $n \geq k_2$. Therefore $u_n v_n < L + \varepsilon$, when $n \geq k$, where k is the larger of the integers k_1 and k_2. But since

$$\overline{\lim_{n \to \infty}}\, u_n v_n = L',$$

we know that $u_n v_n > L' - \varepsilon$, for an infinite number of values of n. Therefore L' cannot be greater than L.

In a similar manner it can be shown that L' is not less than L. Hence $L' = L$.

PROBLEM 75. Given a sequence $(a_n)_{n=1}^{\infty}$ of nonnegative real numbers such that $a_{n+m} \leq a_n a_m$ for all pairs of positive integers, m and n, show that $(\sqrt[n]{a_n})_{n=1}^{\infty}$ converges.

Solution. Since $a_{n+1} \leq a_n a_1$, it follows by induction that $a_n \leq a_1^n$ for all n, and hence that $0 \leq a_n^{1/n} \leq a_1$ for all n. The sequence $(a_n^{1/n})_{n=1}^{\infty}$ is thus bounded. Taking the limit superior, let

$$L = \overline{\lim_{n \to \infty}}\, \sqrt[n]{a_n}.$$

We have $0 \leq L \leq a_1$. We shall show that $a_n^{1/n} \geq L$ for all n. Once this is established, it will follow that the limit inferior

$$\underline{\lim_{n \to \infty}}\, \sqrt[n]{a_n} \geq L = \overline{\lim_{n \to \infty}}\, \sqrt[n]{a_n},$$

which implies the existence of $\lim_{n\to\infty} a_n^{1/n}$.

Let $m_1 < m_2 < m_3 < \cdots$ be such that

$$\lim_{k\to\infty} \sqrt[m_k]{a_{m_k}} = L.$$

Let n be any positive integer, and let the rational number m_k/n be given by $q_k + d_k$ where q_k is an integer and $0 \le d_k < 1$. We observe that $q_k/m_k = (1/n) - d_k/m_k$, so that as $k \to \infty$, $q_k/m_k \to 1/n$, since $d_k/m_k < 1/m_k \to 0$. Now by the given relation

$$a_{m_k} = a_{nq_k+nd_k} \le a_n^{q_k} a_1^{nd_k}.$$

Hence

$$a_{m_k}^{1/m_k} \le a_n^{q_k/m_k} a_1^{nd_k/m_k},$$

and letting $k \to \infty$ we obtain $L \le a_n^{1/n}$, since $q_k/m_k \to 1/n$ and $nd_k/m_k \to 0$. It follows that $\lim_{n\to\infty} a_n^{1/n}$ exists.

Remark. The sequence $(a_n^{1/n})_{n=1}^{\infty}$ need not be monotonic. As an example let

$$a_n = e^{-n} \qquad \text{if n is even,}$$

$$= e^{-n+1} \qquad \text{if n is odd.}$$

For this sequence $(a_n^{1/n})_{n=1}^{\infty}$ is an oscillating sequence converging to 1/e.

PROBLEM 76. Let $u_1 + u_2 + u_3 + \cdots$ be a given series and put $s_n = u_1 + \cdots + u_n$ and

$$\sigma_n = \frac{s_1 + \cdots + s_n}{n}$$

By Problem 40, if $s_n \to s$ as $n \to \infty$, then $\sigma_n \to s$ as $n \to \infty$. Show the following theorem due to Hardy and Landau: If $\sigma_n \to \sigma$ as $n \to \infty$ and either $n(s_n - s_{n+1}) < K$ or $n(s_{n+1} - s_n) < K$, where K is some positive constant, then $s_n \to \sigma$ as $n \to \infty$.

Solution. We may without loss of generality take

$$\sigma = 0, \quad K = 1, \quad \text{and} \quad n(s_n - s_{n+1}) < 1.$$

For if we put

$$U_1 = \frac{u_1 - \sigma}{K}, \quad U_2 = \frac{u_2}{K},$$

etc., and

$$S_n = U_1 + \cdots + U_n,$$

we have

$$S_n = \frac{s_n - \sigma}{K}, \quad \text{and} \quad \frac{1}{n}(S_1 + \cdots + S_n) = \frac{\sigma_n - \sigma}{K}.$$

Thus

$$n(S_n - S_{n+1}) < 1, \quad \text{if} \quad n(s_n - s_{n+1}) < K.$$

In the other case we put

$$U_1 = \frac{\sigma - u_1}{K}, \quad U_2 = \frac{u_2}{K}, \quad \text{etc.}$$

Suppose then we are given $\sigma_n \to 0$ and $n(s_n - s_{n+1}) < 1$. However, it is clear that $\lim_{n\to\infty} s_n$ is not equal to $+\infty$ (or $-\infty$), because if it were we would have $\lim_{n\to\infty} \sigma_n = +\infty$ (or $-\infty$).

(a) If possible, let

$$\overline{\lim_{n\to\infty}}\, s_n = \Lambda,$$

where Λ is $+\infty$, or a finite positive number. Then, if A is any positive number less than Λ,

$$s_n > A, \quad \text{for infinitely many of the values of } n,$$

say $M_1, M_2, \ldots$ But to the arbitrary positive number ε, there corresponds a positive integer μ, such that

$$|\sigma_n| < \varepsilon, \quad \text{when} \quad n \geq \mu.$$

Let M be the first of the sequence $M_1, M_2, \ldots$ which is greater than μ, and such that $MA \geq$ an even positive integer.

Let $2p$ be the largest positive integer not greater than MA. Then $2p \leq MA < 4p$. Also

$$s_{M=1} > s_M - \frac{1}{M},$$

$$s_{M+2} > s_{M+1} - \frac{1}{M+1} > s_M - \left(\frac{1}{M} + \frac{1}{M+1}\right) > s_M - \frac{2}{M},$$

$$s_{M+3} > \cdots > s_M - \frac{3}{M},$$

$$\cdots$$

$$s_{M+p} > \cdots > s_M - \frac{p}{M},$$

and each of these is greater than $A - p/M$. But

$$\sigma_{M+p} = \frac{M\sigma_M + (s_{M+1} + s_{M+2} + \cdots + s_{M+p})}{M + p}$$

$$> \frac{M}{M+p}\sigma_M + \frac{p}{M+p}\left(A - \frac{p}{M}\right)$$

$$> \frac{a}{a+1}(A - a) - \varepsilon, \quad \text{where } a = \frac{p}{M} \le \frac{A}{2}.$$

Now $A - a \ge A/2$ and

$$\frac{a}{a+1} = \frac{A/4}{A/4 + 1},$$

since $p/M = a > A/4$. Therefore

$$\frac{a}{a+1}(A - a) > \frac{A^2}{2(A+4)}, \quad \text{and} \quad \sigma_{M+p} > \frac{A^2}{2(A+4)} - \varepsilon.$$

If

$$\varepsilon = \frac{A^2}{4(A+4)},$$

we have $\sigma_{M+p} > \varepsilon$, which is impossible. Thus

$$\overline{\lim_{n\to\infty}}\, s_n \le 0.$$

(b) We shall now show that

$$\underline{\lim}_{n\to\infty}\, s_n \ge 0.$$

For, if possible, let

$$\underline{\lim}_{n\to\infty}\, s_n = \lambda < 0.$$

Take B any positive number less than $-\lambda$. Then

$$s_n < -B, \quad \text{for an infinite number of values of } n,$$

say $H_1, H_2, \ldots,$ and

$$|\sigma_n| < \varepsilon, \quad \text{when } n \geq \nu.$$

Let H be the first of the sequence $H_1, H_2, \ldots$ such that

$$\frac{H}{1 + \frac{1}{2}B} > \nu$$

and such that between

$$\frac{H}{1 + \frac{1}{2}B} \quad \text{and} \quad \frac{H}{1 + \frac{1}{4}B}$$

there shall be at least one positive integer. Let the integer next above $H/(1 + \frac{1}{2}B)$ be q, and write $H/q = 1 + b$. Then since

$$\frac{H}{1 + \frac{1}{2}B} < q < \frac{H}{1 + \frac{1}{4}B},$$

we have $\frac{1}{4}B < b < \frac{1}{2}B$. Also as before

$$\sigma_H = \frac{q\sigma_q + s_{q+1} + \cdots + s_H}{H}.$$

But

$$s_{H-1} < s_H + \frac{1}{H - 1},$$

$$s_{H-2} < s_{H-1} + \frac{1}{H - 2} < s_H + \left(\frac{1}{H - 1} + \frac{1}{H - 2}\right),$$

$$\cdots$$

$$s_{q+1} < \cdots < s_H + \left(\frac{1}{H - 1} + \frac{1}{H - 2} + \cdots + \frac{1}{q + 1} + \frac{1}{q}\right).$$

Therefore each of these is less than $-B + (H - q)/q < -B + b$. Also

$$\sigma_H < \frac{q}{H}\sigma_q + \frac{H - q}{H}(-B + b) < -\frac{b}{1 + b}(B - b) + \varepsilon.$$

But

$$B - b > \frac{1}{2}B \quad \text{and} \quad \frac{b}{1 + b} > \frac{B}{B + 4}.$$

Therefore

$$\frac{b}{1+b}(B - b) < \frac{B^2}{2(B+4)} \quad \text{and} \quad \sigma_H < -\frac{B^2}{2(B+4)} + \varepsilon.$$

If we take

$$\varepsilon = \frac{B^2}{4(B+4)}.$$

we have $\sigma_H < -\varepsilon$, which is impossible. Thus

$$\underline{\lim}_{n\to\infty} s_n \geq 0.$$

But we have seen that

$$\overline{\lim}_{n\to\infty} s_n \leq 0.$$

It follows that

$$\lim_{n\to\infty} s_n = 0.$$

Remark. From Problem 76 we easily see the validity of the following statement: Let the sequence of arithmetic means σ_n of the series $u_1 + u_2 + u_3 + \cdots$ converge to σ as $n \to \infty$. If a positive integer n_0 exists such that

$$|u_n| < \frac{K}{n}, \quad \text{when} \quad n \geq n_0,$$

where K is a positive number independent of n, then the series $u_1 + u_2 + u_3 + \cdots$ converges and its sum is σ.

PROBLEM 77. Show that if a, b > 0, then

$$\lim_{n\to\infty} \left\{\frac{\sqrt[n]{a} + \sqrt[n]{b}}{2}\right\}^n = \sqrt{ab}.$$

Solution. Let $f(x) = a^x$. Then $f'(x) = a^x \log a$ and $f'(0) = \log a$. Thus

$$\lim_{n\to\infty} \frac{f\left(\frac{1}{n}\right) - f(0)}{\frac{1}{n}} = \lim_{n\to\infty} n(\sqrt[n]{a} - 1) = \log a.$$

Putting

$$x_n = n\left[\frac{\sqrt[n]{a} + \sqrt[n]{b}}{2} - 1\right] = \frac{1}{2}[n(\sqrt[n]{a} - 1) + n(\sqrt[n]{b} -)]$$

we see that

$$x_n \to \frac{1}{2}(\log a + \log b) = \log \sqrt{ab} \quad \text{as } n \to \infty.$$

But

$$\frac{\sqrt[n]{a} + \sqrt[n]{b}}{2} = 1 + \frac{x_n}{n}$$

and

$$\left[1 + \frac{x_n}{n}\right]^n = \left\{\left[1 + \frac{x_n}{n}\right]^{\frac{n}{x_n}}\right\}^{x_n} \to e^{\log\sqrt{ab}} = \sqrt{ab}.$$

Remark. In the foregoing reasoning we made use of the fact that

$$\lim_{\alpha \to 0} (1 + \alpha)^{1/\alpha} = e;$$

this fact is the basis for most applications of the number e.

PROBLEM 78. Let $a_1, a_2, \ldots, a_k$ be given constants and put

$$S_n = \sqrt[k]{(n - a_1)(n - a_2) \cdots (n - a_k)} - n.$$

Evaluate $\lim_{n \to \infty} S_n$.

Solution. In the identity

$$y - z = \frac{y^k - z^k}{y^{k-1} + y^{k-2}z + \cdots + z^{k-1}}$$

we put

$$y = \sqrt[k]{(n - a_1) \cdots (n - a_k)} \quad \text{and} \quad z = n.$$

Then

$$S_n = \frac{(n + a_1) \cdots (n + a_k) - n^k}{(\sqrt[k]{\cdots})^{k-1} + n(\sqrt[k]{\cdots})^{k-2} + \cdots + n^{k-1}}$$

$$= \frac{(a_1 + \cdots + a_k) + \dfrac{a_1a_2 + \cdots + a_{k-1}a_k}{n} + \cdots + \dfrac{a_1a_2 \cdots a_k}{n^{k-1}}}{\left\{\sqrt[k]{\left(1 + \dfrac{a_1}{n}\right) \cdots \left(1 + \dfrac{a_k}{n}\right)}\right\}^{k-1} + \cdots + 1}$$

and so

$$\lim_{n \to \infty} S_n = \frac{a_1 + a_2 + \cdots + a_k}{k}.$$

PROBLEM 79. Let $(x_n)_{n=1}$ be a sequence of real numbers such that $x_n \neq 0$ for all n and $\lim_{n \to \infty} x_n = 0$. Without using differentiation, show that

$$\lim_{n \to \infty} \frac{(1 + x_n)^{1/m} - 1}{x_n} = \frac{1}{m},$$

where m is a positive integer.

Solution. We set

$$(1 + x_n)^{1/m} - 1 = y_n;$$

we get $x_n = (1 + y_n)^m - 1$. But

$$1 - |t| < \sqrt[m]{1 + t} < 1 + |t| \quad \text{for } |t| < 1$$

and so

$$\lim_{n \to \infty} (1 + x_n)^{1/m} = 1,$$

implying that $\lim_{n \to \infty} y_n = 0$. Consequently

$$\lim_{n \to \infty} \frac{(1 + x_n)^{1/m} - 1}{x_n} = \lim_{n \to \infty} \frac{y_n}{(1 + y_n)^m - 1} = \lim_{n \to \infty} \frac{1}{m + \binom{m}{2}y_n + \cdots + y_n^{m-1}}$$

and the latter limit is seen to be equal to 1/m.

PROBLEM 80. Let $b > a > 0$, $d > 0$, and

$$q_n = \frac{a(a + d)(a + 2d) \cdots (a + nd)}{b(b + d)(b + 2d) \cdots (b + nd)}.$$

Show that $\lim_{n \to \infty} q_n = 0$.

Solution. Let $y > x > 0$ and put $u = y - x$. By Problem 62 of Chapter 2, if γ and δ are positive integers larger than 1, then

$$\left(\frac{y}{x}\right)^{\gamma} > 1 + \gamma\frac{u}{x} \quad \text{and} \quad \left(\frac{y}{x}\right)^{1/\delta} < 1 + \frac{u}{\delta x}.$$

Thus

$$\left(1 + \frac{\gamma u}{x}\right)^{1/\gamma} < \frac{y}{x} < \left(1 + \frac{u}{\delta x}\right)^{\delta}$$

or

$$\left(\frac{x}{x + u/\delta}\right)^{\delta} < \frac{x}{y} < \left(\frac{x}{x + \gamma u}\right)^{1/\gamma}.$$

We take for γ and δ only those values for which

$$u/\delta < d \quad \text{and} \quad \gamma y > d.$$

Then

$$\left(\frac{x}{x + d}\right)^{\delta} < \left(\frac{x}{x + u/\delta}\right)^{\delta} < \frac{x}{y} < \left(\frac{x}{x + \gamma u}\right)^{\gamma} < \left(\frac{x}{x + d}\right)^{\gamma}.$$

We therefore have the inequality

$$\left(\frac{x}{x + d}\right)^{\delta} < \frac{x}{y} < \left(\frac{x}{x + d}\right)^{1/\gamma} \tag{E.15}$$

wedging in the proper positive fraction x/y and holding for all integers γ and δ which exceed the largest of the three numbers $(y - x)/d$, $d/(y - x)$, and 1. If in (E.15) we set successively

$$x = a, \quad a + d, \quad a + 2d, \quad \ldots, \quad a + nd,$$

$$y = b, \quad b + d, \quad b + 2d, \quad \ldots, \quad b + nd,$$

we obtain the system of inequalities

$$\left(\frac{a}{a + d}\right)^{\delta} < \frac{a}{b} < \left(\frac{a}{a + d}\right)^{1/\gamma},$$

$$\left(\frac{a + d}{a + 2d}\right)^{\delta} < \frac{a + d}{b + d} < \left(\frac{a + d}{a + 2d}\right)^{1/\gamma},$$

$$\cdots$$

$$\left(\frac{a + nd}{a + md}\right)^{\delta} < \frac{a + nd}{b + nd} < \left(\frac{a + nd}{a + md}\right)^{1/\gamma} \quad \text{with } m = n + 1.$$

Multiplication of the inequalities of the system gives

$$\left(\frac{a}{a + md}\right)^{\delta} < q_n < \left(\frac{a}{a + md}\right)^{1/\gamma} \quad \text{with } m = n + 1.$$

Letting $n \to \infty$, we see that both

$$\left(\frac{a}{a + md}\right)^{\delta} \quad \text{and} \quad \left(\frac{a}{a + md}\right)^{1/\gamma}$$

tend to zero; thus $\lim_{n\to\infty} q_n = 0$.

PROBLEM 81. For $n = 2,3,\ldots,$ let

$$P_n = \left\{1 - \frac{1}{2^2}\right\}\left\{1 - \frac{1}{3^2}\right\} \cdots \left\{1 - \frac{1}{n^2}\right\}.$$

Compute $\lim_{n\to\infty} P_n$.

Solution. We can easily see that

$$P_n = \frac{1\cdot 3}{2^2}\,\frac{2\cdot 4}{3^2}\,\frac{3\cdot 5}{4^2}\,\frac{4\cdot 6}{5^2} \cdots \frac{(n - 1)(n + 1)}{n^2} = \frac{1}{2}\,\frac{n + 1}{n}$$

and so $\lim_{n\to\infty} P_n = 1/2$.

PROBLEM 82. Find the sum of the series

$$\frac{\frac{1}{2^2}}{1 - \frac{1}{2^2}} + \frac{\frac{1}{3^2}}{\left[1 - \frac{1}{2^2}\right]\left[1 - \frac{1}{3^2}\right]} + \frac{\frac{1}{4^2}}{\left[1 - \frac{1}{2^2}\right]\left[1 - \frac{1}{3^2}\right]\left[1 - \frac{1}{4^2}\right]} + \cdots \tag{E.16}$$

Solution. From the Solution of Problem 47 in Chapter 1 we know that

$$\frac{a_1 - 1}{a_1} + \frac{a_2 - 1}{a_1 a_2} + \frac{a_3 - 1}{a_1 a_2 a_3} + \cdots + \frac{a_n - 1}{a_1 a_2 \cdots a_n} = 1 - \frac{1}{a_1 a_2 \cdots a_n}.$$

Thus, if $\lim_{n\to\infty} (a_1 a_2 \cdots a_n)$ exists and is equal to L, then the series

$$\frac{a_1 - 1}{a_1} + \frac{a_2 - 1}{a_1 a_2} + \frac{a_3 - 1}{a_1 a_2 a_3} + \cdots = 1 - \frac{1}{L}.$$

We now set

$$a_1 = 1 \quad \text{and} \quad a_n = 1 - \frac{1}{n^2} \quad \text{for } n = 2,3,\ldots$$

By Problem 81, $(a_1 a_2 \cdots a_n)$ tends to 1/2 as $n \to \infty$; thus $L = 1/2$. Moreover, it is clear that the series (E.16) tends to $(1/L) - 1 = 1$.

PROBLEM 83. Show that

$$\frac{1}{10} + \frac{1}{10 \cdot 98} + \frac{1}{10 \cdot 98 \cdot 9602} + \cdots = 5 - \sqrt{24}$$

(each factor in the denominator is equal to the square of the preceding factor diminished by 2).

Solution. The series to be summed is of the form

$$\frac{1}{y_0} + \frac{1}{y_0 y_1} + \frac{1}{y_0 y_1 y_2} + \cdots,$$

where $y_0 = 10$, $y_1 = 98$, $y_3 = 9602$, ..., and, in general,

$$y_n = y_{n-1}^2 - 2.$$

But from the Solution of Problem 6 we may infer that

$$\frac{1}{y_0} + \frac{1}{y_0 y_1} + \frac{1}{y_0 y_1 y_2} + \cdots = \frac{y_0 - \sqrt{y_0^2 - 4}}{2}$$

when $y_0 = 10$ and $y_n = y_{n-1}^2 - 2$ for $n = 1,2,3,\ldots$

PROBLEM 84. For $n = 1,2,3,\ldots$ define

$$x_n = \left\{1 + \frac{1}{n^2}\right\}\left\{1 + \frac{2}{n^2}\right\}\left\{1 + \frac{3}{n^2}\right\} \cdots \left\{1 + \frac{n}{n^2}\right\}.$$

Find the limit $\lim_{n\to\infty} x_n$.

Solution. We put

$$y_n = \log x_n = \sum_{k=1}^{n} \log\left\{1 + \frac{k}{n^2}\right\}$$

and note that, for $x > 0$,

$$x - \frac{x^2}{2} < \log(1 + x) < x.$$

Thus

$$\frac{k}{n^2} - \frac{k^2}{n^4} < \log\left(1 + \frac{k}{n^2}\right) < \frac{k}{n^2}$$

or

$$\sum_{k=1}^{n} \frac{k}{n^2} - \sum_{k=1}^{n} \frac{k^2}{n^4} < \sum_{k=1}^{n} \log\left(1 + \frac{k}{n^2}\right) < \sum_{k=1}^{n} \frac{k}{n^2}.$$

But

$$\sum_{k=1}^{n} k = \frac{n(n + 1)}{2} \quad \text{and} \quad \sum_{k=1}^{n} k^2 = \frac{n(n + 1)(2n + 1)}{6}$$

(see Problem 15 of Chapter 1). Thus

$$\sum_{k=1}^{n} \frac{k}{n^2} = \frac{n + 1}{2n} \to \frac{1}{2} \quad \text{and} \quad \sum_{k=1}^{n} \frac{k^2}{n^4} \le \frac{n^3}{n^4} = \frac{1}{n} \to 0 \quad \text{as } n \to \infty.$$

This shows that $\lim_{n\to\infty} y_n = 1/2$; therefore $\lim_{n\to\infty} x_n = \sqrt{e}$.

PROBLEM 85. Let x be positive. The sequence

$$x_n = x^{1/2^n}$$

is monotonical (increasing for $x < 1$ and decreasing for $x > 1$) and bounded. Moreover,

$$x_{n+1}^2 = x_n$$

and it is easy to see that $\lim_{n\to\infty} x_n = 1$.

Setting $x_n = x^{1/2^n}$, show that the two sequences

$$a_n = 2^n(x_n - 1) \quad \text{and} \quad b_n = 2^n\left(1 - \frac{1}{x_n}\right)$$

converge to the same limit, the sequence (a_n) decreasing monotonically and the sequence (b_n) increasing monotonically.

Solution. If x and with it x_n is larger than 1, then

$$(x_n + 1)(x_n - 1) = x_n^2 - 1 = x_{n-1} - 1,$$

and thus

$$2(x_n - 1) < x_{n-1} - 1,$$

that is,

$$2^n(x_n - 1) < 2^{n-1}(x_{n-1} - 1).$$

Hence

$$a_n \le a_{n-1}.$$

This shows that the sequence (a_n) decreases monotonically. The same result is found for $0 < x \le 1$. Similar arguments show that (b_n) increases monotonically. Moreover, we have (whether the x_n are all greater than 1 or less than or equal to 1)

$$x_n - 1 \le x_n(x_n - 1),$$

that is

$$1 - \frac{1}{x_n} \le x_n - 1$$

and

$$2^n\left(1 - \frac{1}{x_n}\right) \le 2^n(x_n - 1).$$

Thus

$$b_n \le a_n.$$

Since

$$x_n b_n = a_n$$

and $\lim_{n\to\infty} x_n = 1$, we have

$$\lim_{n\to\infty} a_n = \lim_{n\to\infty} b_n.$$

Remarks. We may define, for positive x,

$$\log x = \lim_{n\to\infty} 2^n(x^{1/2^n} - 1) = \lim_{n\to\infty} 2^n\left(1 - \frac{1}{x^{1/2^n}}\right).$$

From this definition of log x it is easy to see that log 1 = 0; for $x \neq 1$ we have

$$1 - \frac{1}{x} < \log x < x - 1, \qquad 2\left(1 - \sqrt{\frac{1}{x}}\right) < \log x < 2(\sqrt{x} - 1).$$

Moreover, for positive x and y,

$$\log (xy) = \log x + \log y.$$

Indeed, let z = xy and

$$x_n = x^{1/2^n}, \qquad y_n = y^{1/2^n}, \qquad z_n = z^{1/2^n};$$

then

$$z_n = x_n y_n.$$

Hence

$$\log z = \lim_{n\to\infty} 2^n(z_n - 1) = \lim_{n\to\infty} 2^n(x_n y_n - 1)$$

$$= \lim_{n\to\infty} \{2^n x_n(y_n - 1) + 2^n(x_n - 1)\}$$

$$= \lim_{n\to\infty} x_n \cdot \lim_{n\to\infty} 2^n(y_n - 1) + \lim_{n\to\infty} 2^n(x_n - 1)$$

$$= \log y + \log x.$$

In a similar way we can verify that, for positive x and y,

$$\log \frac{x}{y} = \log x - \log y \quad \text{and} \quad \log \frac{1}{x} = - \log x.$$

Noting that log x > 0 for x > 1, this in turn can be used to show that the function f(x) = log x increases monotonically with x.

PROBLEM 86. Starting with any real number x, consider the sequence

$$x_0 = x, \qquad x_{n+1} = \frac{x_n}{1 + \sqrt{1 + x_n^2}} \tag{E.17}$$

For $x < 0$ this sequence increases monotonically, for $x \geq 0$ it decreases monotonically; in either case it converges to 0.

We use this sequence $(x_n)_{n=0}^{\infty}$ to define the two sequences

$$a_n = 2^n x_n \quad \text{and} \quad b_n = \frac{2^n x_n}{\sqrt{1 + x_n^2}}.$$

Verify that $a_n \leq a_{n+1}$, $b_n \geq b_{n+1}$, and $\lim_{n \to \infty} a_n = \lim_{n \to \infty} b_n$.

Solution. We limit ourselves to the case $x > 0$; the considerations for $x \leq 0$ are similar. We have

$$a_{n+1} = 2^{n+1} x_{n+1} = \frac{2^{n+1} x_n}{1 + \sqrt{1 + x_n^2}} < 2^n x_n = a_n.$$

To see that (b_n) increases monotonically, we note that for all $n = 1,2,\ldots,$

$$\sqrt{1 + x_n^2} < 1 + x_n^2.$$

Hence,

$$\tfrac{1}{4}\{(1 + \sqrt{1 + x_n^2})^2 + x_n^2\} < 1 + x_n^2$$

or

$$\frac{2x_n}{(1 + \sqrt{1 + x_n^2})\sqrt{1 + x_n^2}} > \frac{x_n}{\sqrt{1 + x_n^2}}$$

or

$$\frac{2^{n+1} x_{n+1}}{\sqrt{1 + x_{n+1}^2}} > \frac{2^n x_n}{\sqrt{1 + x_n^2}},$$

that is, $b_{n+1} > b_n$.

Clearly

$$\sqrt{1 + z^2} = \frac{\sqrt{1 + x^2} \cdot \sqrt{1 + y^2}}{1 - xy}$$

and therefore

$$z_1 = \frac{x + y}{1 - xy + \sqrt{1 + x^2}\cdot\sqrt{1 + y^2}}.$$

From (E.15) we get

$$x = \frac{2x_1}{1 - x_1^2} \quad \text{and} \quad y = \frac{2y_1}{1 - y_1^2}.$$

Substitution yields

$$z_1 = \frac{x_1 + y_1}{1 - x_1 y_1}.$$

Using induction, one sees that

$$z_n = \frac{x_n + y_n}{1 - x_n y_n}, \quad x_n y_n < 1$$

or

$$2^n z_n = \frac{2^n x_n}{1 - x_n y_n} + \frac{2^n y_n}{1 - x_n y_n}.$$

Passage to the limit as $n \to \infty$ gives

$$\text{arc tan } z = \text{arc tan } x + \text{arc tan } y$$

and so, for any real numbers x and y satisfying $xy < 1$,

$$\text{arc tan } x + \text{arc tan } y = \text{arc tan } \frac{x + y}{1 - xy}.$$

It is also easy to see from the way we defined arc tan x that

$$|\text{arc tan } x| < \frac{2|x|}{1 + \sqrt{1 + x^2}} < 2,$$

$$b_n < a_n.$$

Thus both sequences converge. Since

$$\sqrt{1 + x_n^2} \to 1 \quad \text{as } n \to \infty,$$

the two sequences converge to the same limit.

Remarks. We may define, for all real x,

$$\text{arc tan } x = \lim_{n\to\infty} 2^n x_n,$$

where

$$x_0 = x, \qquad x_{n+1} = \frac{x_n}{1 + \sqrt{1 + x_n^2}}.$$

We note from this definition: If $x > 0$, then

$$\frac{2x}{\sqrt{1 + x^2}\left(1 + \sqrt{1 + x^2}\right)} < \text{arc tan } x < \frac{2x}{1 + \sqrt{1 + x^2}}.$$

For $x < 0$,

$$\frac{2x}{1 + \sqrt{1 + x^2}} < \text{arc tan } x < \frac{2x}{\sqrt{1 + x^2}\left(1 + \sqrt{1 + x^2}\right)}.$$

Obviously,

$$\text{arc tan } 0 = 0.$$

We next check the functional equation of the arc tangent. Let x and y be any numbers satisfying $xy < 1$. We then set

$$z = \frac{x + y}{1 - xy}$$

and write out the numerical sequences (x_n), (y_n), and (z_n) according to (E.17), starting with the numbers x, y, and z.

PROBLEM 87. Let n range over all positive integers ($n = 1,2,3,\ldots$) and p over all prime numbers ($p = 2,3,5,7,11,\ldots$). Show that, for $x > 1$,

$$\sum_n \frac{1}{n^x} = \prod_p \frac{1}{1 - \frac{1}{p^x}}.$$

This is the *Euler Product Formula.*

Solution. Let p_k denote the k-th prime number; then

$$\frac{1}{1 - \frac{1}{p_k^x}} = 1 + \frac{1}{p_k^x} + \frac{1}{(p_k^2)^x} + \frac{1}{(p_k^3)^x} + \cdots$$

Multiplying the finite number of series that correspond to primes smaller than a certain positive integer N, the partial product

$$P_x^{(N)} = \prod_{p_k \le N} \frac{1}{1 - \frac{1}{p_k^x}} = \sum_{n=1}^{\infty}{}' \frac{1}{n^x} = \sum_{n=1}^{N} \frac{1}{n^x} + \sum_{n=N+1}^{\infty}{}' \frac{1}{n^x}, \qquad \text{(E.18)}$$

where the symbol Σ' is to indicate that the summation does not extend over all positive integers, but only those (1 is excluded), which in their prime factorization contain solely prime numbers $\le N$ (the first N prime numbers do, of course, have this property). From (E.18) we get

$$0 < P_x^{(N)} - \sum_{n=1}^{N} \frac{1}{n^x} < \sum_{n=N+1}^{\infty} \frac{1}{n^x};$$

but $\Sigma_{n=1}^{\infty} (1/n^x)$ is convergent and so

$$\lim_{N \to \infty} \left\{ \sum_{n=N+1}^{\infty} \frac{1}{n^x} \right\} = 0,$$

and we obtain the desired result.

PROBLEM 88. Let p range over all prime numbers ($p = 2,3,5,7,11,\ldots$) and let p_k denote the k-th prime number. Using the Euler Product Formula (see Problem 87), show that the set of all prime numbers is infinite and that the series

$$\sum_{k=1}^{\infty} \frac{1}{p_k}$$

diverges.

Solution. The relation (E.16) also holds for $x = 1$ and hence

$$P_1^{(N)} = \prod_{p_k \le N} \frac{1}{1 - \frac{1}{p_k}} > \sum_{n=1}^{N} \frac{1}{n},$$

implying that $P_1^{(N)} \to +\infty$ for $N \to \infty$. Thus

$$\prod_{k=1}^{\infty} \frac{1}{1 - \frac{1}{p_k}}$$

diverges to $+\infty$. If, however, the set of all primes was finite, this product would have to have a finite value. We can also see that

$$\left(1 - \frac{1}{2}\right)\left(1 - \frac{1}{3}\right)\left(1 - \frac{1}{5}\right) \cdots \left(1 - \frac{1}{p_k}\right) \cdots = \prod_{k=1}^{\infty} \left(1 - \frac{1}{p_k}\right) = 0.$$

But

$$\lim_{n \to \infty} \frac{\log\left(1 - \frac{1}{p_k}\right)}{-\frac{1}{p_k}} = 1$$

and so

$$\frac{1}{2} + \frac{1}{3} + \frac{1}{5} + \cdots + \frac{1}{p_k} + \cdots$$

is divergent.

Remarks. The following is another proof of the fact that the infinite series

$$\sum_{k=1}^{\infty} \frac{1}{p_k}$$

diverges.

We assume that the series converges and obtain a contradiction. If the series converges, there is an integer n such that

$$\sum_{m=n+1}^{\infty} \frac{1}{p_m} < \frac{1}{2}.$$

Let $Q = p_1 p_2 \cdots p_n$, and consider the numbers $1 + kQ$ for $k = 1,2,\ldots$ None of these is divisible by any of the primes $p_1, p_2, \ldots, p_n$. Therefore, all the prime factors of $1 + kQ$ occur among the primes $p_{n+1}, p_{n+2}, \ldots$ Thus for each $r \geq 1$ we have

$$\sum_{k=1}^{r} \frac{1}{1 + kQ} \leq \sum_{t=1}^{\infty} \left(\sum_{m=n+1}^{\infty} \frac{1}{p_k}\right)^t,$$

since the sum on the right includes among its terms all the terms on the left. But the right-hand side of this inequality is dominated by the convergent geometric series $\Sigma_{t=1}^{\infty} 2^{-t}$. Therefore the series $\Sigma_{k=1}^{\infty} 1/(1 + kQ)$ has bounded partial sums and hence converges. But this is a contradiction because this series diverges.

PROBLEM 89. Let $a > 0$ and $b > 0$. Putting

$$A(a,b) = \frac{2}{\pi}\int_0^{\pi/2} \log(a \sin^2 x + b \cos^2 x)dx, \tag{E.19}$$

show that

$$A(a,b) = \frac{1}{2} A\{(\frac{a+b}{2})^2, ab\}. \tag{E.20}$$

Moreover, verify that

$$A(a,b) = 2 \log \frac{\sqrt{a} + \sqrt{b}}{2}. \tag{E.21}$$

Solution. Put $x = t/2$ in (E.19) and breaking the interval of integration into two halves, we get

$$A(a,b) = \frac{1}{\pi}\int_0^{\pi/2} \log\left(\frac{a+b}{2} - \frac{a-b}{2}\cos t\right)dt$$

$$+ \frac{1}{\pi}\int_{\pi/2}^{\pi} \log\left(\frac{a+b}{2} - \frac{a-b}{2}\cos t\right)dt.$$

Replacing t by $\pi - t$ in the second integral and recombining, we find

$$A(a,b) = \frac{1}{\pi}\int_0^{\pi/2} \log\left\{\left(\frac{a+b}{2}\right)^2 - \left(\frac{a-b}{2}\right)^2 \cos^2 t\right\}dt$$

$$= \frac{1}{2}\int_0^{\pi/2} \log\left\{\left(\frac{a+b}{2}\right)^2 \sin^2 t + ab \cos^2 t\right\}dt.$$

This proves the invariance property (E.20). The arguments on the right-hand side of (E.20) are the squares of the arithmetic and geometric means of a and b.

To evaluate the integral, that is, to establish (E.21), let

$$a_0 = a, \quad b_0 = b,$$
$$a_{n+1} = \left\{\frac{a_n + b_n}{2}\right\}^2, \quad b_{n+1} = a_n b_n, \tag{E.22}$$

where $n = 0,1,2,3,\ldots$ Define also

$$\alpha_n = \frac{1}{2}(\sqrt{a_n} + \sqrt{b_n}), \quad \beta_n = \frac{1}{2}(\sqrt{a_n} - \sqrt{b_n}), \quad \delta_n = \frac{\beta_n}{\alpha_n}, \tag{E.23}$$

for $n = 0,1,2,3,\ldots$ The recurrence relations (E.22) imply

$$\alpha_{n+1} = \alpha_n^2, \quad \beta_{n+1} = \beta_n^2, \quad \delta_{n+1} = \delta_n^2. \tag{E.24}$$

It follows from $\delta_1 = \beta_0^2/\alpha_0^2$ that $0 \le \delta_1 < 1$ and hence, as $n \to \infty$,

$$\delta_n \to 0, \quad \frac{b_n}{a_n} = \left\{\frac{1 - \delta_n}{1 + \delta_n}\right\}^2 \to 1, \quad \frac{\alpha_n^2}{a_n} \to 1. \tag{E.25}$$

Successive applications of (E.20) show that

$$A(a,b) = 2^{-n} A(a_n,b_n) = 2^{-n} \log a_n + 2^{-n} A(1,b_n/a_n), \tag{E.26}$$

For strictly positive a and b, the integrand of (E.19) is jointly continuous in a, b, x and $A(a,b)$ is continuous in a and b. Since $A(1,1) = 0$ it follows from (E.26) and (E.25) that

$$A(a,b) = \lim_{n\to\infty} 2^{-n} \log a_n = 2 \lim_{n\to\infty} 2^{-n} \log \alpha_n.$$

Finally, because $2^{-n} \log \alpha_n$ is independent of n by (E.24).

$$A(a,b) = 2 \log \alpha_0 = 2 \log \frac{\sqrt{a} + \sqrt{b}}{2},$$

establishing (E.21).

PROBLEM 90. Let

$$S(k) = 1 + \frac{1}{2} + \frac{1}{3} + \cdots + \frac{1}{k}$$

and define k_n to be the least integer k such that $S(k) \ge n$. For example, $k_1 = 1$, $k_2 = 4$, $k_3 = 11$, $k_4 = 31$, $k_5 = 83$, $k_6 = 227$, $k_7 = 616$, ... Find

$$\lim_{n\to\infty} \frac{k_{n+1}}{k_n}.$$

Solution. It is clear that, as $n \to \infty$,

$$0 \le S(k_n) - n < \frac{1}{k_n} \to 0;$$

but $S(k_n) - \log k_n \to c$ (Euler's Constant, see Problem 14). Hence we conclude that $n - \log k_n \to c$. This implies that

$$n + 1 - \log k_{n+1} \to c \quad \text{as } n \to \infty,$$

and subtracting these two limits gives

$$1 - \log \frac{k_{n+1}}{k_n} \to 0 \quad \text{as } n \to \infty.$$

Hence

$$\lim_{n\to\infty} \frac{k_{n+1}}{k_n} = e.$$

PROBLEM 91. Let a_0 be arbitrary but fixed and define

$$a_{n+1} = \sin a_n \quad \text{for } n = 0,1,2,3,\ldots$$

Show that the sequence (na_n^2) is convergent and find

$$\lim_{n\to\infty} n\, a_n^2 = L(a_0).$$

Solution. Suppose first that $0 < a_0 < \pi$. Then clearly $a_n > 0$ for all n, and if we let $f(x) = x - \sin x$, then $f'(x) = 1 - \cos x > 0$ for $0 < x < \pi$; since $f(0) = 0$, it follows that

$$a_n - a_{n+1} = a_n - \sin a_n = f(a_n) > 0.$$

Hence the sequence (a_n) is strictly monotonically decreasing and bounded below by 0, so if $L = \lim_{n\to\infty} a_n$, then $L = \sin L$, implying that $L = 0$. The same can be said for any a_0 which is congruent (mod 2π) to some $a \in (0,\pi)$: $a_1 > a_2 > \cdots > 0$ and $a_n \to 0$. Similarly, if a_0 is congruent (mod 2π) to some $a \in (-\pi,0)$, then $a_1 < a_2 < \cdots < 0$ and $a_n \to 0$. If a_0 is an integral multiple of π, then $a_1 = a_2 = \cdots = 0$.

Consider then for any $a_0 \neq k\pi$ the limit

$$\lim_{n\to\infty} \left(\frac{1}{a_{n+1}^2} - \frac{1}{a_n^2}\right) = \lim_{n\to\infty} \left(\csc^2 a_n - \frac{1}{a_n^2}\right) = \lim_{x\to\infty} \frac{x^2 - \sin^2 x}{x^2 \sin^2 x}$$

which is equal to 1/3 by four applications of L'Hospital's Rule.

If $y_n = a_n^{-2} - a_{n-1}^{-2}$, then (y_n) has 1/3 as its limit, hence the sequence

$$\bar{y}_n = \frac{y_1 + y_2 + \cdots + y_n}{n} \to \frac{1}{3} \quad \text{as } n \to \infty,$$

by Problem 40. Thus

$$\frac{1}{3} = \lim_{n\to\infty} \bar{y}_n = \lim_{n\to\infty} \frac{1}{n} \sum_{k=1}^{n} (a_k^{-2} - a_{k-1}^{-2}) = \lim_{n\to\infty} \frac{1}{n}(a_k^{-2} - a_{k-1}^{-2}) = \lim_{n\to\infty} \frac{1}{na_n^2}.$$

It follows that

$$\lim_{n\to\infty} n\, a_n^2 = 3 \quad \text{for any } a_0 \neq k\pi,$$

$$= 0 \quad \text{for any } a_0 = k\pi.$$

PROBLEM 92. Let (a_n) and (b_n) be sequences of positive numbers such that $a_n^n \to a$ and $b_n^n \to b$ as $n \to \infty$ with $0 < a,b < \infty$. Let p and q be nonnegative numbers such that $p + q = 1$. Show that

$$\lim_{n\to\infty} (pa_n + qb_n)^n = a^p b^q.$$

Solution. Let $0 < x < \infty$; then $x_n^n \to x$ if and only if

$$n(x_n - 1) \to \log x.$$

Both conditions imply $x_n \to 1$, so we can assume that $x_n > 0$ and hence $x_n^n \to x$ if and only if $n \log x_n \to \log x$. If we define $y_n = 1$ if $x_n = 1$ and

$$y_n = \frac{\log x_n}{x_n - 1} = \frac{\log x_n - \log 1}{x_n - 1} \quad \text{if } x_n \neq 1,$$

then $n \log x_n = n(x_n - 1)y_n$ and since $y_n \to 1$, the equivalence is shown.

The limiting behavior of $(pa_n + qb_n)^n$ follows immediately; from $a_n^n \to a$ and $b_n^n \to b$, we conclude that

$$n(a_n - 1) \to \log a \quad \text{and} \quad n(b_n - 1) \to \log b.$$

Letting $x_n = pa_n + qb_n$, we have $n(x_n - 1) = n(pa_n + qb^n - 1) = pn(a_n - 1) + qn(b_n - 1) \to p \log a + q \log b = \log a^p b^q$ and we have the desired result.

PROBLEM 93. Show that, for any fixed $m \geq 2$, the series

$$1 + \frac{1}{2} + \cdots + \frac{1}{m-1} - \frac{x}{m} + \frac{1}{m+1} + \frac{1}{m+2} + \cdots + \frac{1}{2m-1} - \frac{x}{2m}$$

$$+ \frac{1}{2m+1} + \frac{1}{2m+2} + \cdots + \frac{1}{3m-1} - \frac{x}{3m} + \cdots$$

is convergent for exactly one value of x and find the sum of the series for this x.

Solution. Let

$$S_n(x) = \sum_{k=1}^{n} \left\{ \frac{1}{(k-1)m+1} + \cdots + \frac{1}{km-1} - \frac{x}{km} \right\}.$$

If the series converges for x and y, then the sequence

$$S_n(x) - S_n(y) = \frac{y-x}{m} \sum_{k=1}^{n} \frac{1}{k}$$

converges. But this is only possible if $x = y$, since $\Sigma\, 1/k$ diverges. Hence the series converges for at most one value of x.

The sequence

$$A_n = 1 + \frac{1}{2} + \cdots + \frac{1}{nm} - \log(nm)$$

is known to converge to Euler's Constant (see Problem 14). Now

$$S_n(m-1) = A_n + \log(nm) - \sum_{k=1}^{n} \frac{1}{km} - \sum_{k=1}^{n} \frac{m-1}{km}$$

$$= A_n + \log m + \left\{ \log n - \sum_{k=1}^{n} \frac{1}{k} \right\} \to c + \log m - c = \log m.$$

Therefore, when $x = m - 1$, the series converges to $\log m$.

PROBLEM 94. Find the maximum value of α and the minimum value of β for which

$$\left(1 + \frac{1}{n}\right)^{n+\alpha} \leq e \leq \left(1 + \frac{1}{n}\right)^{n+\beta}$$

for all positive integers n.

Solution. On taking logarithms we obtain

$$\alpha_{max} = \inf_n \left\{\frac{1}{\log(1 + 1/n)} - n\right\},$$

$$\beta_{min} = \sup_n \left\{\frac{1}{\log(1 + 1/n)} - n\right\}.$$

We now show that the function

$$F(x) = \frac{1}{\log(1 + 1/x)} - x$$

is monotonically increasing for $x > 0$ by showing its derivative is positive:

$$F'(x) = \frac{1}{x(x + 1)[\log(1 + 1/x)]^2} - 1 = \frac{\sinh^2 u}{u^2} - 1 > 0,$$

where $e^{2u} = 1 + 1/x$. Thus

$$\alpha_{max} = \frac{1}{\log 2} - 1 = 0.4426950\ldots \quad \text{and} \quad \beta_{min} = \lim_{n\to\infty} F(n).$$

By expanding $\log(1 + x)$ in a Maclaurin series,

$$\log(1 + x) = x - \frac{1}{2}x^2 + \frac{1}{3}x^3 - \frac{1}{4}x^4 + \cdots,$$

we have

$$F(n) = \left\{\frac{1}{n} - \frac{1}{2n^2} + O\left(\frac{1}{n^3}\right)\right\}^{-1} - n,$$

where $O(x_n)$ with $x_n > 0$ signifies a quantity that divided by x_n remains bounded; it follows that

$$\beta_{min} = \lim_{n\to\infty} F(n) = \frac{1}{2}.$$

PROBLEM 95. If $r > 1$ is an integer and x is real, define

$$f(x) = \sum_{k=0}^{\infty} \sum_{j=1}^{r-1} \left[\frac{x + jr^k}{r^{k+1}}\right].$$

where the brackets denote the greatest integer function. Show that

$$\begin{aligned} f(x) &= [x] && \text{if } x \ge 0 \\ &= [x + 1] && \text{if } x < 0. \end{aligned}$$

Solution. By Problem 27 of Chapter 1,

$$f(x) = \sum_{k=0}^{\infty} \sum_{j=1}^{r-1} \left[\frac{x}{r^{k+1}} + \frac{j}{r}\right] = \sum_{k=0}^{\infty} \left\{\left[\frac{x}{r^k}\right] - \left[\frac{x}{r^{k+1}}\right]\right\}.$$

Letting S_n represent the n-th partial sum, we have

$$S_n = [x] - \left[\frac{x}{r^{n+1}}\right].$$

Since $r > 1$, there is a positive integer n_0 such that for all $n > n_0$,

$$\left|\frac{x}{r^{n+1}}\right| < 1.$$

Therefore, for fixed x and $n > n_0$, the sequence of partial sums is constant. For $x \geq 0$ and $n > n_0$, the sequence is $[x]$ and for $x < 0$, the sequence is $[x] + 1 = [x + 1]$.

PROBLEM 96. Show that the integer nearest to $n!/e$ is a multiple of $n - 1$.

Solution. The error made in stopping the expansion of e^{-1} with the term $(-1)^n/n!$ is less than $1/(n + 1)!$. Hence the integer nearest to $n!/e$ is

$$P_n = n!\{1 - \frac{1}{1!} + \frac{1}{2!} - \cdots + (-1)^n \frac{1}{n!}\}.$$

The divisibility property in question can be verified as follows:

$$\begin{aligned} P_n &= nP_{n-1} + (-1)^n \\ &= (n - 1)P_{n-1} + P_{n-1} + (-1)^n \\ &= (n - 1)P_{n-1} + (n - 1)P_{n-2} + (-1)^{n-1} + (-1)^n \\ &= (n - 1)(P_{n-1} + P_{n-2}). \end{aligned}$$

PROBLEM 97. Prove that a necessary and sufficient condition for the rationality of

$$R = \sqrt[3]{a + \sqrt[3]{a + \cdots}},$$

where a is a positive integer, is that $a = N(N + 1)(N + 2)$, the product of three consecutive integers. In that case find R.

Solution. Define $R_1 = \sqrt[3]{a}$, $R_n = \sqrt[3]{a + R_{n-1}}$. Now $R_2 > R_1$, and $R_k^3 - R_{k-1}^3 = R_{k-1} - R_{k-2}$, so that by induction (R_n) is monotone increasing, Moreover, $R_1 < 1 + \sqrt[3]{a}$, and $R_{k-1} < 1 + \sqrt[3]{a}$ implies that $R_k^3 < a + 1 + \sqrt[3]{a} < (1 + \sqrt[3]{a})^3$, so that by induction (R_n) is bounded. It follows that (R_n) converges to a limit R. But then $R^3 - R - a = 0$. If R is rational and a integral, then R is integral, and $a = (R - 1)R(R + 1)$, the product of three consecutive integers. Hence the condition is necessary. It is also sufficient, since $R = N + 1$ satisfies the equation $R^3 - R - N(N + 1)(N + 2) = 0$, and, as it is the only real root, it is the value of the radical.

Cardan's Formula yields the explicit expression

$$R = \left\{a/2 + \sqrt{a^2/4 - 1/27}\right\}^{1/3} + \left\{a/2 - \sqrt{a^2/4 - 1/27}\right\}^{1/3}.$$

Remark. The result in Problem 97 can be generalized as follows: a necessary and sufficient condition for the rationality of

$$R = \sqrt[n]{a + \sqrt[n]{a + \cdots}},$$

where a is a positive integer, is that $a = N(N^{n-1} - 1)$.

PROBLEM 98. Show that the total number of permutations of n things is $[n!e]$, where $[x]$ denotes the greatest integer in x.

Solution. The total number P of permutations of n things is the sum of the number of permutations taken n, n - 1, n - 2, ..., 2, 1, 0 at a time. Hence

$$P = \sum_{r=0}^{n} \frac{n!}{r!} = n!e - \sum_{r=n+1}^{\infty} \frac{n!}{r!}.$$

But

$$0 < \sum_{r=n+1}^{\infty} \frac{n!}{r!} < \sum_{r=1}^{\infty} \frac{1}{(n + 1)^r} = \frac{1}{n} \le 1,$$

whence the result.

PROBLEM 99. Let

$$P_n = \sum_{k=0}^{n} \log\binom{pn}{pk}, \qquad S_n = \frac{1}{n^2} P_n,$$

and

$$Q_n = \sum_{k=0}^{n} (-1)^k \log\binom{pn}{pk}, \qquad T_n = \frac{1}{n} Q_n,$$

where p is a positive integer. Show that

$$\lim_{n\to\infty} S_n = \frac{p}{2}, \tag{E.27}$$

and

$$\lim_{n\to\infty} T_{2n} = 0. \tag{E.28}$$

It is clear that $T_{2n-1} = 0$ for all positive integers n.

Solution. We shall make use of the binomial coefficient identity

$$\binom{pn}{pk} = \binom{pn - p}{pk - p}\binom{pn}{p}\binom{pk}{p}^{-1}. \tag{E.29}$$

First we establish (E.27). By means of (E.29) we have, when $n \;\; 1$,

$$P_n = \sum_{k=1}^{n} \log\binom{pn}{pk} = \sum_{k=1}^{n} \log\binom{pn - p}{pk - p} + \sum_{k=1}^{n} \log\binom{pn}{p} - \sum_{k=1}^{n} \log\binom{pk}{p},$$

or

$$P_n - P_{n-1} = \log\left\{\binom{pn}{p}^n \prod_{k=1}^{n} \binom{pk}{p}^{-1}\right\} = \log\frac{(pn)!^n}{(pn - p)!^n (pn)!}.$$

By Problem 39, if b_n increases steadily to ∞ then

$$\lim_{n\to\infty} \frac{P_n}{b_n} = \lim_{n\to\infty} \frac{P_n - P_{n-1}}{b_n - b_{n-1}},$$

provided that the second limit exists. Here we choose $b_n = n^2$, so that $b_n - b_{n-1} = 2n - 1 = n(2 - 1/n)$. Then we have

$$\frac{P_n - P_{n-1}}{b_n - b_{n-1}} = \frac{1}{2 - 1/n} \log\frac{(pn)!}{(pn - p)!(pn)!^{1/n}}.$$

Since the outside factor tends to 1/2 we shall have finished the proof of (E.27) if we can show that the ratio of factorials tends to e^p. To show this we need the fact that

$$\lim_{r\to\infty} \frac{r}{r!^{1/r}} = e$$

(see either Remarks to Problem 16 or Remarks to Problem 41, Part (iv)). Indeed we have

$$\lim_{n\to\infty} \frac{(pn)!}{(pn - p)!(pn)!^{1/n}} = \lim_{n\to\infty} \frac{pn(pn - 1) \cdots (pn - p + 1)}{[(pn)!^{1/pn}]^p}$$

$$= \lim_{r\to\infty} \frac{r(r - 1) \cdots (r - p + 1)}{(r!^{1/r})^p}$$

$$= \lim_{r\to\infty} \left\{\frac{r}{r!^{1/r}}\right\}^p \cdot \left(1 - \frac{1}{r}\right)\left(1 - \frac{2}{r}\right) \cdots \left(1 - \frac{p - 1}{r}\right) = e^p,$$

for $p \geq 2$, and it is clearly also correct when $p = 1$.

In the case of Q_n we find by means of (E.29) that

$$Q_n = \sum_{k=1}^{n} (-1)^k \log\binom{pn - p}{pk - p} + \sum_{k=1}^{n} (-1)^k \log\binom{pn}{p} - \sum_{k=1}^{n} (-1)^k \log\binom{pk}{p},$$

or

$$Q_n + Q_{n-1} = -\frac{1 - (-1)^n}{2} \log\binom{pn}{p} - \sum_{k=1}^{n} (-1)^k \log\binom{pk}{p}, \qquad n \geq 1.$$

Since $Q_n = 0$ when n is odd we have then

$$Q_{2n} = -\sum_{k=1}^{2n} (-1)^k \log\binom{pk}{p} = \sum_{k=1}^{n} \log\binom{(2k - 1)p}{p} - \sum_{k=1}^{n} \log\binom{2kp}{p}$$

$$= \log \prod_{k=1}^{n} \binom{2kp - p}{p}\binom{2kp}{p}^{-1} = \log \prod_{k=1}^{n} \prod_{j=1}^{p} \frac{2pk - p - j + 1}{2pk - j + 1}.$$

Now to show that T_{2n} tends to 0 as $n \to \infty$ we should have to show that the n-th root of the product tends to 1. For example, when $p = 3$ we should have to show that

$$\lim_{n\to\infty} \left\{\prod_{k=1}^{n} \frac{(6k - 3)(6k - 4)(ak - 5)}{6k(6k - 1)(6k - 2)}\right\}^{1/n} = 1.$$

But in general, if $a > b > c \geq 0$, then

$$\lim_{n\to\infty}\left\{\prod_{k=1}^{n}\frac{ak - b}{ak - c}\right\}^{1/n} = 1. \tag{E.30}$$

This is immediate from Problem 42: $\lim_{n\to\infty} a_n^{1/n}$ exists and has the same value as $\lim_{n\to\infty} a_{n+1}/a_n$ provided the latter limit exists. Thus by (E.30) any product of a finite number of such factors has the same property, so we see that T_{2n} tends to zero as n increases indefinitely.

PROBLEM 100. Evaluate

$$\frac{1 - 2^{-2} + 4^{-2} - 5^{-2} + 7^{-2} - 8^{-2} + \cdots}{1 + 2^{-2} - 4^{-2} - 5^{-2} + 7^{-2} + 8^{-2} - \cdots}.$$

Solution. Subtracting term by term the denominator series D from the numerator series N (both unconditionally convergent) we find $N - D = -(1/2)N$. Hence $N/D = 2/3$.

PROBLEM 101. Let $a_1 = 1$, $a_n = n(a_{n-1} + 1)$, and define

$$P_n = \left(1 + \frac{1}{a_1}\right)\left(1 + \frac{1}{a_2}\right)\cdots\left(1 + \frac{1}{a_n}\right)$$

Find $\lim_{n\to\infty} P_n$.

Solution. We have

$$\begin{aligned} P_n &= \{(a_1 + 1)/a_1\}\{(a_2 + 1)/a_2\} \cdots \{(a_n + 1)/a_n\} \\ &= \{(a_1 + 1)/a_2\}\{(a_2 + 1)/a_3\} \cdots \{(a_{n-1} + 1)/a_n\}(a_n + 1) \\ &= (a_n + 1)/n!. \end{aligned}$$

Now

$$\begin{aligned} P_n - P_{n-1} &= \{(a_n + 1)/n!\} - \{(a_{n-1} + 1)/(n - 1)!\} \\ &= \{(a_n + 1)/n!\} - \{a_n/n!\} = 1/n!. \end{aligned}$$

Hence

$$P_n = P_1 + 1/2! + 1/3! + \cdots + 1/n!$$

$$= 1 + 1/1! + 1/2! + 1/3! + \cdots + 1/n!$$

and $\lim_{n\to\infty} P_n = e$.

PROBLEM 102. There are given $p_n = [en!] + 1$ points in space. Each pair of these points is connected by a line, and each line is colored with one of n different colors. Show that there is at least one triangle all of whose sides are of the same color.

Solution. Define a sequence (b_n) inductively by $b_1 = 2$ and also by $b_{n+1} = (n + 1)b_n + 1$. We will prove:

(i) When the segments connecting a set of $b_n + 1$ points are colored with n colors, at least one single-color triangle results;

(ii) $p_n \geq b_n + 1$ for all n.

Statement (i) is clear for n = 1. Suppose it is true for n = k, and let a set of $b_{k+1} + 1$ points be given. Starting at any point A in this set, there are b_{k+1} segments joining A to the remaining points; since $b_{k+1} > (k + 1)b_k$, one of the k + 1 colors (call it "blue") must be used at least $b_k + 1$ times in coloring those segments. Thus we have a subset B consisting of $b_k + 1$ points, each joined to A by a blue segment. If any segment joining two points of B is blue, they will form with A an all-blue triangle; otherwise the segments of B are all colored with the k remaining colors, and the induction hypothesis assures us that a monochromatic triangle exists in this case also.

Now if $a_n = b_n/n!$ for each n, we see that $a_1 = 2$ and $a_{n+1} = a_n + 1/(n + 1)!$. Thus (a_n) is the (increasing) sequence of partial sums of the usual series for e. Therefore, for all n, $a_n < e$ and consequently $b_n < en!$, $b_n \leq [en!]$, and finally $b_n + 1 \leq p_n$.

PROBLEM 103. Let

$$\zeta(s) = \sum_{n=1}^{\infty} \frac{1}{n^s} \quad (s > 1).$$

Show that, for n = 2,3,4,...,

$$\zeta(2)\zeta(2n - 2) + \zeta(4)\zeta(2n - 4) + \cdots + \zeta(2n - 2)\zeta(2) = (n + \tfrac{1}{2})\zeta(2n).$$

Solution. The left-hand side, written out at length, is the limit, as $N \to \infty$, of

$$\sum_{j=1}^{N} \sum_{k=1}^{N} \left\{ \frac{1}{k^2} \cdot \frac{1}{j^{2n-2}} + \frac{1}{k^4} \cdot \frac{1}{j^{2n-4}} + \cdots + \frac{1}{k^{2n-2}} \cdot \frac{1}{j^2} \right\} \tag{E.31}$$

(N a positive integer). Summing the expression within braces, and taking note of the exceptional case $k = j$, this becomes

$$\sum_j \left\{ \sum_k{}' \frac{j^{2-2n} - k^{2-2n}}{k^2 - j^2} + (n - 1) \frac{1}{j^{2n}} \right\}. \tag{E.32}$$

Throughout the present discussion, all sums run from 1 to N, unless otherwise indicated, and an accent on an inner Σ indicates that the index (in this case, k) does not take on the value of the index of the outer sum (j, here). Ignoring the term on the far right, (E.32) is equal to

$$\sum_j \sum_k{}' \frac{j^{2-2n}}{k^2 - j^2} + \sum_j \sum_k{}' \frac{k^{2-2n}}{j^2 - k^2}. \tag{E.33}$$

Inverting the order of summation in the second double sum, and noting that the condition $k \neq j$ is the same as the condition $j \neq k$, (E.33) may be written as

$$\begin{aligned} &\sum_j \frac{1}{j^{2n-2}} \sum_k{}' \frac{1}{k^2 - j^2} + \sum_k \frac{1}{k^{2n-2}} \sum_j{}' \frac{1}{j^2 - k^2} \\ &= 2 \sum_j \frac{1}{j^{2n-2}} \sum_k{}' \frac{1}{k^2 - j^2}, \end{aligned} \tag{E.34}$$

the latter form arising out of an interchange of the dummy indices in the second sum. Combining (E.31) - (E.34), we find that

$$\begin{aligned} &\sum_j \sum_k \left\{ \frac{1}{k^2} \cdot \frac{1}{j^{2n-2}} + \cdots + \frac{1}{k^{2n-2}} \cdot \frac{1}{j^2} \right\} \\ &= (n - 1) \sum_j \frac{1}{j^{2n}} + 2 \sum_j \frac{1}{j^{2n-2}} \sum_k{}' \frac{1}{k^2 - j^2}. \end{aligned} \tag{E.35}$$

Now,

$$\begin{aligned} 2j \sum_k{}' \frac{1}{k^2 - j^2} &= \sum_k{}' \frac{1}{k - j} - \sum_k{}' \frac{1}{k + j} \\ &= \sum_{k=1}^{j-1} \frac{1}{k - j} + \sum_{k=j+1}^{N} \frac{1}{k - j} - \sum_{k=1}^{N} \frac{1}{k + j} + \frac{1}{2j} \end{aligned}$$

$$= -\sum_{k=1}^{j-1}\frac{1}{k} + \sum_{k=1}^{N-j}\frac{1}{k} - \sum_{k=j+1}^{N+j}\frac{1}{k} + \frac{1}{2j}$$

$$= -\sum_{k=1}^{N+j}\frac{1}{k} + \frac{1}{j} + \sum_{k=1}^{N-j}\frac{1}{k} + \frac{1}{2j}$$

$$= \frac{3}{2j} - \left\{\frac{1}{N - j + 1} + \frac{1}{N - j + 2} + \cdots + \frac{1}{N + j}\right\}.$$

When we substitute this into (E.35), we get

$$\sum_k \frac{1}{k^2}\sum_j \frac{1}{j^{2n-2}} + \cdots + \sum_k \frac{1}{k^{2n-2}}\sum_j \frac{1}{j^2}$$
$$= (n + \tfrac{1}{2})\sum_j \frac{1}{j^{2n}} - \sum_j \frac{1}{j^{2n-1}}\left\{\frac{1}{N - j + 1} + \frac{1}{N - j + 2} + \cdots + \frac{1}{N + j}\right\}. \tag{E.36}$$

Finally,

$$0 < \frac{1}{N - j + 1} + \frac{1}{N - j + 2} + \cdots + \frac{1}{N + j} < \frac{2j}{N - j + 1}$$

and so

$$0 < \sum_j \frac{1}{j^{2n-1}}\left\{\frac{1}{N - j + 1} + \cdots + \frac{1}{N + j}\right\}$$

$$< 2\sum_j \frac{1}{j^{2n-2}}\,\frac{1}{N - j + 1} \le 2\sum_j \frac{1}{j(N - j + 1)} \qquad \left(n \ge \frac{3}{2}\right)$$

$$= \frac{2}{N + 1}\sum_j \left\{\frac{1}{j} + \frac{1}{N - j + 1}\right\} = \frac{4}{N + 1}\sum_j \frac{1}{j}$$
$$< \frac{4}{N + 1}(1 + \log N) \to 0 \quad \text{as } N \to \infty. \tag{E.37}$$

Statements (E.36) and (E.37), taken together, complete the proof.

Remarks. By successive applications of the result in Problem 103, one can express $\zeta(2n)$ as a rational multiple of $\{\zeta(2)\}^n$. From Problem 7 we know that $\zeta(2) = \pi^2/6$; hence $\zeta(4) = \pi^4/90$ and so forth. Problem 103 tells us nothing, however, about $\zeta(2)$, itself. In Problem 104 we will circumvent this difficulty.

PROBLEM 104. For $s > 0$, let $\xi(s) = \Sigma_{j=0}^{\infty} (-1)^j (2j + 1)^{-s}$. Show that

$$\xi(1)\xi(2n-1) + \xi(3)\xi(2n-3) + \cdots + \xi(2n-1)\xi(1)$$

$$= (n - \tfrac{1}{2}) \sum_{j=0}^{\infty} \frac{1}{(2j+1)^{2n}} \quad (n = 1,2,3,\ldots).$$

Solution. The solution proceeds along the same lines as the solution of Problem 103. In the first place, we have, as in (E.31) - (E.35),

$$\sum_j \sum_k (-1)^{j+k} \left\{ \frac{1}{2k+1} \cdot \frac{1}{(2j+1)^{2n-1}} + \cdots + \frac{1}{(2k+1)^{2n-1}} \cdot \frac{1}{2j+1} \right\}$$

$$= n \sum_j \frac{1}{(2j+1)^{2n}} \tag{E.38}$$

$$+ 2 \sum_j \frac{1}{(2j+1)^{2n-1}} \sum_k{}' (-1)^{j+k} \frac{2k+1}{(2k+1)^2 - (2j+1)^2}.$$

Here, all sums run from 0 to N, and the accent has the same significance as in the Solution of Problem 103. Again,

$$4 \sum_k{}' (-1)^{j+k} \frac{2k+1}{(2k+1)^2 - (2j+1)^2} = \sum_k{}' (-1)^{j+k} \frac{2k+1}{(k+j+1)(k-j)}$$

$$= \sum_k{}' (-1)^{j+k} \frac{1}{k-j} + \sum_k{}' (-1)^{j+k} \frac{1}{k+j+1} \tag{E.39}$$

$$= -\frac{1}{2j+1} + (-1)^{N-j} \left\{ \frac{1}{N-j+1} - \frac{1}{N-j+2} + - \cdots + \frac{1}{N+j+1} \right\},$$

the last step involving reasoning analogous to the corresponding calculations in the Solution of Problem 103. Equations (E.38) and (E.39) now yield

$$\sum_k (-1)^k \frac{1}{2k+1} \sum_j (-1)^j \frac{1}{(2j+1)^{2n-1}}$$

$$+ \cdots + \sum_k (-1)^k \frac{1}{(2k+1)^{2n-1}} \sum_j (-1)^j \frac{1}{2j+1}$$

$$= (n - \tfrac{1}{2}) \sum_j \frac{1}{(2j+1)^{2n}} \tag{E.40}$$

$$+ \frac{1}{2} \sum_j (-1)^{N-j} \frac{1}{(2j+1)^{2n-1}} \left\{ \frac{1}{N-j+1} - \frac{1}{N-j+2} + - \cdots + \frac{1}{N+j+1} \right\}.$$

But

$$0 < \frac{1}{N-j+1} - \frac{1}{N-j+2} + - \cdots + \frac{1}{N+j+1} \le \frac{1}{N-j+1},$$

and so the absolute value of the last sum in (E.40) is

$$\le \sum_j \frac{1}{(2j+1)^{2n-1}} \cdot \frac{1}{N-j+1} \le \sum_j \frac{1}{(2j+1)(N-j+1)} \qquad (n \ge 1)$$

$$= \frac{2}{2N+3} \sum_j \left\{\frac{1}{2j+1} + \frac{1}{2(N-j+1)}\right\}$$

$$= \frac{2}{2N+3}\left\{1 + \frac{1}{2} + \frac{1}{3} + \cdots + \frac{1}{2N+1} + \frac{1}{2N+2}\right\} \to 0 \quad \text{as } N \to \infty, \tag{E.41}$$

as with (E.37). From (E.40) and (E.41) the desired result follows at once.

Remarks. In the Remark to Problem 10 we noted the fact that

$$\xi(1) = 1 - \frac{1}{3} + \frac{1}{5} - + \cdots = \text{arc tan } 1 = \frac{\pi}{4}.$$

Hence, taking n = 1 in Problem 104, we find

$$\sum_{j=0}^{\infty} \frac{1}{(2j+1)^2} = 2\{\xi(1)\}^2 = \frac{\pi^2}{8}.$$

Coupling this with the simple identity

$$\sum_{j=0}^{\infty} \frac{1}{(2j+1)^s} = (1 - 2^{-s})\zeta(s) \qquad (s > 1),$$

the familiar value (see Problem 7), $\zeta(2) = \pi^2/6$, emerges at once.

PROBLEM 105. Show that

$$1 + \left(\frac{1 + \frac{1}{2}}{2}\right)^2 + \left(\frac{1 + \frac{1}{2} + \frac{1}{3}}{3}\right)^2 + \left(\frac{1 + \frac{1}{2} + \frac{1}{3} + \frac{1}{4}}{4}\right)^2 + \cdots = \frac{17\pi^4}{360}.$$

Solution. Observing that

$$\frac{1}{m} = \left(\frac{1}{m} - \frac{1}{m+n}\right) + \left(\frac{1}{m+n} - \frac{1}{m+2n}\right) + \left(\frac{1}{m+2n} - \frac{1}{m+3m}\right) + \cdots$$

$$= \sum_{j=0}^{\infty} \frac{n}{(m+jn)(m+jn+n)},$$

we get

$$\frac{1}{n}\sum_{m=1}^{n} \frac{1}{m} = \sum_{m=1}^{n} \sum_{j=0}^{\infty} \frac{1}{(m+jn)(m+jn+n)} = \sum_{k=1}^{\infty} \frac{1}{k(k+n)},$$

because k = m + jn takes all values 1,2,3,... if k takes the values of 0,1, 2,... and m the values 1,2,...,n.

In the following all summations run from 1 to infinity unless indicated differently. If we denote the sum in question by S, then we have

$$S = \sum_n \left\{\frac{1}{n}\sum_{m=1}^{n}\frac{1}{m}\right\}^2 = \sum_n\left\{\sum_k \frac{1}{k(k+n)}\right\}^2 = \sum_{k,j,n}\frac{1}{jk(k+n)(j+n)}$$

$$= \sum_{k<j}\frac{1}{jk(k+n)(j+n)} + \sum_{k=j}\frac{1}{jk(k+n)(j+n)} + \sum_{k>j}\frac{1}{jk(k+n)(j+n)}$$

$$= \sum\frac{1}{k^2(k+n)^2} + 2\sum_{k<j}\frac{1}{jk(k+n)(j+n)}$$

$$= \sum_{k<j}\frac{1}{k^2j^2} + 2\sum\frac{1}{k(k+m)(k+n)(k+m+n)}$$

$$= \frac{1}{2}\left\{\sum\frac{1}{k^2j^2} - \sum_{k=j}\frac{1}{k^2j^2}\right\} + 2\sum\left\{\frac{1}{kmn(k+m+n)} - \frac{1}{mn(k+m)(k+n)}\right\}$$

$$= \frac{1}{2}\left\{\left(\sum\frac{1}{k^2}\right)^2 - \sum\frac{1}{k^4}\right\} + 2\sum\frac{1}{kmn(k+m+n)} - 2S$$

according to the third expression for S. Since

$$\sum\frac{1}{k^2} = \frac{\pi^2}{6} \quad\text{and}\quad \sum\frac{1}{k^4} = \frac{\pi^4}{90}$$

(see Problem 7 and Remarks to Problem 103), we obtain

$$\frac{1}{2}\left(3S - \frac{\pi^4}{120}\right) = \sum\frac{1}{kmn(k+m+n)} = \sum\int_0^1 \frac{x^{k+m+n-1}}{kmn}\,dx = \int_0^1\left(\sum\frac{x^k}{k}\right)^3\frac{dx}{x}$$

$$= -\int_0^1 \{\log(1-x)\}^3\frac{dx}{x} = -\int_0^1 (\log x)^3\frac{dx}{1-x}$$

$$= -\sum\int_0^1 x^{n-1}(\log x)^3\,dx = -\sum\int_{-\infty}^0 e^{nt}\,t^3\,dt = \sum\frac{6}{n^4} = \frac{\pi^4}{15},$$

whence $S = 17\pi^4/360$, as required.

PROBLEM 106. Any positive integer may be written in the form

$$n = 2^k(2j+1).$$

Let $a_n = e^{-k}$ and $b_n = a_1 a_2 \cdots a_n$. Show that $\Sigma_{n=1}^{\infty} b_n$ converges.

Solution. The problem means that every positive integer n can be written uniquely in the form $2^k(2j + 1)$, where k and j are nonnegative integers. Thus $k = 0$ when n is odd, and k is a positive integer when n is even. Hence $a_n = 1$ when n id odd, and $a_n \le e^{-1}$ when n is even. Therefore, $b_{2n} = b_{2n+1}$ and the series is equivalent to

$$b_1 + 2b_2 + 2b_4 + \cdots$$

But since $b_{2n+2}/b_{2n} = a_{2n+1}a_{2n+2} \le e^{-1}$, the latter series converges. Since the terms approach zero, the grouping of the given series does not affect convergence and the given series converges.

PROBLEM 107. Let the entries t_{nm} $(1 \le m \le n)$ of an infinite triangular matrix

$$\begin{matrix} t_{11} \\ t_{21} & t_{22} \\ t_{31} & t_{32} & t_{33} \\ \cdots \end{matrix}$$

satisfy the following conditions:

(i) The elements of each column tend to zero, i.e.,

$$t_{nm} \to 0 \quad \text{as } n \to \infty \quad (m \text{ fixed});$$

(ii) the sum of the absolute values of all elements of any row do not exceed a constant K, i.e., for all $n = 1,2,3,\ldots,$

$$|t_{n1}| + |t_{n2}| + \cdots + |t_{nn}| \le K < \infty.$$

If $(x_n)_{n=1}^{\infty}$ is a sequence that converges to zero, show that the sequence

$$x'_n = t_{n1}x_1 + t_{n2}x_2 + \cdots + t_{nn}x_n$$

converges to zero also.

(b) Suppose that the coefficients t_{nm} satisfy in addition to conditions (i) and (ii) of part (a) also the condition

(iii) $T_n = t_{n1} + t_{n2} + \cdots + t_{nn} \to 1$ as $n \to \infty$.

If $(y_n)_{n=1}^{\infty}$ is a sequence that converges to a (finite) real number a, show that

$$y'_n = t_{n1}y_1 + t_{n2}y_2 + \cdots + t_{nn}y_n \to a \quad \text{as } n \to \infty.$$

Solution. (a) Let $\varepsilon > 0$. Then there exists an integer m such that $|x_n| < \varepsilon/2K$ for $n \geq m$; for these n we have, by condition (ii),

$$|x'_n| < |t_{n1}x_1 + \cdots + t_{nm}x_m| + \frac{\varepsilon}{2}.$$

Since m is kept fixed, there exists by condition (i) an $n_0 \geq m$ such that for $n \geq n_0$ we have $|t_{n1}x_1 + \cdots + t_{nm}x_m| < \varepsilon/2$. Thus $|x'_n| < \varepsilon$.

(b) Since

$$y'_n = t_{n1}(y_1 - a) + t_{n2}(y_2 - a) + \cdots + t_{nn}(y_n - a) + T_n a,$$

we may apply part (a) to the sequence $(y_n - a)_{n=1}^{\infty}$ and the proof is complete.

Remarks. The result in Problem 107 is due to O. Toeplitz and has many applications. For example, setting

$$t_{n1} = t_{n2} = \cdots = t_{nn} = \frac{1}{n},$$

we see that conditions (i), (ii), and (iii) of Problem 107 are satisfied. We can now deduce immediately the result in Problem 40.

Consider the result in Problem 39 and for simplicity assume that the sequence $(y_n)_{n=1}^{\infty}$ is strictly increasing and suppose that

$$\frac{x_n - x_{n-1}}{y_n - y_{n-1}} \quad (n = 1,2,\ldots;\ x_0 = y_0 = 0)$$

tends to a; putting

$$t_{nm} = \frac{y_m - y_{m-1}}{y_n},$$

we can verify that conditions (i), (ii), and (iii) of Problem 107 are satisfied. We may then conclude that

$$\frac{x_n}{y_n} = \sum_{m=1}^{n} t_{nm} \frac{x_m - x_{m-1}}{y_m - y_{m-1}}$$

converges to a, thereby having another solution for Problem 39.

PROBLEM 108. Let $(x_n)_{n=1}^{\infty}$ and $(y_n)_{n=1}^{\infty}$ both tend to zero and

$|y_1| + |y_2| + \cdots + |y_n| \le K \quad (n = 1,2,\ldots;\ K \text{ constant}).$

Show that

$$z_n = x_1y_n + x_2y_{n-1} + \cdots + x_ny_1$$

tends to zero also.

Solution. Set $t_{nm} = y_{n-m+1}$ and apply part (a) of Problem 107.

PROBLEM 109. Let $x_n \to a$ and $y_n \to b$ as $n \to \infty$. Show that

$$z_n = \frac{x_1y_n + x_2y_{n-1} + \cdots + x_ny_1}{n} \to ab \quad \text{as } n \to \infty.$$

Solution. Assume first that $a = 0$. If we replace y_n by y_n/n in Problem 108, we get $z_n \to 0$ as $n \to \infty$. The condition imposed on y_n in Problem 108 is not violated by this since the y_n's are bounded, $|y_n| \le K$.

Turning now to the general case, we note that

$$z_n = \frac{(x_1 - a)y_n + \cdots + (x_n - a)y_1}{n} + \frac{y_1 + \cdots + y_n}{n}\, a.$$

But we have already shown that

$$\frac{(x_1 - a)y_n + \cdots + (x_n - a)y_1}{n} \to 0 \quad \text{as } n \to \infty$$

and, by Problem 40,

$$a\,\frac{y_1 + \cdots + y_n}{n} \to ab \quad \text{as } n \to \infty.$$

PROBLEM 110. Let $x_n \to a$ as $n \to \infty$. Show that

$$x_n' = \frac{1\cdot x_0 + \binom{n}{1}x_1 + \cdots + \binom{n}{n}x_n}{2^n} \to a \quad \text{as } n \to \infty.$$

Solution. We set

$$t_{nm} = \frac{\binom{n}{m}}{2^n}$$

and apply part (b) of Problem 107. Since $\binom{n}{m} < n^m$ and $(n^m)/(2^n) \to 0$ as $n \to \infty$, we see that condition (i) of Problem 107 is satisfied. Since

$$\sum_{m=0}^{n} \binom{n}{m} = (1 + 1)^n = 2^n,$$

conditions (ii) and (iii) of Problem 107 are satisfied too and the desired result follows.

Remark. Let $x_n \to a$ as $n \to \infty$ and $z > 0$. One can show in analogy to the solution of Problem 110 that the sequences

$$x'_n = \frac{1 \cdot x_0 + \binom{n}{1} z x_1 + \cdots + \binom{n}{n} x^n z_n}{(1 + z)^n}$$

and

$$x''_n = \frac{z^n x_0 + \binom{n}{1} z^{n-1} x_1 + \cdots + \binom{n}{n} x_n}{(1 + z)^n}$$

tend to a as $n \to \infty$.

PROBLEM 111. Evaluate the convergent infinite series

$$S = \sum_{k=1}^{\infty} \left\{ \sum_{p=1}^{2k-1} \frac{1}{p} \right\} \frac{1}{2k(2k + 2)}.$$

Solution. We have

$$4S = \sum_{k=1}^{\infty} \left\{ \sum_{p=1}^{2k-1} \frac{1}{p} \right\} \frac{1}{k(k + 1)}$$

$$= \sum_{k=1}^{\infty} 1/k(k + 1) + (1/2 + 1/3) \sum_{k=2}^{\infty} 1/k(k + 1)$$

$$+ (1/4 + 1/5) \sum_{k=3}^{\infty} 1/k(k + 1) + \cdots.$$

But

$$\sum_{k=m}^{\infty} 1/k(k + 1) = 1/m.$$

Therefore

$$4S = 1 + \sum_{k=2}^{\infty} [1/(2k - 2) + 1/(2k - 1)]/k$$

$$= 1 + (1/2) \sum_{k=2}^{\infty} 1/k(k - 1) + \sum_{k=2}^{\infty} 1/k(2k - 1)$$

$$= 1 + 1/2 + 2 \sum_{k=2}^{\infty} [1/(2k - 1) - 1/2k]$$

$$= 3/2 + 2(1/3 - 1/4 + 1/5 - 1/6 + \cdots)$$

$$= 3/2 + 2[\log 2 - (1 - 1/2)] \qquad \text{(see Problem 15)}$$

$$= 1/2 + 2 \log 2$$

and

$$S = \frac{1}{8} + \frac{\log 2}{2}.$$

PROBLEM 112. Show that

$$\sum_{m=0}^{\infty} \sum_{n=0}^{\infty} \frac{m!n!}{(m + n + 2)!} = \frac{\pi^2}{6}.$$

Solution. Multiplying each term by $(m + n + 2) - (n + 1)$ and dividing by $(m + 1)$, the summation over n can be written as a telescoping series, and the double sum becomes

$$\sum_{m=0}^{\infty} \frac{m!}{m + 1} \sum_{n=0}^{\infty} \left(\frac{n!}{(m + n + 1)!} - \frac{(n + 1)!}{(m + n + 2)!}\right)$$

$$= \sum_{m=0}^{\infty} \frac{m!}{m + 1} \frac{0!}{(m + 1)!} = \sum_{m=0}^{\infty} \frac{1}{(m + 1)^2} = \frac{\pi^2}{6}. \qquad \text{(see Problem 7)}$$

PROBLEM 113. Euler's Constant c is by definition

$$c = \lim_{m\to\infty} \left\{ \sum_{k=1}^{m} \frac{1}{k} - \log m \right\}$$

(see Solution to Problem 14).

Derive the following series representations for Euler's Constant c:

$$c = 1 - \sum_{n=1}^{\infty} \sum_{m=2^{n-1}+1}^{2^n} \frac{n}{(2m - 1)(2m)}, \tag{E.42}$$

$$c = \frac{1}{2} + \sum_{n=1}^{\infty} \sum_{m=2^{n-1}}^{2^n - 1} \frac{n}{(2m)(2m + 1)(2m + 2)}. \tag{E.43}$$

Solution. Let

$$s_n = \sum_{k=1}^{2^n} \frac{1}{k}, \qquad \sigma_n = \sum_{k=1}^{2^n} \frac{(-1)^{k+1}}{k}, \qquad (n = 0,1,2,\ldots).$$

Then, since

$$\sum_{k=1}^{2^n} \frac{(-1)^{k+1}}{k} = \sum_{k=1}^{2^n} \frac{1}{k} - 2 \sum_{k=1}^{2^{n-1}} \frac{1}{2k},$$

we have the relation

$$\sigma_n = s_n - s_{n-1}, \qquad (n = 1,2,\ldots). \tag{E.44}$$

We start from the definition

$$c = \lim_{m\to\infty} \left\{ \sum_{k=1}^{m} \frac{1}{k} - \log m \right\},$$

and consider the subsequence in the right member corresponding to $m = 2^n$, $(n = 1,2,\ldots)$. Then, also,

$$c = \lim_{n\to\infty} (s_n - n \log 2). \tag{E.45}$$

But

$$\log 2 = 1 - \frac{1}{2} + \frac{1}{3} - \frac{1}{4} + \cdots + (-1)^{n+1} \frac{1}{n} + \cdots$$

(see Problem 15) and so

$$\log 2 = \sigma_n + r_n, \qquad (n = 1,2,\ldots),$$

where

$$r_n = \sum_{k=2^n+1}^{\infty} \frac{(-1)^{k+1}}{k}.$$

Since $0 < r_n < 1/2^n$, $\lim_{n\to\infty} (nr_n) = 0$, and (E.45) becomes

$$c = \lim_{n\to\infty} (s_n - n\sigma_n). \tag{E.46}$$

The sequence in (E.46) is now converted into a series in the usual way, and we have

$$c = (s_1 - \sigma_1) + \sum_{n=1}^{\infty} \{s_{n+1} - (n + 1)\sigma_{n+1} - (s_n - n\sigma_n)\}.$$

Using (E.44), this simplifies to

$$c = 1 - \sum_{n=1}^{\infty} n(\sigma_{n+1} - \sigma_n). \tag{E.47}$$

Equivalently, on replacing the σ_n by their values

$$c = 1 - \sum_{n=1}^{\infty} n\left(\frac{1}{2^n + 1} - \frac{1}{2^n + 2} + \cdots - \frac{1}{2^{n+1}}\right). \tag{E.48}$$

By combining consecutive pairs of terms in the parenthesis in (E.48), we get (E.42).

To show the equivalence of (E.43) and (E.42), expand the general term in (E.43) into partial fractions. Then

$$c = \frac{1}{2} + \frac{1}{2} \sum_{n=1}^{\infty} \sum_{m=2^{n-1}}^{2^n-1} n\left(\frac{1}{2m} - \frac{2}{2m + 1} + \frac{1}{2m + 2}\right)$$

$$= \frac{1}{2} + \sum_{n=1}^{\infty} n\left(\sigma_n - \sigma_{n+1} + \frac{1}{2^{n+2}}\right).$$

Since $\Sigma_{n=1}^{\infty} n/2^{n+2} = 1/2$, the equation

$$c = \frac{1}{2} + \sum_{n=1}^{\infty} n\left(\sigma_n - \sigma_{n+1} + \frac{1}{2^{n+2}}\right)$$

is equivalent with (E.47).

Remarks. It follows readily that the series (E.48) converges also with the parenthesis removed, in which case the resulting representation has the form

$$c = 1 + \sum_{t=3}^{\infty} (-1)^t \frac{[\log(t - 1)/\log 2]}{t}. \tag{E.49}$$

(Here and in the sequel a square bracket will denote the greatest integer function.) For choose any integer $k > 3$ and set $a = [\log(t - 1)/\log 2]$. Then we have

$$-\frac{a}{2^a + 1} \le \sum_{t=2^a+1}^{k} (-1)^t \frac{[\log(t - 1)/\log 2]}{t} < 0.$$

Thus as $k \to \infty$, the series in this inequality will approach zero as a limit. As this series represents the difference between the partial sums of the series in (E.49) and corresponding partial sums in (E.48), the convergence of (E.49) is established.

Another representation of the form (E.49) is obtained by modifying (E.47) as follows. Since $\Sigma_{n=1}^{\infty} n/2^{n+1} = 1$, we have from (E.47) that

$$c = -\sum_{n=1}^{\infty} n\left(\sigma_{n+1} - \sigma_n - \frac{1}{2^{n+1}}\right).$$

As before, it may be shown that the parenthesis in this series can be removed. The resulting series representation is

$$c = \sum_{t=1}^{\infty} (-1)^t \frac{[\log t/\log 2]}{t}.$$

PROBLEM 114. Let $S_n = 1 + 1/2 + \cdots + 1/n$. Show that

$$c < S_p + S_q - S_{pq} \le 1,$$

where c is Euler's Constant (see Problem 113).

Solution. If we set $(p,q) = S_p + S_q - S_{pq}$, then

$$(p,q) - (p-1,q) = \frac{1}{p} - \frac{1}{pq - q + 1} - \frac{1}{pq - q + 2} - \cdots - \frac{1}{pq}$$

$$< \frac{1}{p} - q\left(\frac{1}{pq}\right) = 0.$$

Thus (p,q) is a decreasing function of p and therefore also of q (by symmetry). The cases $p \to \infty$, $q \to \infty$ and $p = 1$, $q = 1$ are the two sides of the inequality in question.

PROBLEM 115. Let $a > 0$ and $b > a + 1$. Show that

$$\frac{a}{b} + \frac{a}{b}\cdot\frac{a+1}{b+1} + \frac{a}{b}\cdot\frac{a+1}{b+1}\cdot\frac{a+2}{b+2} + \frac{a}{b}\cdot\frac{a+1}{b+1}\cdot\frac{a+2}{b+2}\cdot\frac{a+3}{b+3} + \cdots = \frac{a}{d},$$

where $d = b - a - 1$.

Solution. Let $q_{-1} = 0$, $Q_{-1} = a$, and, for $n = 0,1,2,3,\ldots,$

$$q_n = \frac{a}{b}\cdot\frac{a+1}{b+1}\cdot\frac{a+2}{b+2}\cdots\frac{a+n}{b+n}, \qquad Q_n = q_n(a + n + 1).$$

Then

$$Q_{n-1} - Q_n = dq_n.$$

Letting $n = 0,1,2,\ldots,s$ in the last expression and then adding, we get

$$d(q_0 + q_1 + \cdots + q_s) = Q_{-1} - Q_s = a - Q_s.$$

But

$$Q_s = aP_s \quad \text{with} \quad P_s = \frac{A(A+1)(A+2)\cdots(A+s)}{B(B+1)(B+2)\cdots(B+s)}$$

and $A = a + 1$, $B = b$. But $P_s \to 0$ as $s \to \infty$ because $B > A > 0$ (see Problem 80). Hence $q_0 + q_1 + q_2 + \cdots = a/d$. But this is the desired result.

PROBLEM 116. Let $S_0 = 1$, $S_1 = 3$, $S_{n+1} = 2S_n^2 - 1$ for $n \geq 1$. Show that

$$\lim_{n\to\infty} \frac{S_n}{2^n S_0 S_1 \cdots S_{n-1}} = \sqrt{2}.$$

Solution. From

$$S_n^2 - 1 = (S_n + 1)(S_n - 1) = 2S_{n-1}^2(2S_{n-1}^2 - 2) = 2^2 S_{n-1}^2(S_{n-1}^2 - 1)$$

it follows that

$$S_n^2 - 1 = 2^{2(n-1)} S_{n-1}^2 S_{n-2}^2 \cdots S_1^2 (S_1^2 - 1);$$

whence

$$\frac{S_n}{2^n S_0 S_1 \cdots S_{n-1}} = \frac{S_n}{2S_0 \sqrt{\dfrac{S_n^2 - 1}{S_1^2 - 1}}} = \frac{\sqrt{S_1^2 - 1}}{2S_0} \sqrt{\frac{S_n^2}{S_n^2 - 1}},$$

which has as limit

$$\sqrt{(S_1^2 - 1)/2S_0} = \sqrt{2}.$$

PROBLEM 117. Evaluate

$$\lim_{n\to\infty} \sum_{j=1}^{n^2} \frac{n}{n^2 + j^2}.$$

Solution. For all positive integers n

$$\int_1^{n^2+1} \frac{n}{n^2 + j^2}\, dj \le \sum_{j=1}^{n^2} \frac{n}{n^2 + j^2} \le \int_0^{n^2} \frac{n}{n^2 + j^2}\, dj$$

and thus it is clear that the limit as $n \to \infty$ is $\pi/2$.

PROBLEM 118. If

$$S_{2m} = \frac{1}{2^{2m}} + \frac{1}{4^{2m}} + \frac{1}{6^{2m}} + \cdots,$$

find the value of

$$\frac{S_2}{2\cdot 3} + \frac{S_4}{4\cdot 5} + \frac{S_6}{6\cdot 7} + \cdots$$

Solution. If we let

$$T_{2m} = \frac{1}{2\cdot 3}\left(\frac{1}{2m}\right)^2 + \frac{1}{4\cdot 5}\left(\frac{1}{2m}\right)^4 + \frac{1}{6\cdot 7}\left(\frac{1}{2m}\right)^6 + \cdots,$$

then the value of the given expression is the same as that of

$$\sum_{m=1}^{\infty} T_{2m}.$$

By integrating $x/(1 - x^2) = x + x^3 + x^5 + \cdots$, $|x| < 1$, from 0 to r twice and then setting $r = 1/2m$, we find

$$T_{2m} = 1 - \frac{2m + 1}{2} \log \frac{2m + 1}{2m} + \frac{2m - 1}{2} \log \frac{2m - 1}{2m},$$

$m = 1,2,3,\ldots$, whence

$$\sum_{m=1}^{n} T_{2m} = n + \log \frac{2^n (n!)}{(2n + 1)^{n+\frac{1}{2}}} = \log \frac{2^n e^n (n!)}{(2n + 1)^{n+\frac{1}{2}}}.$$

Using Stirling's Formula (see Problem 34),

$$\lim_{n \to \infty} \frac{e^n n!}{n^{n+\frac{1}{2}}} = \sqrt{2\pi},$$

we obtain

$$\sum_{m=1}^{\infty} T_{2m} = \frac{1}{2} \log \frac{\pi}{e}.$$

PROBLEM 119. Verify the following identity, due to Gauss:

$$\prod_{s=1}^{\infty} \frac{1 - x^{2s}}{1 - x^{2s-1}} = \sum_{s=1}^{\infty} x^{s(s-1)/2}, \qquad |x| < 1. \tag{E.50}$$

Solution. Let $P_0 = 1$ and

$$P_n = \prod_{s=1}^{n} \frac{1 - x^{2s}}{1 - x^{2s-1}} \quad \text{for } n = 1,2,3,\ldots$$

We shall show first that

$$A_n = \sum_{s=0}^{n-1} \frac{P_n}{P_s} x^{s(2n+1)} = \sum_{s=1}^{2n} x^{s(s-1)/2} = S_n. \tag{E.51}$$

Indeed, we readily verify that

$$(1 - x^{2n-1})x^{s(2n-1)} + (1 - x^{2s+1})x^{(s+1)(2n-1)} - (1 - x^{2s})x^{s(2n-1)}$$

$$= (1 - x^{2n})x^{s(2n+1)},$$

and multiplying by

$$\frac{P_{n-1}}{P_s(1 - x^{2n-1})}$$

we find

$$\frac{P_n}{P_s} x^{s(2n+1)} = \frac{P_{n-1}}{P_s} x^{s(2[n-1]+1)} + \alpha_{s,n} - \beta_{s,n} \tag{E.52}$$

where

$$\alpha_{s,n} = \frac{1 - x^{2s+1}}{1 - x^{2n-1}} x^{(s+1)(2n-1)}$$

and

$$\beta_{s,n} = \frac{1 - x^{2s}}{1 - x^{2n-1}} \frac{P_{n-1}}{P_s} x^{s(2n-1)}.$$

Now

$$\beta_{s+1,n} = \alpha_{s,n} \qquad (\text{for } s = 0,1,2,\ldots,n - 2)$$

and since, further,

$$\beta_{0,n} = 0 \quad \text{and} \quad \alpha_{n-1,n} = x^{n(2n-1)},$$

by summing (E.52) from $s = 0$ to $s = n - 1$ we obtain:

$$A_n = A_{n-1} + x^{(n-1)(2n-1)} + x^{n(2n-1)}.$$

But this may be written $A_n - A_{n-1} = S_n - S_{n-1}$, and by induction

$$A_n - S_n = A_1 - S_1 = \frac{1 - x^2}{1 - x} - (1 + x) = 0.$$

This proves (E.51).

From (E.51) we now readily obtain (E.50). The leading term in A_n (that is $s = 0$ in the left side of equation (E.51)) is P_n. Since the remaining terms ($s = 1, 2, \ldots, n = 1$) are of order x^{2n+1} and higher, the power series of the function $P_n(x)$ must agree with that of $S_n(x)$ at least to terms of order x^{2n}. By induction the function P_∞ must have power series S_∞ which proves (E.50).

PROBLEM 120. Show that

$$\log(1.1010010001\ldots) = \frac{1}{11} + \frac{1}{2}\cdot\frac{1}{101} + \frac{1}{3}\cdot\frac{1}{1001} + \cdots.$$

Solution. By the result in Problem 119,

$$\log(1 + x + x^3 + x^6 + \cdots) = \sum_{n=1}^{\infty} (-1)^n \log(1 - x^n)$$

$$= \sum_{n=1}^{\infty} (-1)^{n+1} \left\{ \sum_{p=1}^{\infty} \frac{x^{pn}}{p} \right\} = \sum_{p=1}^{\infty} \sum_{n=1}^{\infty} (-1)^{n+1} \left(\frac{x^{pn}}{p}\right) = \sum_{p=1}^{\infty} \frac{x^p}{p(1 + x^p)},$$

where the operations are justified by the absolute convergence. The desired formula results from placing $x = 0.1$.

PROBLEM 121. Find the sum of

$$1 + \frac{1}{2}\cdot\frac{1}{3^2} + \frac{1\cdot 3}{2\cdot 4}\cdot\frac{1}{5^2} + \frac{1\cdot 3\cdot 5}{2\cdot 4\cdot 6}\cdot\frac{1}{7^2} + \cdots.$$

Solution. Consider the integral

$$\int_0^x \frac{\text{arc sin } z}{z}\, dz$$

and expand the integrand in a power series in z. Then

$$\int_0^x \frac{\text{arc sin } z}{z}\, dz = \int_0^x \left\{1 + \frac{1}{2}\cdot\frac{z^2}{3} + \frac{1\cdot 3}{2\cdot 4}\cdot\frac{z^4}{5} + \frac{1\cdot 3\cdot 5}{2\cdot 4\cdot 6}\cdot\frac{z^6}{7} + \cdots\right\} dz$$

$$= x + \frac{1}{2}\cdot\frac{x^3}{3^2} + \frac{1\cdot 3}{2\cdot 4}\cdot\frac{x^5}{5^2} + \frac{1\cdot 3\cdot 5}{2\cdot 4\cdot 6}\cdot\frac{x^7}{7^2} + \cdots.$$

Inasmuch as the series in the integrand is convergent for $|z| < 1$, the resulting power series in x is convergent for $|x| < 1$. Moreover, this series converges for $x = 1$ and is then just the series whose sum, S, is to be found. By Abel's theorem,

$$S = \int_0^1 \frac{\text{arc sin } z}{z}\, dz.$$

The substitution $z = \sin t$ yields the result in terms of a standard improper integral

$$S = -\int_0^{\pi/2} \log \sin t \, dt$$

which is easily evaluated by replacing sin t by 2 sin ½t cos ½t. One obtains $S = \frac{1}{2}\pi \log 2$. Indeed, let $J = -S$ and $t = 2x$. Then

$$J = 2\int_0^{\pi/4} \log \sin 2x \, dx = \frac{\pi}{2} \log 2 + 2\int_0^{\pi/4} \log \sin x \, dx$$

$$+ 2\int_0^{\pi/4} \log \cos x \, dx.$$

But, under the substitution $x = \frac{\pi}{2} - u$, we obtain

$$\int_0^{\pi/4} \log \cos x \, dx = \int_{\pi/4}^{\pi/2} \log \sin u \, du$$

and so $J = \frac{\pi}{2} \log 2 + 2J$ or $J = -\frac{\pi}{2} \log 2$.

Remarks. Denoting by x_s the roots of $x^{2n} = 1$, we get

$$x_s = \cos \frac{2s\pi}{n} + i \sin \frac{2s\pi}{n} \quad (s = 1,2,\ldots,2n).$$

Thus

$$x^{2n} - 1 = \prod_{s=1}^{2n} (x - x_s) = \prod_{s=1}^{n-1} (x - x_s) \prod_{s=n+1}^{2n-1} (x - x_s)\cdot(x^2 - 1),$$

since $x_n = -1$, $x_{2n} = 1$. But $x_{2n-s} = \bar{x}_s$ with $\bar{x}_s$ being the complex conjugate of x_s and so

$$x^{2n} - 1 = (x^2 - 1) \prod_{s=1}^{n-1} (x - x_s)(x - \bar{x}_s)$$

$$= (x^2 - 1) \prod_{s=1}^{n-1} \left(x^2 - 2x \cos \frac{s\pi}{n} + 1\right)$$

or

$$x^{2n-2} + x^{2n-4} + \cdots + x^2 + 1 = \prod_{s=1}^{n-1} \left(x^2 - 2x \cos \frac{s\pi}{n} + 1\right).$$

Letting x = 1, we get

$$n = \prod_{s=1}^{n-1} \left(2 - 2\cos\frac{s\pi}{n}\right) = \prod_{s=1}^{n-1} 4\sin^2\frac{s\pi}{2n}$$

$$= 2^{2(n-1)} \sin^2\frac{\pi}{2n} \sin^2\frac{2\pi}{2n} \cdots \sin^2\frac{(n-1)\pi}{2n}$$

or

$$\sin\frac{\pi}{2n} \sin\frac{2\pi}{2n} \cdots \sin\frac{(n-1)\pi}{2n} = \frac{\sqrt{n}}{2^{n-1}}.$$

But this identity leads to the evaluation of J directly:

$$\int_0^{\pi/2} \log \sin x \, dx = \lim_{n\to\infty} \frac{\pi}{2} \frac{\frac{1}{2}\log n - (n-1)\log 2}{n} = -\frac{\pi}{2}\log 2.$$

PROBLEM 122. Show that

$$\left(1 + \frac{1}{n}\right)\left(1 + \frac{2}{n}\right)^{1/2}\left(1 + \frac{3}{n}\right)^{1/3} \cdots \left(1 + \frac{n}{n}\right)^{1/n}$$

tends to $\exp(\pi^2/12)$ as $n \to \infty$.

Solution. On taking logarithms we see that

$$\log\left(1 + \frac{1}{n}\right) + \frac{1}{2}\log\left(1 + \frac{2}{n}\right) + \frac{1}{3}\log\left(1 + \frac{3}{n}\right) + \cdots + \frac{1}{n}\log\left(1 + \frac{n}{n}\right)$$

$$= \frac{1}{n}\left[n\log\left(1 + \frac{1}{n}\right) + \frac{n}{2}\log\left(1 + \frac{2}{n}\right) + \cdots + \frac{n}{n}\log\left(1 + \frac{n}{n}\right)\right]$$

$$\to \int_0^1 \frac{\log(1 + x)}{x}\, dx \quad \text{as } n \to \infty.$$

But, for $0 \le x \le 1$,

$$\frac{\log(1 + x)}{x} = 1 - \frac{1}{2}x + \frac{1}{3}x^2 - \cdots + (-1)^{n-1}\frac{1}{n}x^{n-1} + \cdots$$

and so

$$\int_0^1 \frac{\log(1 + x)}{x}\, dx = \sum_{n=1}^{\infty} (-1)^{n-1}\frac{1}{n^2} = \sum_{n=1}^{\infty} \frac{1}{n^2} - 2\sum_{n=1}^{\infty} \frac{1}{(2n)^2} = \frac{\pi^2}{12}$$

because $\Sigma_{n=1}^{\infty} n^{-2} = \pi^2/6$ (see Problem 7).

PROBLEM 123. Show that

$$\left(\left\{1 + \frac{1^2}{n^2}\right\}\left\{1 + \frac{2^2}{n^2}\right\}\left\{1 + \frac{3^2}{n^2}\right\} \cdots \left\{1 + \frac{n^2}{n^2}\right\}\right)^{1/n}$$

tends to $2 \exp\left(\frac{\pi - 4}{2}\right)$ as $n \to \infty$.

Solution. On taking logarithms we see that

$$\frac{1}{n}\left[\log\left\{1 + \frac{1^2}{n^2}\right\} + \log\left\{1 + \frac{2^2}{n^2}\right\} + \cdots + \log\left\{1 + \frac{n^2}{n^2}\right\}\right]$$

$$\to \int_0^1 \log(1 + x^2)\, dx \qquad \text{as } n \to \infty.$$

But

$$\int_0^1 \log(1 + x^2)\, dx = x \log(1 + x^2) \Big|_0^1 - 2\int_0^1 \frac{x^2}{1 + x^2}\, dx$$

$$= \log 2 - 2\int_0^1 \left(1 - \frac{1}{1 + x^2}\right) dx = \log 2 - 2 + 2\,\frac{\pi}{4}$$

$$= \frac{\pi}{2} + \log 2 - 2 = \log 2 + \frac{\pi - 4}{2}$$

and the desired result follows.

Remarks. In an entirely similar way we can find the limit of

$$2 \log n - \log(1 + n^2)^{1/n}(2^2 + n^2)^{1/n} \cdots (2n^2)^{1/n}$$

as $n \to \infty$. Indeed, the given expression can be rewritten as

$$-\frac{1}{n}\left[\log\left\{\frac{1^2}{n^2} + 1\right\} + \log\left\{\frac{2^2}{n^2} + 1\right\} + \cdots + \log\left\{\frac{n^2}{n^2} + 1\right\}\right]$$

and tends to $-\int_0^1 \log(1 + x^2)\, dx = 2 - \log 2 - \pi/2$ as $n \to \infty$.

PROBLEM 124. Let A_n and G_n denote the arithmetic and geometric mean of the binomial coefficients $\binom{n}{0}, \binom{n}{1}, \binom{n}{2}, \ldots, \binom{n}{n}$, respectively. Show that

$$\lim_{n\to\infty} \sqrt[n]{A_n} = 2 \quad \text{and} \quad \lim_{n\to\infty} \sqrt[n]{G_n} = \sqrt{e}.$$

Solution. We have $A_n = 2^n/(n + 1)$. But

$$\sqrt[n]{\frac{1}{2}\cdot\frac{2}{3}\cdot\frac{3}{4}\cdot\frac{4}{5} \cdots \frac{n}{n + 1}} \to 1 \quad \text{as } n \to \infty$$

(by Problem 41) and so $\lim_{n\to\infty} \sqrt[n]{A_n} = 2$. Moreover

$$\binom{n}{0}\binom{n}{1}\binom{n}{2} \cdots \binom{n}{n} = \frac{(n!)^{n+1}}{(1!\ 2!\ 3! \cdots n!)^2} = \prod_{k=1}^{n} (n + 1 - k)^{n+1-2k}$$

$$= \prod_{k=1}^{n} \left(\frac{n + 1 - k}{n + 1}\right)^{n+1-2k},$$

because

$$\sum_{k=1}^{n} (n + 1 - 2k) = 0.$$

But

$$\lim_{n\to\infty} \frac{1}{n} \log G_n = \lim_{n\to\infty} \frac{1}{n} \sum_{k=1}^{n} \left(1 - \frac{2k}{n + 1}\right) \log\left(1 - \frac{k}{n + 1}\right)$$

$$= \int_0^1 (1 - 2x) \log(1 - x)\, dx = \frac{1}{2}$$

and so $\lim_{n\to\infty} \sqrt[n]{G_n} = \sqrt{e}$.

PROBLEM 125. Find the limit of S_n as $n \to \infty$, where

$$S_n = \frac{1^2}{n^3 + 1^3} + \frac{2^2}{n^3 + 2^3} + \frac{3^2}{n^3 + 3^3} + \cdots + \frac{n^2}{n^3 + n^3}.$$

Solution. Evidently

$$S_n = \sum_{k=1}^{n} \frac{1}{n} \frac{k^2/n^2}{1 + k^3/n^3} \to \int_0^1 \frac{x^2}{1 + x^3}\, dx = \frac{1}{3} \log 2 \quad \text{as } n \to \infty.$$

PROBLEM 126. Let $x > 0$. Show that

$$\frac{1}{x} = \frac{1}{x+1} + \sum_{n=1}^{\infty} \frac{n!}{(x+1)(x+2)\cdots(x+n+1)}.$$

Solution. We easily see that

$$\frac{1}{x} - \frac{1}{x+1} = \frac{1}{x(x+1)}, \quad \frac{1}{x} - \frac{1}{x+1} - \frac{1}{(x+1)(x+2)} = \frac{2}{x(x+1)(x+2)},$$

and, more generally, that

$$\frac{1}{x} - \frac{1}{x+1} - \sum_{k=1}^{n-1} \frac{k!}{(x+1)(x+2)\cdots(x+n+1)} = \frac{n!}{x(x+1)\cdots(x+n)}.$$

But, for $x > 0$,

$$\lim_{n\to\infty} \frac{n!}{(x+1)\cdots(x+n)} = 0$$

by the result in Problem 80.

PROBLEM 127. Let

$$T_n = \frac{1}{n}\left(\sin\frac{t}{n} + \sin\frac{2t}{n} + \cdots + \sin\frac{(n-1)t}{n}\right).$$

Show that

$$\lim_{n\to\infty} T_n = \frac{1 - \cos t}{t}.$$

Solution. Dividing the interval $[0,t]$ into n equal parts, we see that

$$\lim_{n\to\infty} \frac{t}{n} \sum_{k=1}^{n} \sin\frac{kt}{n} = \int_0^t \sin x\, dx = 1 - \cos t$$

and so

$$\lim_{n\to\infty} \frac{1}{n} \sum_{k=1}^{n-1} \sin\frac{kt}{n} = \frac{1 - \cos t}{t}$$

because

$$\lim_{n\to\infty} \frac{\sin nt}{n} = 0.$$

PROBLEM 128. Let

$$P_n = \frac{\sqrt[n]{(n+1)(n+2)\cdots(n+n)}}{n}.$$

Show that $\lim_{n\to\infty} P_n = 4/e$.

Solution. We have

$$P_n = \left\{\left(1+\frac{1}{n}\right)\left(1+\frac{2}{n}\right)\cdots\left(1+\frac{n}{n}\right)\right\}^{1/n}$$

and so, as $n \to \infty$,

$$\log P_n = \frac{1}{n}\left\{\log\left(1+\frac{1}{n}\right) + \log\left(1+\frac{2}{n}\right) + \cdots + \log\left(1+\frac{n}{n}\right)\right\}$$

$$\to \int_0^1 \log(1+x)\,dx = \log 4 - 1.$$

PROBLEM 129. Let

$$S_n = \frac{n}{2n+1} + \sum_{k=1}^{n} \frac{1}{(2k)^3 - 2k}.$$

Show that $\lim_{n\to\infty} S_n = \log 2$.

Solution. By the result in Problem 105 in Chapter 1 we have

$$S_n = \frac{1}{n+1} + \frac{1}{n+2} + \cdots + \frac{1}{n+n}.$$

Thus, as $n \to \infty$,

$$S_n = \frac{1}{n}\left[\frac{1}{1+\frac{1}{n}} + \frac{1}{1+\frac{2}{n}} + \cdots + \frac{1}{1+\frac{n}{n}}\right] \to \int_0^1 \frac{dx}{1+x} = \log 2.$$

Remark. From the result in Problem 129 we can easily see that

$$\sum_{k=1}^{\infty} \frac{1}{(2k)^3 - 2k} = -\frac{1}{2} + \log 2.$$

PROBLEM 130. Show that $1 - 1/4 + 1/7 - 1/10 + \cdots = (1/3)(\pi/\sqrt{3} - \log 2)$.

Solution. Noting that

$$\frac{1}{a} - \frac{1}{a + b} + \frac{1}{a + 2b} - \frac{1}{a + 3b} + \cdots = \int_0^1 \frac{t^{a-1}}{1 + t^b}\, dt \qquad (a,b > 0),$$

we see that we merely have to evaluate

$$\int_0^1 \frac{dt}{1 + t^3}.$$

But

$$\int \frac{dt}{1 + t^3} = \frac{1}{6}\left\{\log \frac{(t + 1)^2}{t^2 - t + 1} + 2\sqrt{3} \operatorname{arc\,tan} \frac{2x - 1}{\sqrt{3}}\right\} + C$$

and the desired result easily follows.

PROBLEM 131. Show that, with m a positive integer,

$$\sum_{n=1}^{\infty} \frac{1}{n(n + m)} = \frac{1}{m}\left(1 + \frac{1}{2} + \frac{1}{3} + \cdots + \frac{1}{m}\right).$$

Solution. For $a > -1$, $b > -1$, and $a \neq b$ we have

$$\sum_{n=1}^{\infty} \frac{1}{(n + a)(n + b)} = \frac{1}{b - a}\int_0^1 \frac{x^a - x^b}{1 - x}\, dx.$$

Thus, taking $a = 0$ and $b = m$, we obtain the desired result.

PROBLEM 132. Let $b - 2 > a > 0$. Show that

$$\frac{a}{b} + 2\cdot\frac{a(a + 1)}{b(b + 1)} + 3\cdot\frac{a(a + 1)(a + 2)}{b(b + 1)(b + 2)} + \cdots = \frac{a(b - 1)}{(b - a - 1)(b - a - 2)}.$$

Solution. By the result in Problem 115, if $b - 1 > a > 0$, then

$$1 + \frac{a}{b} + \frac{a(a + 1)}{b(b + 1)} + \frac{a(a + 1)(a + 2)}{b(b + 1)(b + 2)} + \cdots = \frac{b - 1}{b - a - 1}. \tag{E.53}$$

Hence, if $b - 2 > a > 0$, then

$$1 + \frac{a + 1}{b} + \frac{(a + 1)(a + 2)}{b(b + 1)} + \frac{(a + 1)(a + 2)(a + 3)}{b(b + 1)(b + 2)} + \cdots = \frac{b - 1}{b - a - 2}. \tag{E.54}$$

Subtracting from the series in (E.54) the series in (E.53), we get

$$\frac{1}{b} + 2\cdot\frac{a + 1}{b(b + 1)} + 3\cdot\frac{(a + 1)(a + 2)}{b(b + 1)(b + 2)} + \cdots = \frac{b - 1}{b - a - 2} - \frac{b - 1}{b - a - 1}$$

$$= \frac{b - 1}{(b - a - 1)(b - a - 2)}$$

and the desired result follows immediately.

PROBLEM 133. Let

$$T_n = \frac{1}{n}\left\{\frac{n}{1} + \frac{n - 1}{2} + \frac{n - 2}{3} + \cdots + \frac{1}{n} - \log(n!)\right\}.$$

Find $\lim_{n\to\infty} T_n$.

Solution. Let

$$a_n = 1 + \frac{1}{2} + \frac{1}{3} + \cdots + \frac{1}{n} - \log n.$$

We have that $a_n \to c$ as $n \to \infty$, where c is Euler's Constant (see Problem 113). But

$$T_n = \frac{a_1 + a_2 + \cdots + a_n}{n}$$

and so, by the result in Problem 40, $\lim_{n\to\infty} T_n = c$.

PROBLEM 134. Show that

$$\frac{1}{3} + \frac{2}{3\cdot 5} + \frac{3}{3\cdot 5\cdot 7} + \frac{5}{3\cdot 5\cdot 7\cdot 9} + \cdots = \frac{1}{2}.$$

Solution. Since

$$\frac{1}{3} + \frac{2}{3\cdot 5} + \frac{3}{3\cdot 5\cdot 7} + \cdots + \frac{n}{3\cdot 5\cdot 7 \cdots (2n + 1)}$$

$$= \frac{1}{2}\left\{1 - \frac{1}{3\cdot 5\cdot 7 \cdots (2n + 1)}\right\},$$

the desired result follows by letting $n \to \infty$.

PROBLEM 135. Show that

$$1 = \frac{a_1}{a_1 + 1} + \frac{a_2}{(a_1 + 1)(a_2 + 1)} + \frac{a_3}{(a_1 + 1)(a_2 + 1)(a_3 + 1)} + \cdots$$

whenever $(a_1 + 1)(a_2 + 1) \cdots (a_n + 1) \to \infty$ as $n \to \infty$.

Solution. By Problem 106 of Chapter 1,

$$\frac{a_1}{a_1 + 1} + \frac{a_2}{(a_1 + 1)(a_2 + 1)} + \frac{a_3}{(a_1 + 1)(a_2 + 1)(a_3 + 1)}$$

$$+ \frac{a_{n-1}}{(a_1 + 1)(a_2 + 1) \cdots (a_{n-1} + 1)}$$

$$= 1 - \frac{1}{(a_1 + 1)(a_2 + 1) \cdots (a_n + 1)}$$

and the desired result follows.

Remarks. The result in Problem 135 easily yields a number of interesting identities. For example, letting

$$a_1 = 0,\ a_2 = 1,\ a_3 = 2,\ a_4 = 3,\ \ldots,\ a_n = n - 1,\ \ldots,$$

we obtain

$$1 = \frac{1}{1 \cdot 2} + \frac{2}{1 \cdot 2 \cdot 3} + \frac{3}{1 \cdot 2 \cdot 3 \cdot 4} + \cdots + \frac{n - 1}{n!} + \cdots.$$

Putting $a_1 = 0$, $a_2 = 2$, $a_3 = 4$, $a_4 = 6$, ..., $a_{n+1} = 2n$, ..., we get

$$1 = \frac{2}{1 \cdot 3} + \frac{4}{1 \cdot 3 \cdot 5} + \frac{6}{1 \cdot 3 \cdot 5 \cdot 7} + \cdots + \frac{2n}{1 \cdot 3 \cdot 5 \cdot 7 \cdots (2n + 1)} + \cdots$$

and, putting $a_1 = 1$, $a_2 = 3$, $a_3 = 5$, $a_4 = 7$, ..., $a_n = 2n - 1$, ...,

$$1 = \frac{1}{2} + \frac{3}{2 \cdot 4} + \frac{5}{2 \cdot 4 \cdot 6} + \frac{7}{2 \cdot 4 \cdot 6 \cdot 8} + \cdots + \frac{2n - 1}{2 \cdot 4 \cdot 6 \cdot 8 \cdots 2n} + \cdots.$$

PROBLEM 136. Show that, for $x > 2$,

$$\frac{1}{x} + \frac{2!}{x(x + 1)} + \frac{3!}{x(x + 1)(x + 2)} + \cdots = \frac{1}{x - 2}.$$

Solution. Since

$$\frac{1}{x} = \frac{1!}{x - 2} - \frac{2!}{(x - 2)x},$$

$$\frac{2!}{x(x + 1)} = \frac{2!}{(x - 2)x} - \frac{3!}{(x - 2)x(x + 1)},$$

$$\frac{3!}{x(x + 1)(x + 2)} = \frac{3!}{(x - 2)x(x + 1)} - \frac{4!}{(x - 2)x(x + 1)(x + 2)},$$

and, in general,

$$\frac{n!}{x(x + 1)(x + 2) \cdots (x + n - 1)}$$

$$= \frac{n!}{(x - 2)x(x + 1)(x + 2) \cdots (x + n - 2)}$$

$$- \frac{(n + 1)!}{(x - 2)x(x + 1)(x + 2) \cdots (x + n - 1)}$$

we see that

$$\frac{1}{x} + \frac{2!}{x(x + 1)} + \frac{3!}{x(x + 1)(x + 2)} + \cdots + \frac{n!}{x(x + 1)(x + 2) \cdots (x + n)}$$

$$= \frac{1}{x - 2} - \frac{(n + 1)!}{(x - 2)x(x + 1)(x + 2) \cdots (x + n - 1)}.$$

But, for $x > 2$,

$$\frac{2 \cdot 3 \cdot 4 \cdots (n + 1)}{x(x + 1)(x + 2) \cdots (x + n - 1)} \to 0 \quad \text{as } n \to \infty$$

by the result in Problem 80 and so our claim follows.

PROBLEM 137. Show that

$$\frac{1}{1 \cdot 2 \cdot 3} + \frac{1}{3 \cdot 4 \cdot 5} + \frac{1}{5 \cdot 6 \cdot 7} + \cdots = \log 2 - \frac{1}{2}.$$

Solution. Using the identity

$$\frac{1}{2!} \int_0^1 t^{n-1} (1 - t)^2 \, dt = \frac{1}{n(n + 1)(n + 2)}$$

and observing that, for $|t| < 1$,

$$(1 - t)^2(t^2 + t^4 + t^6 + \cdots) = (1 - t)^2 \left(\frac{1}{1 - t^2} - 1\right)$$

$$= - t^2 + 2t - 2 + \frac{2}{t + 1}$$

we see that

$$\frac{1}{1\cdot 2\cdot 3} + \frac{1}{3\cdot 4\cdot 5} + \frac{1}{5\cdot 6\cdot 7} + \cdots$$

$$= \frac{1}{1\cdot 2\cdot 3} + \frac{1}{2}\int_0^1 \left(-\,t^2 + 2t - 2 + \frac{2}{t+1}\right) dt = \log 2 - \frac{1}{2}.$$

PROBLEM 138. Show that

$$\lim_{n\to\infty}\left\{\frac{\sqrt{n-1}}{n} + \frac{\sqrt{2n-1}}{2n} + \cdots + \frac{\sqrt{n^2-1}}{n^2}\right\} = 2.$$

Solution. We have, for $n \to \infty$,

$$\sum_{k=1}^{n} \frac{\sqrt{kn-1}}{kn} = \frac{1}{n}\sum_{k=1}^{n} \frac{n}{k}\sqrt{\frac{k}{n} - \frac{1}{n^2}} \to \int_0^1 \frac{\sqrt{x}}{x}\,dx = \int_0^1 x^{-\frac{1}{2}}\,dx = 2.$$

Remark. In a completely similar manner we can verify that

$$\lim_{n\to\infty}\sum_{k=1}^{n} \frac{\sqrt{kn-a}}{kn-c} = 2.$$

PROBLEM 139. Show that

$$\lim_{n\to\infty}\sum_{k=1}^{n} \frac{(k^2 n - a)^{1/3}}{kn} = \frac{3}{2}.$$

Solution. We have, for $n \to \infty$,

$$\sum_{k=1}^{n} \frac{(k^2 n - a)^{3/2}}{kn} = \frac{1}{n}\sum_{k=1}^{n} \frac{n}{k}\left(\left(\frac{k}{n}\right)^2 - \frac{a}{n^3}\right)^{1/3} \to \int_0^1 \frac{x^{2/3}}{x}\,dx = \frac{3}{2}.$$

PROBLEM 140. Show that

$$\lim_{n\to\infty}\sum_{k=1}^{n-1} \frac{n^2}{(n^2 + k^2)^{3/2}} = \frac{1}{\sqrt{2}}.$$

Solution. We have, for $n \to \infty$,

$$\sum_{k=1}^{n-1} \frac{n^2}{(n^2 + k^2)^{3/2}} = \frac{1}{n} \sum_{k=1}^{n-1} \frac{1}{\{1 + (\frac{k}{n})\}^{3/2}} \to \int_0^1 \frac{dx}{(1 + x^2)^{3/2}}$$

$$= \frac{x}{\sqrt{1 + x^2}} \Bigg|_0^1 = 2^{-\frac{1}{2}}.$$

PROBLEM 141. Show that

$$\lim_{n \to \infty} \sum_{k=1}^{n} \frac{1}{\sqrt{2a^2kn - 1}} = \frac{\sqrt{2}}{a}.$$

Solution. We have, for $n \to \infty$,

$$\sum_{k=1}^{n} \frac{1}{\sqrt{2a^2kn - 1}} = \frac{1}{n} \sum_{k=1}^{n} \frac{1}{\sqrt{2a^2k/n - 1/n^2}} \to \int_0^1 \frac{dx}{\sqrt{2a^2x}} = \frac{\sqrt{2}}{a}.$$

PROBLEM 142. Let $x \neq 0$ and put $u_1 = x$ and

$$u_{n+1} = \log \frac{e^{u_n} - 1}{u_n} \qquad \text{for } n = 1,2,3,\ldots$$

Show that

$$e^x - 1 = u_1 + u_1u_2 + u_1u_2u_3 + \cdots$$

Solution. Since the law of formation of the sequence is

$$e^{u_n} - 1 = u_n\, e^{u_{n+1}},$$

we can write successively

$$e^{u_{n-1}} - 1 = u_{n-1} + u_{n-1}u_n\, e^{u_{n+1}},$$

$$\cdots$$

$$e^x - 1 = u_1 + u_1u_2 + u_1u_2u_3 + \cdots + u_1u_2 \cdots u_n\, e^{u_{n+1}}.$$

The last term of this expansion, that is,

$$u_1 u_2 \cdots u_n e^{u_{n+1}},$$

is the remainder of the series when we stop with the n-th term. We shall show that $\lim_{n\to\infty} u_n = 0$.

The inequalities

$$x > \log \frac{e^x - 1}{x} > 0, \quad x > 0; \qquad x < \log \frac{e^x - 1}{x} < 0, \quad x < 0$$

imply that the sequence u_n is steadily decreasing in the first case, $u_n > 0$, and increasing in the second case, $u_n < 0$. We have $\lim_{n\to\infty} u_n = u = 0$ because

$$u > \log \frac{e^u - 1}{u} \quad \text{for } u > 0$$

and

$$u < \log \frac{e^u - 1}{u} \quad \text{for } u < 0.$$

PROBLEM 143. Show that

$$\frac{1}{4} = \frac{1}{5} + \frac{4}{5\cdot 17} + \frac{4\cdot 16}{5\cdot 17\cdot 257} + \frac{4\cdot 16\cdot 256}{5\cdot 17\cdot 257\cdot 65537} + \cdots$$

Solution. Evidently it is enough to show that

$$\frac{1}{4} = \frac{4}{17} + \frac{4\cdot 16}{17\cdot 257} + \frac{4\cdot 16\cdot 256}{17\cdot 257\cdot 65537} + \cdots$$

But the latter series is of the form

$$\frac{1}{y_0} + \frac{1}{y_0 y_1} + \frac{1}{y_0 y_1 y_2} + \cdots,$$

where $y_0 = \frac{17}{4}$, $y_1 = \frac{256}{16}$, $y_2 = \frac{65537}{256}$, ..., and, in general,

$$y_n = y_{n-1}^2 - 2.$$

Now from the Solution of Problem 6 we see that

$$\frac{1}{y_0} + \frac{1}{y_0 y_1} + \frac{1}{y_0 y_1 y_2} + \cdots = \frac{y_0 - \sqrt{y_0^2 - 4}}{2}.$$

Here

$$y_0 = 4 + \frac{1}{4} \quad \text{and} \quad y_0^2 - 4 = \left(4 - \frac{1}{4}\right)^2.$$

Remark. In Problem 83 a series was summed by the same method as the one used in the foregoing Solution.

PROBLEM 144. Show that

$$(1 + 2^{-1})(1 + 2^{-2})(1 + 2^{-3})(1 + 2^{-4}) \cdots$$

$$= 1 + \frac{1}{1} + \frac{1}{1\cdot 3} + \frac{1}{1\cdot 3\cdot 7} + \frac{1}{1\cdot 3\cdot 7\cdot 15} + \frac{1}{1\cdot 3\cdot 7\cdot 15\cdot 31} + \cdots.$$

Solution. Let

$$z = (1 + xz)(1 + x^2z)(1 + x^3z) \cdots (1 + x^nz)$$

$$= 1 + P_1z + P_2z^2 + P_3z^3 + \cdots + P_nz^n + \cdots,$$

where P_1, P_2, ..., P_n, ... are functions of x only which are to be determined; changing z to xz, we see that

$$\frac{z}{1 + xz} = 1 + P_1xz + P_2x^2z^2 + \cdots + P_nx^nz^n + \cdots.$$

Hence

$$1 + P_1z + P_2z^2 + \cdots = (1 + xz)(1 + P_1xz + P_2x^2z^2 + \cdots)$$

and so, by comparison of coefficients, we obtain

$$P_1 = \frac{x}{1 - x}, \quad P_2 = \frac{x^3}{(1 - x)(1 - x^2)}, \quad P_3 = \frac{x^6}{(1 - x)(1 - x^2)(1 - x^3)},$$

$$\ldots, \quad P_n = \frac{x^{\frac{n(n+1)}{2}}}{(1 - x)(1 - x^2) \cdots (1 - x^n)}, \quad \ldots.$$

In other words

$$1 + \frac{x}{1 - x}z + \frac{x^3}{(1 - x)(1 - x^2)} + \cdots + \frac{x^{\frac{n(n+1)}{2}}}{(1 - x)(1 - x^2) \cdots (1 - x^n)}z^n + \cdots$$

$$= (1 + xz)(1 + x^2z)(1 + x^3z) \cdots (1 + x^nz) \cdots$$

Putting $x = 1/2$ and $z = 1$, we obtain the desired result.

Remarks. Note that

$$(1 + 2^{-1})(1 + 2^{-2})(1 + 2^{-3})(1 + 2^{-4}) \cdots = \frac{3 \cdot 5 \cdot 9 \cdot 17 \cdots}{2\ 4\ 8\ 16 \cdots}.$$

The result in Problem 144 shows that

$$\frac{3 \cdot 5 \cdot 9 \cdot 17 \cdots}{2\ 4\ 8\ 16 \cdots}$$

is between 2 and 12/5. Indeed,

$$\frac{1}{1 \cdot 3} + \frac{1}{1 \cdot 3 \cdot 7} + \frac{1}{1 \cdot 3 \cdot 7 \cdot 15} + \cdots$$

$$< \frac{1}{3}(1 + 1/6 + 1/6^2 + \cdots) = \frac{1}{3}\frac{1}{1 - 1/6} = \frac{2}{5}.$$

From the Formula of Wallis (see Problem 6) we know that the product of all even numbers divided by the product of all odd numbers gives the value $\sqrt{\pi/2}$. If in the quotient of the product of all even numbers divided by the product of all odd numbers we omit all even numbers which are powers of 2 and all odd numbers which are powers of 2 augmented by unity, then the resulting quotient will be between $2(\sqrt{\pi/2})$ and $(12/5)(\sqrt{\pi/2})$.

PROBLEM 145. Let k be a positive integer. Show that

$$\lim_{n \to \infty} \frac{2 \cdot 4 \cdot 6 \cdots 2kn}{1 \cdot 3 \cdot 5 \cdots (2kn - 1)} \frac{1 \cdot 3 \cdot 5 \cdots (2n - 1)}{2 \cdot 4 \cdot 6 \cdots 2n} = \sqrt{k}.$$

Solution. We have

$$\frac{2 \cdot 4 \cdot 6 \cdots 2kn}{1 \cdot 3 \cdot 5 \cdots (2kn - 1)} \frac{1 \cdot 3 \cdot 5 \cdots (2n - 1)}{2 \cdot 4 \cdot 6 \cdots 2n} = \frac{2^{2kn}[(kn)!]^2}{(2kn)!} \frac{(2n)!}{2^{2n}(n!)^2}.$$

But, for large m, $m! = \sqrt{2\pi m}\ m^m\ e^{-m}$ approximately (see Problem 34) and the desired result follows.

PROBLEM 146. Find

$$I_{2m} = \int \frac{dx}{(x^2 + a^2)^m},$$

m being a positive integer. In particular, show that

$$\int_0^\infty \frac{dx}{(x^2 + a^2)^m} = \frac{(2m - 3)(2m - 5) \cdots 1}{(2m - 2)(2m - 4) \cdots 2} \frac{1}{a^{2m-1}} \frac{\pi}{2}.$$

Then use this identity together with the Formula of Wallis (see Problem 8) to verify that

$$\int_0^\infty e^{-x^2}\, dx = \frac{\sqrt{\pi}}{2}.$$

Solution. Let

$$P_{2m} = \frac{x}{(x^2 + a^2)^{m-1}}.$$

Since

$$I_{2m} = \frac{P_{2m}}{(2m - 2)a^2} + \frac{2m - 3}{2m - 2} \frac{1}{a^2} I_{2(m-1)},$$

$$I_{2(m-1)} = \frac{P_{2(m-1)}}{(2m - 4)a^2} + \frac{2m - 5}{2m - 4} \frac{1}{a^2} I_{2(m-2)}$$

and so forth, we see that

$$I_{2m} = \frac{1}{2m - 2} \frac{P_{2m}}{a^2} + \frac{2m - 3}{(2m - 2)(2m - 4)} \frac{P_{2(m-1)}}{a^4}$$

$$+ \frac{(2m - 3)(2m - 5)}{(2m - 2)(2m - 4)(2m - 6)} \frac{P_{2(m-2)}}{a^6}$$

$$+ \cdots + \frac{(2m - 3)(2m - 5) \cdots 1}{(2m - 2)(2m - 4)(2m - 6) \cdots 2} \text{arc tan } \frac{x}{a}.$$

Hence

$$\int_0^\infty \frac{dx}{(x^2 + a^2)^m} = \frac{(2m - 3)(2m - 5) \cdots 1}{(2m - 2)(2m - 4) \cdots 2} \frac{1}{a^{2m-1}} \frac{\pi}{2}.$$

Putting $a = 1$ and $x = z/\sqrt{m}$, we get

$$\frac{1}{\sqrt{m}} \int_0^\infty \frac{dz}{(1 + z^2/m)^m} = \frac{(2m - 3)(2m - 5) \cdots 1}{(2m - 2)(2m - 4) \cdots 2} \frac{\pi}{2}.$$

Since

$$\lim_{m\to\infty} (1 + z^2/m)^m = e^{z^2},$$

we therefore obtain

$$\int_0^\infty e^{-z^2}\, dz = \frac{\pi}{2} \lim_{m\to\infty} \frac{1\cdot3\cdot5 \cdots (2m-3)}{2\cdot4\cdot6 \cdots (2m-2)} \sqrt{m}.$$

By the Formula of Wallis (see Problem 8),

$$\frac{2\cdot4\cdot6 \cdots (2m-2)}{1\cdot3\cdot5 \cdots (2m-3)} \quad \text{and} \quad \sqrt{\frac{\pi}{2}(2m-1)}$$

become infinite in a ratio of equality. Therefore

$$\frac{\pi}{2} \lim_{m\to\infty} \frac{1\cdot3\cdot5 \cdots (2m-3)}{2\cdot4\cdot6 \cdots (2m-2)} \sqrt{m} = \frac{\pi}{2} \lim_{m\to\infty} \frac{\sqrt{m}}{\sqrt{\frac{\pi}{2}(2m-1)}} = \frac{\sqrt{\pi}}{2}$$

and so

$$\int_0^\infty e^{-z^2}\, dz = \frac{\sqrt{\pi}}{2}.$$

PROBLEM 147. Show that

$$s_n = \frac{1}{\sqrt{n^2}} + \frac{1}{\sqrt{n^2+1}} + \cdots + \frac{1}{\sqrt{n^2+2n}} \to 2 \quad \text{as } n \to \infty.$$

Solution. Since

$$\frac{1}{\sqrt{n^2+k-1}} \ge \int_{k-1}^{k} \frac{dx}{\sqrt{n^2+x}} \ge \frac{1}{\sqrt{n^2+k}},$$

we see that

$$s_n - \frac{1}{\sqrt{n^2+2n}} \ge \int_0^{2n} \frac{dx}{\sqrt{n^2+x}} \ge s_n - \frac{1}{\sqrt{n^2}}.$$

But the integral

$$\int_0^{2n} \frac{dx}{\sqrt{n^2+x}} = \int_0^2 \frac{dx}{\sqrt{1+t/n}}$$

lies between

$$\int_0^2 dt = 2 \quad \text{and} \quad \left(1 + \frac{2}{n}\right)^{-\frac{1}{2}} \int_0^2 dt$$

and so $s_n \to 2$ as $n \to \infty$.

PROBLEM 148. Let $\Sigma_{n=1}^{\infty} a_n$ be a divergent series of positive decreasing terms. Show that

$$\frac{a_1 + a_3 + \cdots + a_{2n-1}}{a_2 + a_4 + \cdots + a_{2n}} \to 1 \quad \text{as } n \to \infty.$$

Solution. Call the numerator and denominator N and D. Clearly $N \geq D$, and $N - a_1 = a_3 + a_5 + \cdots + a_{2n-1} \leq D$. Also

$$2\,D \geq D + N - a_1 = \sum_{k=2}^{n} a_k \to \infty \quad \text{as } n \to \infty.$$

Hence

$$\left|\frac{N}{D} - 1\right| = \frac{|N - D|}{D} \leq \frac{a_1}{D} \to 0.$$

PROBLEM 149. Let $a_1, a_2, a_3, \ldots$ be any numbers such that

$$0 < a_n < 1, \quad (1 - a_n)a_{n+1} > \frac{1}{4} \quad (n = 1,2,3,\ldots).$$

Show that $\lim_{n\to\infty} a_n = 1/2$.

Solution. For any real number x we have $x(1 - x) \leq 1/4$. Hence

$$(1 - a_n)a_{n+1} > (1 - a_n)a_n,$$

that is, the a's are monotone increasing. Since they are bounded they approach a limit L which must satisfy $(1 - L)L \geq 1/4$. Hence $L = 1/2$.

PROBLEM 150. Find $\lim_{n\to\infty} n \sin(2\pi e n!)$.

Solution. Using the familiar series for e and sin x and the periodicity of the latter, we have

$$n \sin(2\pi e n!) = n \sin\left\{2\pi n! \left[\sum_{k=0}^{\infty} 1/k!\right]\right\} = n \sin(2\pi R_n)$$

where

$$R_n = \sum_{k=1}^{\infty} 1/(n+1)(n+2) \cdots (n+k).$$

But

$$1/(n+1) < R_n < \sum_{k=1}^{\infty} 1/(n+1)^k = 1/n.$$

Therefore

$$\lim_{n\to\infty} R_n = 0 \quad \text{and} \quad \lim_{n\to\infty} nR_n = 1$$

and so

$$\lim_{n\to\infty} n \sin(2\pi e n!) = \lim_{n\to\infty} n \sin(2\pi R_n)$$

$$= \lim_{n\to\infty} 2\pi n R_n \frac{\sin(2\pi R_n)}{2\pi R_n} = 2\pi.$$

PROBLEM 151. Show that

$$\pi = \sum_{n=0}^{\infty} \frac{(n!)^2\, 2^{n+1}}{(2n+1)!}.$$

Solution. Using the fact that

$$\int_0^{\pi/2} \cos^{2n+1} x\, dx = \frac{(n!)^2\, 2^{2n}}{(2n+1)!},$$

we obtain

$$\frac{(n!)^2\, 2^{n+1}}{(2n+1)!} = 2\int_0^{\pi/2} 2^{-n} \cos^{2n+1} x\, dx$$

$$= 2\int_0^{\pi/2} (\cos x)(\tfrac{1}{2}\cos^2 x)^n\, dx,$$

and

$$\sum_{n=0}^{\infty} \frac{(n!)^2\, 2^{n+1}}{(2n+1)!} = 2\int_0^{\pi/2} (\cos x) \left\{ \sum_{n=0}^{\infty} (\tfrac{1}{2}\cos^2 x)^n \right\} dx$$

$$= \int_0^{\pi/2} (\cos x)(1 - \tfrac{1}{2}\cos^2 x)^{-1}\, dx = \pi.$$

PROBLEM 152. Evaluate

$$\int_0^{\infty} \frac{e^{-x^2}\, dx}{(x^2 + \tfrac{1}{2})^2}.$$

Solution. Applying integration by parts twice, we get

$$\int \frac{e^{-x^2}\, dx}{(x^2 + \tfrac{1}{2})^2} = \frac{e^{-x^2}}{2x} \frac{-1}{x^2 + \tfrac{1}{2}} - \int \frac{e^{-x^2}}{x^2}\, dx = \frac{-e^{-x^2}}{2x(x^2 + \tfrac{1}{2})} + \frac{e^{-x^2}}{x} + 2\int e^{-x^2}\, dx$$

$$= \frac{xe^{-x^2}}{x^2 + \tfrac{1}{2}} + 2\int e^{-x^2}\, dx.$$

The value of the definite integral corresponding to the last term is known to us from Problem 146, and thus

$$\int_0^{\infty} \frac{e^{-x^2}\, dx}{(x^2 + \tfrac{1}{2})^2} = \sqrt{\pi}.$$

CHAPTER 4

REAL FUNCTIONS

PROBLEM 1. The *Mean Value Theorem* states: If f is a continuous function in a closed interval [a,b] and is differentiable in the open interval (a,b), then there exists a point $c \in (a,b)$ such that

$$\frac{f(b) - f(a)}{b - a} = f'(c).$$

We assume the Mean Value Theorem as known and turn to the problem of obtaining a mean value theorem for higher-order differences. Suppose that f is defined in [a,b] and for x and x + h in [a,b] we define

$$\Delta_h f(x) = f(x + h) - f(x).$$

If x + 2h is also in [a,b], then we set

$$\Delta_h^2 f(x) = \Delta_h\{\Delta_h f(x)\} = f(x + 2h) - 2f(x + h) + f(x).$$

If n is a positive integer so that $x + nh \in [a,b]$, we define $\Delta_h^n f(x)$ inductively by means of the formula

$$\Delta_h^n f(x) = \Delta_h\{\Delta_h^{n-1} f(x)\}.$$

Prove the following *Mean Value Theorem for Higher-Order Differences*: If a function f is continuous in a closed interval [a,b] and is n times differentiable in the open interval (a,b) and if with $x \in [a,b]$ we have $x + nh \in [a,b]$ for $h \neq 0$, then there exists some θ so that $0 < \theta < 1$ and

$$\Delta_h^n f(x) = f^{(n)}(x + n\theta h)h^n,$$

where $f^{(n)}$ denotes the n-th order derivative of f.

Solution. For the sake of argument we shall suppose that $h > 0$ and we choose to denote by P(n) the statement of our claim and to proceed by induction. Clearly, the statement P(1) is a mere reformulation of the Mean Value Theorem and hence is true. Assume now that P(n - 1) is true for $n > 1$ and set

$$g(x) = \frac{\Delta_h f(x)}{h}.$$

The function g is defined and continuous in $[a,b - h]$ and is $n - 1$ times differentiable in $(a,b - h)$. If $x + nh \in [a,b]$, then $x + (n - 1)h \in [a,b]$. Consequently, we may apply P(n - 1) to g and find some θ_1 so that

$$\Delta_h^{n-1} g(x) = g^{(n-1)}(x + \{n - 1\}\theta_1 h)h^{n-1}, \quad 0 < \theta_1 < 1. \tag{E.1}$$

Now

$$g^{(n-1)}(x + \{n - 1\}\theta_1 h)$$

$$= \frac{1}{h}\{f^{(n-1)}(x + \{n - 1\}\theta_1 h + h) - f^{(n-1)}(x + \{n - 1\}\theta_1 h)\}.$$

Applying the Mean Value Theorem to the right-hand side expression, we get

$$\Delta_h^{n-1} g(x) = f^{(n)}(x + \{n - 1\}\theta_1 h + \theta_2 h)h^{n-1}, \quad 0 < \theta_2 < 1.$$

If we set

$$\theta = \frac{(n - 1)\theta_1 + \theta_2}{n},$$

then it is clear that $0 < \theta < 1$. Further, since

$$\Delta_h^{n-1} g(x) = \Delta_h^{n-1}\{\Delta_h f(x)/h\} = \{\Delta_h^n f(x)\}/h, \tag{E.2}$$

it follows from (E.1) and (E.2) that

$$\Delta_h^n f(x) = f^{(n)}(x + n\theta h)h^n. \tag{E.3}$$

If $h < 0$, a similar argument will lead to the same conclusion (E.3). Thus we have shown that P(n - 1) implies P(n) and the rest follows by the principle of mathematical induction. This completes the proof of the Mean Value Theorem for Higher-Order Differences.

Remarks. Since $\Delta_h x = x + h - x = h$, from formula (E.3) we get, if $f^{(n)}$ is continuous at x,

$$f^{(n)}(x) = \lim_{\Delta_h x \to 0} \frac{\Delta_h^n f(x)}{(\Delta_h x)^n}; \tag{E.4}$$

this justifies the otherwise surprising notation

$$\frac{d^n f}{dx^n}$$

for higher-order derivatives. It does not, however, afford a means of defining the n-th order derivative; the limit on the right-hand side of (E.4) may well exist even if $f^{(n)}$ does not exist. For example, let

$$f(0) = 0, \quad f(x) = x^3 \sin \frac{1}{x} \quad \text{for } x \neq 0.$$

Here $f''(0)$ does not exist since

$$\frac{f'(0 + \Delta_h x) - f'(0)}{\Delta_h x} = 3\Delta_h x \cdot \sin \frac{1}{\Delta_h x} - \cos \frac{1}{\Delta_h x};$$

but

$$\frac{\Delta_h^2 f(0)}{(\Delta_h x)^2} = 8\Delta_h x \cdot \sin \frac{1}{2\Delta_h x} - 2\Delta_h x \cdot \sin \frac{1}{\Delta_h x}$$

tends to 0 as $\Delta_h x \to 0$.

By Problem 1, if f is k times differentiable in (a,b) and continuous in [a,b], then there exists a t, a < t < b such that

$$\Delta_h^k f^{(k)}(t) = \Delta_h^k f(a),$$

where $h = (b - a)/k$ and

$$\Delta_h^k f(a) = \sum_{j=0}^{k} (-1)^{k-j} \binom{k}{j} f(a + jh).$$

PROBLEM 2. Let c be a real number such that n^c is an integer for every positive integer n. Show that c is a nonnegative integer.

Solution. The case n = 2 shows that c is nonnegative. If the Mean Value

Theorem (see Problem 1) is applied to x^c in the interval $[u,u + 1]$ there is a t with $u < t < u + 1$ such that

$$ct^{c-1} = (u + 1)^c - u^c.$$

For any positive integer u the right-hand side is a positive integer. Now, in case $0 < c < 1$, u could be taken large enough so $u^{c-1} < 1/c$ and so $ct^{c-1} < 1$. Thus the Mean Value Theorem eliminates all c with $0 < c < 1$.

By the Mean Value Theorem for Higher-Order Differences (see Problem 1), if f is a k times differentiable function in $[a,b]$, then there exists a t, $a < t < b$, such that

$$\Delta_h^k f^{(k)}(t) = \Delta_h^k f(a),$$

where $h = (b - a)/k$ and

$$\Delta_h^k f(a) = \sum_{j=0}^{k} (-1)^{k-j} \binom{k}{j} f(a + jh).$$

We apply this fact to the interval $[u,u + k]$, where k is the unique integer such that $k - 1 \le c < k$, and obtain that there is a t with $u < t < u + k$ such that

$$c(c - 1)(c - 2) \cdots (c - k + 1)\, t^{c-k} = \Delta_h^k f(u), \tag{E.5}$$

with u a positive integer. The right-hand side of (E.5) is an integer, and by taking u sufficiently large t^{c-k} becomes sufficiently small so that the left-hand side of (E.5), though nonnegative, is less than 1. Hence

$$c(c - 1)(c - 2) \cdots (c - k + 1) = 0$$

and so $c = k - 1$.

PROBLEM 3. Let f be a nonzero differentiable function in $[a,b]$ and $f(a) = f(b) = 0$. Show that there is a point t in the interval such that

$$|f'(t)| > \frac{1}{(b - a)^2} \int_a^b f(x)\, dx.$$

Solution. Let

$$M = \sup_{a \le x \le b} |f'(x)|.$$

Then, by the Mean Value Theorem (see Problem 1),

$$f(x) = f'(t)(x - a) \le M(x - a) \qquad \text{for } a \le x \le \frac{a + b}{2},$$

$$f(x) = f'(x)(x - b) \le M(b - x) \qquad \text{for } \frac{a + b}{2} \le x \le b,$$

$$a < t < x, \qquad x < s < b.$$

The function $M(x - a)$ for $a \le x \le (a + b)/2$ and $M(b - x)$ for $(a + b)/2 \le x \le b$ is not differentiable at $x = (a + b)/2$. Hence we can not have that $f(x) = M(x - a)$ for $a \le x \le (a + b)/2$ or $f(x) = M(b - x)$ for $(a + b)/2 \le x \le b$ simultaneously. Thus, setting $m = (a + b)/2$,

$$\int_a^b f(x)\, dx < M \int_a^m (x - a)\, dx + M \int_m^b (b - x)\, dx = M\, \frac{(b - a)^2}{4}$$

or

$$M > \frac{4}{(b - a)^2} \int_a^b f(x)\, dx.$$

PROBLEM 4. Let f have a continuous derivative in [a,b]. Are there two points x_1, x_2 with $x_1 < x_2$ for every point t with x_1, x_2, and t in [a,b] such that

$$\frac{f(x_2) - f(x_1)}{x_2 - x_1} = f'(t)?$$

Solution. The answer is no in general. Take, for example,

$$f(x) = x^3 \quad \text{and} \quad t = 0.$$

PROBLEM 5. Let n be a positive integer. Show that $f^{(n)}(x) = 0$, where $f(x) = (x^2 - 1)^n$, has exactly n distinct real roots between -1 and 1.

Solution. Rolle's Theorem states: If f is a continuous function in a closed interval [a,b] and is differentiable in the open interval (a,b), and if $f(a) = f(b)$, then there is a point t in (a,b) such that $f'(t) = 0$.

We note that $f(x) = (x^2 - 1)^n = (x - 1)^n(x + 1)^n$ and its first $n - 1$

consecutive derivatives vanish at $x = \pm 1$. Thus, by Rolle's Theorem f' has at least one real root between -1 and 1; again by Rolle's Theorem, f'' has at least two distinct real roots between -1 and 1, etc.; $f^{(n-1)}$ has at least $n - 1$ distinct real roots between -1 and 1. Finally, applying Rolle's Theorem to $f^{(n-1)}$ we obtain the desired result (noting that $f^{(n)}$ is a polynomial of degree n).

PROBLEM 6. Let f be as in Problem 5. Show that $f^{(n)}(x)$ is positive for $x \geq 1$.

Solution. By a theorem of Leibniz (easily verified by induction), if g and h have derivatives of order n at a point x, then so does $f = gh$ and

$$f^{(n)}(x) = \sum_{j=0}^{n} \binom{n}{j} g^{(j)}(x)\, h^{(n-j)}(x).$$

Taking $g(x) = (x - 1)^n$ and $h(x) = (x + 1)^n$, we get the desired result.

PROBLEM 7. Show the following theorem of Hermite: The number e, where

$$e = \lim_{n\to\infty} \left(1 + \frac{1}{n}\right)^n,$$

satisfies no relation of the form

$$c_0 + c_1 e + c_2 e^2 + \cdots + c_m e^m = 0 \tag{E.6}$$

having integral coefficients not all zero. (Stated otherwise, e is a transcendental number.)

Solution. Suppose that e is a root of the polynomial

$$c_0 + c_1 x + c_2 x^2 + \cdots + c_m x^m$$

with integral coefficients $c_0, c_1, \ldots, c_m$ which are not all zero; we shall show that this assumption leads to a contradiction.

Let f be a polynomial of degree n. Then $f^{(n+1)} = 0$ and repeated integration by parts yields

$$\int_0^b f(x)\, e^{-x}\, dx = -e^{-x}\{f(x) + f'(x) + \cdots + f^{(n)}(x)\}\Big|_0^b .$$

We put

$$F(x) = f(x) + f'(x) + \cdots + f^{(n)}(x).$$

Then

$$e^b F(0) = F(b) + e^b \int_0^b f(x)\, e^{-x}\, dx. \tag{E.7}$$

In (E.7) we let $b = 0,1,2,\ldots,m$, successively, and multiply the resulting equations by c_0, c_1, c_2, ..., c_m, respectively; finally, we add up the resulting relations and obtain, by (E.6),

$$0 = c_0F(0) + c_1F(1) + \cdots + c_mF(m) + \sum_{j=1}^{m} c_j\, e^j \int_0^b f(x)\, e^{-x}\, dx, \tag{E.8}$$

which, according to our assumption, must hold for *every* polynomial f (with integral coefficients). We shall show that there is in fact an f for which (E.8) is false; this will, of course, prove our claim. To this end we put

$$f(x) = \frac{1}{(p-1)!}\, x^{p-1}(x-1)^p(x-2)^p \cdots (x-m)^p,$$

where p is an odd prime larger than both m and $|c_0|$. The derivatives of order p and larger of this polynomial have integral coefficients which are divisible by p; this is an immediate consequence of the fact that the product of p consecutive integers is divisible by p! (see Remark to Problem 22 in Chapter 1). Since the polynomial f and its first p - 1 derivatives vanish for $x = 1,2,\ldots,m$, we have $F(1)$, $F(2)$, ..., $F(m)$ are integral multiples of p. But the situation is different for $F(0)$. For $x = 0$, besides $f(x)$ only the first p - 2 derivatives vanish and so

$$F(0) = f^{(p-1)}(0) + f^{(p)}(0) + \cdots$$

holds. All summands beginning with the second summand are integral multiples of p; but, since

$$f^{(p-1)}(0) = [(-1)^m m!]^p,$$

$F(0)$ is not divisible by p. Since p is a prime number both larger than m and $|c_0|$, it follows in particular that c_0 is not divisible by p. We therefore see that the first sum in (E.8), namely,

$$c_0F(0) + c_1F(1) + \cdots + c_mF(m),$$

is not divisible by p and thus can *not* be equal to zero. We now turn our attention to the second sum in (E.8). On the interval [0,m] we obviously have

$$|f(x)| < \frac{1}{(p-1)!} m^{p-1} m^p m^p \cdots = \frac{m^{mp+p-1}}{(p-1)!}.$$

Thus

$$\left|\int_0^j f(x)\, e^{-x}\, dx\right| < \frac{m^{mp+p-1}}{(p-1)!}\int_0^j e^{-x}\, dx < \frac{m^{mp+p-1}}{(p-1)!}$$

and, on setting

$$C = |c_0| + |c_1| + \cdots + |c_m|,$$

we obtain

$$\left|\sum_{j=1}^{m} c_j\, e^j \int_0^j f(x)\, e^{-x}\, dx\right| < C\, e^m\, \frac{m^{mp+p-1}}{(p-1)!} = C\, e^m\, \frac{(m^{m+1})^{p-1}}{(p-1)!}.$$

By Problem 2 in Chapter 3, the last factor tends to zero as $p \to \infty$. Thus, by taking p large enough, the absolute value of the second sum in (E.8) can be made as small as we please. In other words, by choosing p suitably large, the total sum on the right-hand side of (E.8) cannot be equal to zero. We have thus arrived at a contradiction.

PROBLEM 8. Give an example of a function f discontinuous at all $x \neq 0$, but differentiable at $x = 0$.

Solution. Let

$$f(x) = x \quad \text{for rational } x, \text{ including } x = 0,$$
$$= x^2 + x \quad \text{for irrational } x.$$

Actually, $f'(0) = 1$.

PROBLEM 9. Let f be a differentiable function in the interval [a,b] and suppose that $f'(a) < y_0 < f'(b)$. Show that there is a point x_0 in the open interval (a,b) such that $f'(x_0) = y_0$.

Solution. We first consider the special case where $f'(a) < 0$, $f'(b)$ and show that there is an x in (a,b) such that $f'(x) = 0$.

We note that since f is differentiable it must be continuous. It accordingly attains its smallest value on $[a,b]$. Since $f'(a) < 0$ there are points $x_1 \in (a,b)$ with $f(x_1) < f(a)$. Since $f'(b) > 0$ there are points $x_2 \in (a,b)$ with $f(x_2) < f(b)$. Thus the least value of f in $[a,b]$ is attained at an $x \in (a,b)$. But then $f'(x) = 0$.

Now suppose f is differentiable and only that $f'(a) < y_0 < f'(b)$. We show that there is an $x_0 \in (a,b)$ such that $f'(x_0) = y_0$.

Consider the auxiliary function $g(t) = f(t) - y_0 t$. Then

$$g'(a) = f'(a) - y_0 < 0 \quad \text{and} \quad g'(b) = f'(b) - y_0 > 0.$$

Since g satisfies the conditions of the special case, there is an $x_0 \in (a,b)$ for which $g'(x_0) = 0$. Now $f'(x_0) = g'(x_0) + y_0 = y_0$, establishing the desired result.

PROBLEM 10. Give an example of a function which is differentiable everywhere but whose derivative fails to be continuous at some point.

Solution. Consider the function

$$f(x) = x^2 \sin \frac{1}{x} \quad \text{for } x \neq 0,$$

$$= 0 \quad \text{for } x = 0.$$

Here

$$f'(x) = 2x \sin \frac{1}{x} - \cos \frac{1}{x} \quad \text{for } x \neq 0,$$

$$= 0 \quad \text{for } x = 0.$$

Since

$$\lim_{x \to \infty} \cos \frac{1}{x}$$

does not exist, f' is not continuous at $x = 0$.

PROBLEM 11. Let $[a,b]$ be a given interval. A *partition* P *of* $[a,b]$ is any finite set of points $x_0, x_1, \ldots, x_n$ such that

$$a = x_0 < x_1 < \cdots < x_n = b.$$

We write, for $1 \le k \le n$,

$$\Delta x_k = x_k - x_{k-1}$$

and denote by

$$|P| = \max\{x_k : 1 \le k \le n\};$$

$|P|$ is called the *mesh of the partition* P.

A partition P* of $[a,b]$ is called a *refinement of a partition* P of $[a,b]$ if every point of P is a point of P*. Given two partitions P_1 and P_2 of $[a,b]$, we call $P^* = P_1 \cup P_2$ their *common refinement*.

Let f be a bounded real-valued function on a bounded closed interval $[a,b]$. Corresponding to each partition P of $[a,b]$ we put

$$M_k = \sup\{f(x) : x_{k-1} \le x \le x_k\},$$

$$m_k = \inf\{f(x) : x_{k-1} \le x \le x_k\},$$

and define the *upper* and the *lower Darboux sums of* f *relative to* P, respectively, by

$$U(p,f) = \sum_{k=1}^{n} M_k \, \Delta x_k \quad \text{and} \quad L(P,f) = \sum_{k=1}^{n} m_k \, \Delta x_k;$$

finally we put

$$\overline{\int_a^b} f(x) \, dx = \inf U(P,f), \tag{E.9}$$

$$\underline{\int_a^b} f(x) \, dx = \sup L(P,f), \tag{E.10}$$

where the infimum and the supremum are taken over all partitions P of $[a,b]$. The left-hand members of (E.9) and (E.10) are called the *upper* and the *lower Riemann integrals of* f *over* $[a,b]$, respectively. If the upper and lower Riemann integrals are equal, we say that f is *Riemann integrable on* $[a,b]$ and we denote the common value of (E.9) and (E.10) by the symbol

$$\int_a^b f(x) \, dx.$$

To see that the upper and lower Riemann integrals exist for every bounded function f on a closed bounded interval [a,b], we observe that the numbers L(P,f) and U(P,f) form a bounded set. Indeed, since f is bounded, there are two numbers, m and M, such that $m \le f(x) \le M$ for $a \le x \le b$; hence, for any partition P, we have

$$m(b - a) \le L(P,f) \le U(P,f) \le M(b - a).$$

Evidently,

$$\underline{\int_a^b} f(x)\, dx \le \overline{\int_a^b} f(x)\, dx$$

and if P* is a refinement of P, then

$$L(P,f) \le L(P^*,f) \quad \text{and} \quad U(P^*,f) \le U(P,f).$$

Show that f is Riemann integrable on [a,b] if and only if for any $\varepsilon > 0$ there is a partition P such that

$$U(P,f) - L(P,f) < \varepsilon. \tag{E.11}$$

Solution. For any P we have

$$L(P,f) \le \underline{\int_a^b} f(x)\, dx \le \overline{\int_a^b} f(x)\, dx \le U(P,f).$$

Thus (E.11) implies

$$0 \le \overline{\int_a^b} f(x)\, dx - \underline{\int_a^b} f(x)\, dx < \varepsilon.$$

Consequently, if (E.11) holds for any $\varepsilon > 0$, then

$$\overline{\int_a^b} f(x)\, dx = \underline{\int_a^b} f(x)\, dx,$$

implying that f is Riemann integrable on [a,b].

Conversely, suppose that f is Riemann integrable on [a,b], and let $\varepsilon > 0$ be given. Then there are partitions P_1 and P_2 of [a,b] such that

$$U(P_2,f) - \int_a^b f(x)\, dx < \frac{\varepsilon}{2} \tag{E.12}$$

and

$$\int_a^b f(x)\, dx - L(P_1,f) < \frac{\varepsilon}{2}. \tag{E.13}$$

Let P be the common refinement of P_1 and P_2. Then, by (E.12) and (E.13), we get

$$U(P,f) \leq U(P_2,f) < \int_a^b f(x)\,dx + \frac{\varepsilon}{2} < L(P_1,f) + \varepsilon \leq L(P,f) + \varepsilon;$$

thus (E.11) holds for the partition $P = P_1 \cup P_2$.

PROBLEM 12, Let f be a bounded real-valued function defined on an interval J; we define

$$\omega(f;J) = \sup\{f(x): x \in J\} - \inf\{f(x): x \in J\}$$

and call $\omega(f;J)$ the *oscillation of* f *over* J.

Let σ and ω be arbitrary positive numbers and f be a bounded function on an interval [a,b] of finite length. Show that f is Riemann integrable on [a,b] if and only if there is a mode of division of [a,b] into subintervals such that the sum of the lengths of the subintervals in which the oscillation of f is greater than or equal to ω is less than σ.

Solution. Let $P = \{a = x_0 < x_1 < x_2 < \cdots < x_n = b\}$ be a partition of [a,b] and consider the sum

$$Z(P,f) = \sum_{k=1}^{n} \omega_k \, \Delta x_k,$$

where $\Delta x_k = x_k - x_{k-1}$ and $\omega_k = M_k - m_k$ with

$$M_k = \sup\{f(x): x \in [x_{k-1},x_k]\}, \quad m_k = \inf\{f(x): x \in [x_{k-1},x_k]\}$$

(that is, ω_k is the oscillation of f in $[x_{k-1},x_k]$). We let

$$\Omega = M - m,$$

where

$$M = \sup\{f(x): x \in [a,b]\}, \quad m = \inf\{f(x): x \in [a,b]\},$$

and denote the length of the interval [a,b] by L.

We now derive bounds for Z(P,f); incidentally, Z(P,f) equals U(P,f) - L(P,f). Let δ be the sum of the lengths of the subintervals obtained by the partition P in which the oscillation of f is greater than or equal to ω. Then

$$Z(P,f) \geq \delta\omega.$$

But in these subintervals the oscillation of f is less than or equal to Ω and in the remaining subintervals (the sum of whose lengths is $L - \delta$) the oscillation of f is less than ω. Thus

$$Z(P,f) \leq \delta\Omega + (L - \delta)\omega.$$

Since $L - \delta \leq L$, we see that

$$\delta\omega \leq Z(P,f) \leq \delta\Omega + L\Omega.$$

If f is Riemann integrable on [a,b], then by Problem 11 there exists a partition P such that for any preassigned positive numbers σ and ω we have

$$Z(P,f) < \omega\sigma.$$

But then $\delta\omega < \omega\sigma$, that is, $\delta < \sigma$.

Conversely, if there exists a partition P for which $\delta < \sigma$, we choose

$$\omega = \frac{\varepsilon}{2L} \quad \text{and} \quad \delta = \frac{\varepsilon}{2\Omega}.$$

Then $Z(P,f) \leq \delta\Omega + L\Omega < \varepsilon/2 + \varepsilon/2 = \varepsilon$.

PROBLEM 13. Let f be Riemann integrable on [a,b]. Show that

$$\lim_{p\to\infty} \int_a^b f(x) \begin{matrix}\sin px \\ \cos px\end{matrix} \, dx = 0.$$

Solution. For any bounded interval $[\alpha,\beta]$ we have

$$\left|\int_\alpha^\beta \sin px \, dx\right| = \left|\frac{\cos p\alpha - \cos p\beta}{p}\right| \leq \frac{2}{p}.$$

Let $|f(x)| \leq A$ on [a,b] and be the usual arbitrary positive number. There is a partition of [a,b], say

$$a = x_0 < x_1 < \cdots < x_n = b,$$

such that $U(P,f) - L(P,f) < \varepsilon/2$. Thus

$$\left|\int_a^b f(x) \sin px \, dx\right| = \left|\sum_{k=1}^{n} \int_a^b [f(x_k) + f(x) - f(x_k)] \sin px \, dx\right|$$

$$\leq \sum_{k=1}^{n} \left\{ |f(x_k)| \left|\int_{x_{k-1}}^{x_k} \sin px \, dx\right| + \int_{x_{k-1}}^{x_k} |f(x) - f(x_k)||\sin px| \, dx \right\}$$

$$< \frac{2nA}{p} + [U(P,f) - L(P,f)] < \frac{2nA}{p} + \frac{\varepsilon}{2} < \varepsilon$$

when $p > (4nA)/\varepsilon$. Hence $\int_a^b f(x) \sin px\, dx$ tends to 0 as $p \to \infty$. In the same way, $\int_a^b f(x) \cos px\, dx$ tends to 0 as $p \to \infty$.

PROBLEM 14. Every rational number x can be written in the form $x = p/q$, where $q > 0$, and p and q are integers without any common divisors. When $x = 0$, we take $q = 1$. Consider the function on the interval $[0,1]$ defined by

$$f(x) = 0 \quad \text{for } x \text{ irrational,}$$

$$= \frac{1}{q} \quad \text{for } x = \frac{p}{q}.$$

Then f is continuous at every irrational point of (0,1) and discontinuous at every rational point of (0,1).

Indeed, let x_0 be any point of (0,1). Given $\varepsilon > 0$, there is only a finite number of positive integers q that are not larger than $1/\varepsilon$ and this means that in (0,1) there are only finitely many rational points p/q for which $f(p/q) = 1/q \geq \varepsilon$. Thus one may construct around the point x_0 a neighborhood $(x_0 - \delta, x_0 + \delta)$ with $\delta > 0$ such that in this neighborhood there is no point x for which $f(x) \geq \varepsilon$ (except possibly the point x_0 itself). Thus, if $0 < |x - x_0| < \delta$, then for both rational and irrational x we have $|f(x)| < \varepsilon$. Letting, for $h > 0$,

$$f(x_0+) = \lim_{h\to\infty} f(x_0 + h) \quad \text{and} \quad f(x_0-) = \lim_{h\to\infty} f(x_0 - h),$$

we get

$$f(x_0+) = f(x_0-) = 0$$

for every point x_0. If x_0 is irrational, then $f(x_0) = 0$, that is, f is continuous at x_0; if x_0 is rational, then $f(x_0) \neq 0$, that is, f is discontinuous at x_0.

Show that f is Riemann integrable on $[0,1]$.

Solution. Let $[0,1]$ be partitioned into subintervals of length $\Delta x_k \leq \lambda$ by a partition P with $|P| = \lambda$. We pick an arbitrary positive integer I. The subintervals fall into two classes:

1. To the first class belong those intervals that contain the points p/q with q at most equal to I; since there are only finitely many such points

p/q, say $k = k_I$, there are only at most $2k_I$ such intervals; their total length does not exceed $2k_I\lambda$.

2. To the second class belong those intervals that do not contain the points mentioned in class 1; over these intervals the oscillation $M_k - m_k$ of the function f is clearly smaller than $1/I$.

We now observe that

$$U(P,f) - L(P,f) = \sum_{k=1}^{n} (M_k - m_k)\Delta x_k < 2k_I\lambda + \frac{1}{I}.$$

Taking $I > 2/\varepsilon$ and then $\lambda < \varepsilon/(4k_I)$, we get

$$\sum_{k=1}^{n} (M_k - m_k)\Delta x_k < \varepsilon.$$

Thus f is Riemann integrable on [0,1] by Problem 11.

Remarks. Noting that there is only a finite number of points at which the functional value of f exceeds an assigned positive real number, it is clear that $\int_0^1 f(x)\, dx = 0$.

We can also see that if

$$g(y) = 1 \quad \text{for } 0 < y \le 1,$$
$$= 0 \quad \text{for } y = 0,$$

then the function

$$h(x) = g[f(x)]$$

for $x \in [0,1]$ is not Riemann integrable over [0,1], because $h(x) = 1$ for rational x and $h(x) = 0$ for irrational x; however, the upper Riemann integral of h over [0,1] equals 1 and the lower Riemann integral of h over [0,1] is equal to 0.

PROBLEM 15. Let f be Riemann integrable on [a,b] and g be a positive-valued, bounded, and nonincreasing function in [a,b]. We denote by $g(a+)$ and $g(b-)$ the limits

$$\lim_{h\to 0} g(a + h) \quad \text{and} \quad \lim_{h\to 0} g(b - h),$$

respectively, where $h > 0$. Show that

$$\int_a^b f(x)\, g(x)\, dx = g(a+) \int_a^t f(x)\, dx \quad \text{for some } t \in (a,b).$$

If g is positive-valued and nondecreasing, the corresponding formula is

$$\int_a^b f(x)\ g(x)\ dx = g(b-) \int_t^b f(x)\ dx,$$

where $a < t < b$.

If g is any monotonic function, there is a number t between a and b such that

$$\int_a^b f(x)\ g(x)\ dx = g(a+) \int_a^t f(x)\ dx + g(b-) \int_t^b f(x)\ dx.$$

Solution. Let ε be a positive number less than $g(a+) - g(b-)$. Then there is a point x_1 such that

$$\begin{aligned} g(a+) - g(x) &< \varepsilon \quad (a < x < x_1) \\ &\geq \varepsilon \quad (x > x_1). \end{aligned}$$

Similarly there are points x_2, x_3, ... such that

$$\begin{aligned} g(x_{k-1}+) - g(x) &< \varepsilon \quad (x_{k-1} < x < x_k) \\ &\geq \varepsilon \quad (x > x_k), \end{aligned}$$

so long as $g(x_{k-1}+) - g(b-) > \varepsilon$. Otherwise we take $x_n = b$. The point b is thus reached in a finite number of steps, since the variation of g in each interval (x_{k-1}, x_k) is at least ε.

Let $h(x) = g(x_k+)$ in each interval $x_k \leq x \leq x_{k+1}$. Then

$$0 \leq h(x) - g(x) < \varepsilon$$

except possibly at the points $a = x_0, x_1, x_2, \ldots, b$, and

$$\int_a^b h(x)\ f(x)\ dx = \sum_{k=0}^{n-1} g(x_k+) \int_{x_k}^{x_{k+1}} f(x)\ dx.$$

Let $F(x) = \int_a^x f(s)\ ds$; then, if m and M are the lower and upper bounds of F, it follows from Abel's Inequality (see Problem 58 in Chapter 2) that

$$m\ g(a+) \leq \int_a^b h(x)\ f(x)\ dx \leq M\ g(a+).$$

But

$$\left|\int_a^b h(x)\ f(x)\ dx - \int_a^b g(x)\ f(x)\ dx\right| \le \varepsilon \int_a^b |f(x)|\ dx,$$

which tends to zero with ε. Hence, making $\varepsilon \to 0$, it follows that

$$m\ g(a+) \le \int_a^b g(x)\ f(x)\ dx \le M\ g(a+).$$

Since F is continuous, it takes every value between m and M, and so, at every $x = t$, say, the value

$$\frac{1}{g(a+)} \int_a^b g(x)\ f(x)\ dx.$$

The rest follows easily.

PROBLEM 16. Let $E_1, E_2, \ldots, E_n$ be n intervals which are situated in the unit interval [0,1]. If each point of [0,1] belongs to at least q of these intervals E_j with $j = 1, \ldots, n$, show that at least one of these intervals must have length $\ge q/n$.

Solution. For $x \in [0,1]$, define $f_j(x) = 1$ if $x \in E_j$ and $f_j(x) = 0$ if $x \notin E_j$; then let

$$f(x) = \sum_{j=1}^{n} f_j(x).$$

Evidently, $f(x) \ge q$ for every x in the interval [0,1] and so

$$q \le \int_0^1 f(x)\ dx = \int_0^1 \sum_{j=1}^{n} f_j(x)\ dx = \sum_{j=1}^{n} \int_0^1 f_j(x)\ dx = \sum_{j=1}^{n} |E_j|,$$

where $|E_j|$ denotes the length of the interval E_j. It is clear that not every summand in the last sum can be less than q/n, for if it were then we would have $q < n(q/n)$.

PROBLEM 17. Let f be a continuous function in the interval [0,1] and suppose that

$$\int_0^1 f(x)\ dx = 0, \quad \int_0^1 x\ f(x)\ dx = 0, \quad \ldots,$$

$$\int_0^1 x^{n-1} f(x)\, dx = 0, \qquad \int_0^1 x^n f(x)\, dx = 1.$$

Show that $|f(x)| \geq 2^n(n + 1)$ on some part of $[0,1]$.

Solution. The conditions imply

$$\int_0^1 (x - \tfrac{1}{2})^n f(x)\, dx = 1.$$

Suppose $|f(x)| < 2^n(n + 1)$ everywhere on $[0,1]$. Then

$$1 = \int_0^1 (x - \tfrac{1}{2})^n f(x)\, dx < 2^n(n + 1) \int_0^1 |x - \tfrac{1}{2}|^n\, dx = 1.$$

But this is a contradiction.

PROBLEM 18. Consider a polynomial $f(x)$ with real coefficients having the property

$$f[g(x)] = g[f(x)]$$

for every polynomial $g(x)$ with real coefficients. Show that $f(x) = x$.

Solution. Let $g(x) = x + h$ so that $f(x + h) = f(x) + h$ and

$$\frac{f(x + h) - f(x)}{h} = 1$$

for all $h \neq 0$. Therefore $f'(x) = 1$ and $f(x) = x + c$. Letting $g(x)$ be the zero polynomial shows that it is necessary that $c = 0$ and thus $f(x) = x$. It is easily seen that $f(x) = x$ is sufficient.

PROBLEM 19. Show that there exists a real-valued, everywhere differentiable function H on the real line R^1 such that H is monotone on no subinterval of R^1 and H' is bounded. In other words, a function may be everywhere oscillating and still have a finite derivative at every point.

Solution. We divide the proof into seven steps.

I. Let r and s be real numbers.

(i) If $r > s > 0$, then $(r - s)/(r^2 - s^2) < 2/r$.

(ii) If $r > 1$ and $s > 1$, then $(r + s - 2)/(r^2 + s^2 - 2) < 2/s$.

Proof. Assertion (i) is obvious. Inequality (ii) is equivalent to

$$(r - s)^2 + (r - 1)(s - 1) + r^2 + r + 3s > 5.$$

But this too is obvious when $r > 1$ and $s > 1$.

II. Let $v(x) = (1 + \ x\ )^{-\frac{1}{2}}$ for $x \in R^1$. Then

$$\frac{1}{b - a}\int_a^b v(x)\ dx < 4 \min\{v(a), v(b)\}$$

whenever a and b are distinct real numbers.

Proof. We may suppose that $a < b$. In case $0 \le a$, we have by step 1 that

$$\frac{1}{b - a}\int_a^b v(x)\ dx = \frac{2(\sqrt{1 + b} - \sqrt{1 + a})}{(1 + b) - (1 + a)} < \frac{4}{\sqrt{1 + b}} = 4 \min\{v(a), v(b)\}.$$

Since $v(-x) = v(x)$, the case $b \le 0$ needs no special consideration. We therefore suppose that $a < 0 < b$. Then step I yields

$$\frac{1}{b - a}\int_a^b v(x)\ dx = \frac{2(\sqrt{1 + b} + \sqrt{1 - a} - 2)}{(1 + b) + (1 - a) - 2} < 4 \min\{v(a), v(b)\}.$$

III. If v is as in step II and w is any function of the form

$$w(x) = \sum_{j=1}^{n} c_j\ v[\lambda_j(x - \alpha_j)],$$

where $c_1, \ldots, c_n$ and $\lambda_1, \ldots, \lambda_n$ are positive real numbers and $\alpha_1, \ldots, \alpha_n$ are any real numbers, then

$$\frac{1}{b - a}\int_a^b w(x)\ dx < 4 \min\{w(a), w(b)\}.$$

Proof. This follows directly from step II and the fact that

$$\frac{1}{b - a}\int_a^b v[\lambda(x - \alpha)]\ dx = \frac{1}{\lambda(b - \alpha) - \lambda(a - \alpha)}\int_{\lambda(a-\alpha)}^{\lambda(b-\alpha)} v(t)\ dt.$$

IV. Let $(w_n)_{n=1}^{\infty}$ be any sequence of functions as in step III. For $x \in R^1$ and each n define

$$W_n(x) = \int_0^x w_n(t)\ dt.$$

Let $\Sigma_{n=1}^{\infty} w_n(a) = s < \infty$ for some $a \in R^1$. Then the series

$$F(x) = \sum_{n=1}^{\infty} W_n(x)$$

converges uniformly on every bounded subset of R^1, the function F is differentiable at a, and $F'(a) = s$. In particular, if

$$\sum_{n=1}^{\infty} w_n(t) = f(t) < \infty$$

for all $t \in R^1$, then F is differentiable everywhere on R^1 and $F' = f$.

Proof. Let $b \in R^1$ satisfy $b \geq |a|$. Then, using step III, $-b \leq x \leq b$ implies

$$|W_n(x)| \leq \left|\int_0^a w_n(t)\,dt\right| + \left|\int_a^x w_n(t)\,dt\right|$$

$$\leq 4\,|a|\,w_n(a) + 4|x - a|\,w_n(a) \leq 12\,w_n(a).$$

Thus, uniform convergence on [-b,b] follows from the M-Test of Weierstrass.

To show that $F'(a) = s$, let $\varepsilon > 0$ be given. Choose n_0 such that

$$10 \sum_{n=n_0+1}^{\infty} w_n(a) < \varepsilon.$$

Since each w_n is continuous at a, there exists some $\delta > 0$ such that

$$\left|\frac{1}{h}\int_a^{a+h} w_n(t)\,dt - w_n(a)\right| < \frac{\varepsilon}{2n_0}$$

whenever $0 < |h| < \delta$ and $1 \leq n \leq n_0$. Therefore, using step III again, $0 < |h| < \delta$ implies that

$$\left|\frac{F(a+h) - F(a)}{h} - s\right| = \left|\sum_{n=1}^{\infty} \frac{1}{h}\int_a^{a+h} w_n(t)\,dt - w_n(a)\right|$$

$$\leq \sum_{n=1}^{n_0}\left|\frac{1}{h}\int_a^{a+h} w_n(t)\,dt - w_n(a)\right| + \sum_{n=n_0+1}^{\infty}\left|\frac{1}{h}\int_a^{a+h} w_n(t)\,dt - w_n(a)\right|$$

$$< \frac{\varepsilon}{2} + \sum_{n=n_0+1}^{\infty} 5\,w_n(a) < \varepsilon.$$

V. Let $I_1, \ldots, I_n$ be disjoint open intervals, let α_j be the midpoint of I_j, and let ε and $y_1, \ldots, y_n$ be positive real numbers. Then there is a function w as in step III such that for each j

(i) $w(\alpha_j) > y_j$,

(ii) $w(x) < y_j + \varepsilon$ if $x \in I_j$,

(iii) $w(x) < \varepsilon$ if $x \notin I_1 \cup \cdots \cup I_n$.

Proof. Choose $c_j = y_j + \varepsilon/2$ and define $v_j(x)$ to be

$$v_j(x) = c_j\, v[\lambda_j(x - \alpha_j)],$$

where λ_j is chosen so large that $v_j(x) < \varepsilon/2n$ if $x \notin I_j$ (one needs only to check this inequality at an endpoint of I_j). Take

$$w = v_1 + \cdots + v_n.$$

Since the I_j's are disjoint and because v_j takes its largest value c_j at α_j, properties (i), (ii), and (iii) are clear.

VI. Let $\{\alpha_j\}_{j=1}^{\infty}$ and $\{\beta_j\}_{j=1}^{\infty}$ be disjoint countable subsets of R^1. Then there exists a real-valued, everywhere differentiable function F on R^1 satisfying $F'(\alpha_j) = 1$, $F'(\beta_j) < 1$ for all j, and $0 < F'(x) \le 1$ for all x.

Proof. We obtain F as in step IV by first constructing

$$F' = f = \sum_{n=1}^{\infty} w_n,$$

or, more precisely, the partial sums

$$f_n = \sum_{k=1}^{n} w_k,$$

in such a way that

$$f_n(\alpha_j) > 1 - \frac{1}{n} \qquad (1 \le j \le n), \tag{A_n}$$

$$f_n(x) < 1 - \frac{1}{n+1} \qquad (x \in R^1), \tag{B_n}$$

$$w_n(\beta_j) < \frac{1}{2n\cdot 2^n} \qquad (1 \le j \le n). \tag{C_n}$$

Suppose that this were done we would have

$$F'(\alpha_j) = \lim_{n \to \infty} f_n(\alpha_j) = 1,$$

$$0 < F'(x) = \lim_{n \to \infty} f_n(x) \le 1,$$

and, picking $n > j$,

$$F'(\beta_j) = f_{n-1}(\beta_j) + \sum_{k=n}^{\infty} w_k(\beta_j) < 1 - \frac{1}{n} + \sum_{k=n}^{\infty} \frac{1}{2k \cdot 2^k}$$

$$< 1 - \frac{1}{n} + \frac{1}{2n} \cdot 1 = 1 - \frac{1}{2n} < 1$$

and thus we would have the desired F.

We proceed inductively. First choose an open interval I with midpoint α_1 such that $\beta_1 \notin I$. Then apply step V with $\varepsilon = y_1 = 1/4$ to obtain $f_1 = w_1$ that satisfies (A_1), (B_1), and (C_1).

Suppose that $n > 1$ and that f_{n-1} and w_{n-1} have been chosen which satisfy (A_{n-1}), (B_{n-1}), and (C_{n-1}). Select disjoint open intervals $I_1, \ldots, I_n$ such that, for each $j \in \{1,\ldots,n\}$, α_j is the midpoint of I_j, I_j being disjoint from $\{\beta_1, \ldots, \beta_n\}$ and

$$f_{n-1}(x) < f_{n-1}(\alpha_j) + \delta$$

for $x \in I_j$, where

$$\delta = \frac{1}{n(n+1)} - \frac{1}{2n \cdot 2^n} > 0.$$

We now apply step V with

$$\varepsilon = \frac{1}{2n \cdot 2^n} \quad \text{and} \quad y_j = 1 - \frac{1}{n} - f_{n-1}(\alpha_j) \quad \text{when } 1 \le j \le n,$$

to obtain w_n. It is clear that (C_n) is satisfied. Also

$$f_n(\alpha_j) = f_{n-1}(\alpha_j) + w_n(\alpha_j) > f_{n-1}(\alpha_j) + y_j = 1 - \frac{1}{n}$$

when $1 \le j \le n$, and so (A_n) is satisfied. To check (B_n), observe that if $x \in I_j$, then

$$f_n(x) = f_{n-1}(x) + w_n(x) < f_{n-1}(\alpha_j) + \delta + y_j + \varepsilon$$

$$= 1 - \frac{1}{n} + \frac{1}{n(n+1)} = 1 - \frac{1}{n+1};$$

while if $x \notin \cup_{j=1}^{n} I_j$, then

$$f_n(x) = f_{n-1}(x) + w_n(x) < 1 - \frac{1}{n} + \varepsilon < 1 - \frac{1}{n+1}.$$

VII. *Proof of the Claim in Problem 19:* Let $\{\alpha_j\}_{j=1}^{\infty}$ and $\{\beta_j\}_{j=1}^{\infty}$ be disjoint subsets of R^1. Apply step VI to obtain everywhere differentiable functions F and G on R^1 such that

$$F'(\alpha_j) = G'(\beta_j) = 1, \quad G'(\alpha_j) < 1, \quad F'(\beta_j) < 1,$$

$$0 < F'(x) \leq 1, \quad 0 < G'(x) \leq 1$$

for all j and x. Now put

$$H = F - G.$$

Then

$$H'(\alpha_j) > 0, \quad H'(\beta_j) < 0, \quad -1 < H'(x) < 1$$

for all j and x. Since $\{\alpha_j\}_{j=1}^{\infty}$ and $\{\beta_j\}_{j=1}^{\infty}$ are both dense in R^1, H cannot be monotone on an interval.

PROBLEM 20. A set G of real numbers is said to be *open* if for every x in G there is an open interval $I = (x - \delta, x + \delta)$ with $\delta > 0$ such that I is contained in G; I is called a *neighborhood of* x.

Show that any open set of real numbers is the union of a countable collection of disjoint open intervals. This representation is unique up to the order in which the terms appear in the union.

Solution. If G is an open set of real numbers, we say that the real numbers x and y are "equivalent" and write $x \sim y$ if and only if there is an open interval (a,b) contained in G such that x and y are points of (a,b). The relation $\sim$ is clearly an equivalence relation on G and the resulting equivalence classes are disjoint open intervals whose union is G. The fact that there can be only countably many follows since each must contain a distinct rational number.

Finally, if there were two distinct representations of G in terms of countable unions of disjoint open intervals, then there would exist a point x in G which would belong to a component interval I in the first representation and to a component interval $J \neq I$ in the second representation. But then one of these component intervals, e.g., J, would extend beyond the other; it would follow from this that one of the endpoints of I belongs to J, which is impossible, inasmuch as the endpoints of I do not belong to G.

PROBLEM 21. A real number x is called a *limit point of a set* F (of real numbers) if every neighborhood of x contains a point $y \neq x$ such that y belongs to F; F is said to be a *closed set* if every limit point of F is a point of F.

Let f be a real-valued function defined on the real line R^1. Let J be any bounded open interval in R^1. We define $\omega(f;J)$, that is, the oscillation of f over J, as in Problem 12; for $a \in R^1$, we put

$$\omega(f;a) = \inf \omega(f;J),$$

where the infimum is taken over all bounded intervals J containing the point a.

One can easily see that if f is continuous at $a \in R^1$, then $\omega(f;a) = 0$ and if f is not continuous at $a \in R^1$, then $\omega(f;a) > 0$.

Suppose that $r > 0$ and let E_r be the set of all $a \in R^1$ such that $\omega(f;a) \geq 1/r$. Show that E_r is a closed set.

Solution. Let x be any limit point of E_r. We have to show that x belongs to E_r, that is, we must verify that $\omega(f;x) \geq 1/r$. To do this, it is enough to show that if J is a bounded open interval containing x, then $\omega(f;J) \geq 1/r$, since $\omega(f;x)$ is the infimum of such $\omega(f;J)$. Since the open interval J contains the limit point x of E_r, J must contain a point y of E_r. But then $\omega(f;J) \geq \omega(f;y) \geq 1/r$ and we have what we set out to do.

Remark. Let D denote the set of all points x at which f is discontinuous. Then, letting $E_{1/n} = \{a \in R^1 : \omega(f;a) \geq 1/n\}$,

$$D = \bigcup_{n=1}^{\infty} E_{1/n}.$$

Indeed, if $x \in D$, then $\omega(f;x) > 0$. For some positive integer n we must have $\omega(f;x) \geq 1/n$. This shows that

$$D \subset \bigcup_{n=1}^{\infty} E_{1/n}.$$

Conversely, if

$$x \in \bigcup_{n=1}^{\infty} E_{1/n},$$

then $\omega(f;x) > 0$ and so $x \in D$. Hence

$$D = \bigcup_{n=1}^{\infty} E_{1/n}$$

and D is seen to be a countable union of closed sets.

PROBLEM 22. A subset E of the real line R^1 is said to have *measure zero* if, for any $\varepsilon > 0$, there is a countable set $\{I_k\}_{k=1}^{\infty}$ of intervals such that

$$\bigcup_{k=1}^{\infty} I_k \supset E \quad \text{and} \quad \sum_{k=1}^{\infty} |I_k| < \varepsilon,$$

where $|I_k|$ denotes the length of the interval I_k.

Clearly, any countable set $\{t_n: n = 1, 2, \ldots\}$ has measure zero, the point t_n can be enclosed in an interval of length $\varepsilon/2^n$, and

$$\sum_{n=1}^{\infty} \frac{\varepsilon}{2^n} = \varepsilon.$$

Show that if

$$E = \bigcup_{n=1}^{\infty} E_n,$$

where each E_n has measure zero, then E has measure zero.

Solution. Given $\varepsilon > 0$, the set E_1 can be enclosed in the union

$$\bigcup_{k=1}^{\infty} I_{1,k},$$

where

$$\sum_{k=1}^{\infty} |I_{1,k}| \le \frac{\varepsilon}{2},$$

and in general, the set E_m can be enclosed in

$$\bigcup_{k=1}^{\infty} I_{m,k},$$

where

$$\sum_{k=1}^{\infty} |I_{m,k}| \le \frac{\varepsilon}{2^m}.$$

The intervals $I_{m,k}$ with m, k = 1,2,... then satisfy

$$\bigcup_{m,k=1}^{\infty} I_{m,k} \supset \bigcup_{m=1}^{\infty} E_m = E, \quad \text{and}$$

$$\sum_{m,k=1}^{\infty} |I_{m,k}| \le \frac{\varepsilon}{2} + \frac{\varepsilon}{2^2} + \cdots + \frac{\varepsilon}{2^m} + \cdots = \varepsilon.$$

PROBLEM 23. The *Cantor Set* C consists of all numbers in the interval [0,1] that admit a ternary development in which the digit 1 does not appear. Show that C has measure zero and is an uncountable set.

Solution. The set C can be constructed by deleting the open middle third of the interval [0,1], then deleting the open middle thirds of each of the intervals [0,1/3] and [2/3,1], and so on. If F_n denotes the union of the 2^n closed intervals of length $1/3^n$ which remain at the n-th stage, then

$$C = \bigcap_{n=1}^{\infty} F_n.$$

F_n (and therefore C) contains no interval of length larger than $1/3^n$. The sum of the lengths of the intervals that compose F_n is $(2/3)^n$, which is less than ε if n is taken sufficiently large. Hence C has measure zero. Finally, each number x in (0,1] has a unique non-terminating binary development

$$x = 0.x_1x_2x_3\ldots$$

If $y_i = 2x_i$, then $0.y_1y_2y_3\ldots$ is the ternary development with $y_i \ne 1$ of some point y of C. This correspondence between x and y, extended by mapping 0 onto itself, defines a one-to-one map of [0,1] onto a (proper) subset of C. It follows that C is uncountable; it has cardinality c (the power of the continuum).

PROBLEM 24. Let C be the Cantor set (see Problem 23). Show that any subinterval $[a,b]$ of $[0,1]$ contains an interval (a',b') free of points of C such that its length satisfies the inequality

$$b' - a' \geq \frac{1}{5}(b - a).$$

Solution. If $[a,b]$ is free of points of C, then the claim is trivially true. If not, then let V be one of those closed intervals of length 3^{-n} that remained at the n-th step of the removal of the open middle thirds such that $V \subset [a,b]$ but with n being as small as possible. If $V = [a,b]$, then the claim is true, for at the $(n + 1)$-th step we remove the open middle third of V. In the other case, that is, in the case when V is a proper subset of $[a,b]$, there are adjoining V (possibly on the left and on the right) intervals of the same length as V which are free of points of C. If one of these (two) intervals has a subinterval of length $\geq (1/5)(b - a)$ in common with (a,b), the claim is once again fulfilled. If this, however, is not the case, then the length of V is $\geq (3/5)(b - a)$. Since the middle third (a',b') of V is free of points of C and $b' - a' \geq (1/5)(b - a)$ holds, the claim follows in general.

PROBLEM 25. Let C be the Cantor set (see Problem 23). We arrange the complementary intervals into groups as follows: The first group contains the interval $(1/3,2/3)$, the second the two intervals $(1/9,2/9)$ and $(7/9,8/9)$, the third group the four intervals $(1/27,2/27)$, $(7/27,8/27)$, $(19/27,20/27)$ and $(25/27,26/27)$, etc. The n-th group contains then 2^{n-1} intervals.

We define the *Cantor Function* g as follows:

$$\begin{aligned} g(x) &= 1/2 \quad \text{for } x \in (1/3,2/3) \\ &= 1/4 \quad \text{for } x \in (1/9,2/9) \\ &= 3/4 \quad \text{for } x \in (7/9,8/9) \end{aligned}$$

In the four intervals of the third group we set the function g consecutively equal to 1/8, 3/8, 5/8, and 7/8. In the 2^{n-1} intervals of the n-th group we set g consecutively equal to

$$\frac{1}{2^n}, \quad \frac{3}{2^n}, \quad \frac{5}{2^n}, \quad \ldots, \quad \frac{2^n - 1}{2^n}.$$

The function g is in this way defined on the open set $[0,1] - C$ and is seen to be constant on each component interval and is also seen to be nondecreasing on $[0,1] - C$.

We extend the domain of definition of g to all of $[0,1]$ by putting

$$g(0) = 0, \quad g(1) = 1,$$

and

$$g(x_0) = \sup\{g(x): x \in [0,1] - C, x < x_0\}.$$

We see that g is now defined on all of $[0,1]$ and is nondecreasing throughout $[0,1]$.

Show that the Cantor function g is continuous on $[0,1]$.

Solution. The claim follows from the fact that the set of values of the function g on the set $[0,1] - C$ is dense in $[0,1]$, that is, every subinterval of $[0,1]$ contains at least one point of the set of values of the function g on $[0,1] - C$. Indeed, since g is a nondecreasing function, any point of discontinuity of g must be a simple jump; hence, if g is discontinuous at x_0, then at least one of the intervals

$$(g(x_0 - 0), g(x_0)) \quad \text{and} \quad (g(x_0), g(x_0 + 0)),$$

where $g(x_0 - 0)$ denotes the limit from the left and $g(x_0 + 0)$ the limit from the right of $g(x)$ at $x = x_0$, must be free of values of g.

PROBLEM 26. Let $P(x)$ denote a statement concerning the point x of the interval $[a,b]$. We say that $P(x)$ holds *almost everywhere* on $[a,b]$ if $P(x)$ holds for every point of $[a,b]$ except for a subset of points of $[a,b]$ having measure zero (see Problem 22).

The Cantor function g (see Problem 25) is an example of a continuous, nondecreasing function on $[0,1]$ having derivative 0 almost everywhere on $[0,1]$. Moreover, $g(0) = 0$ and $g(1) = 1$.

Give an example of a strictly increasing continuous function on the interval $[0,1]$ whose derivative is zero almost everywhere on $[0,1]$.

Solution. Let g denote the Cantor function and let us set $g(x) = 0$ for $x < 0$ and $g(x) = 1$ for $x > 1$. If we have the interval $[a,b]$, we define the corresponding Cantor function in a similar way, namely, we set it equal to

$$g\left(\frac{x - a}{b - a}\right)$$

for $a \le x \le b$.

Now, let I_1, I_2, ... be the intervals

$$[0,1], \quad [0,1/2], \quad [1/2,1], \quad [0,1/4], \quad \ldots, \quad [0,1/8], \quad \ldots$$

and let g_n be the Cantor function corresponding to I_n. Then

$$f(x) = \sum_{n=1}^{\infty} \frac{g_n(x)}{2^n}$$

is continuous and strictly increasing on $[0,1]$; moreover, $f'(x) = 0$ almost everywhere on $[0,1]$.

PROBLEM 27. Prove the following characterization of the Riemann integral, due to Lebesgue: A bounded function f on $[a,b]$ is Riemann integrable if and only if the set D of discontinuities of f has measure zero.

Solution. Let $E_n = \{t \in [a,b]: \omega(t) \ge 1/n\}$, where $\omega(t) = \omega(f;t)$ with $\omega(f;t)$ as in Problem 21. Clearly, f is continuous at t if and only if $\omega(t) = 0$. The set E_n is closed (by Problem 21), and the set D of points where f is discontinuous equals $\cup_{n=1}^{\infty} E_n$.

Now assume that f is integrable, so that for any $\varepsilon > 0$ there is a partition P_ε such that $U(f,P_\varepsilon) - L(f,P_\varepsilon) < \varepsilon$ (see Problem 11). If

$$P = \{a = t_0 < t_1 < \cdots < t_{n-1} < t_n = b\},$$

then the set of intervals

$$I_1 = (t_0,t_1), \quad \ldots, \quad I_n = (t_{n-1},t_n)$$

can be split into two groups, where the intervals in the first group have nonempty intersection with E_m, and those of the second group do not meet E_m (m fixed). Then

$$U(f,P_\varepsilon) - L(f,P_\varepsilon) = \sum{}' (M_k - m_k)(t_k - t_{k-1}) + \sum{}'' (M_k - m_k)(t_k - t_{k-1}) < \varepsilon,$$

where the prime (resp. double prime) indicates summation over intervals of the first group (resp. second group). On the primed intervals, $M_k - m_k \ge 1/m$, so that

$$\frac{1}{m}\sum{}'(t_k - t_{k-1}) \le \sum{}'(M_k - m_k)(t_k - t_{k-1}) < \varepsilon$$

and $\Sigma'(t_k - t_{k-1}) < m\varepsilon$. This E_m can be enclosed in finitely many intervals of length less than $m\varepsilon$, and since ε is arbitrary, we see that E_m has measure zero. It now follows from Problem 22 that D has measure zero.

Conversely, assume that $D = \cup_{n=1}^{\infty} E_n$ has measure zero and, hence, that E_n has measure zero for all n. Choose K so that $1/K < \varepsilon$. The set $E_K^c = [a,b] - E_k$ is open; therefore, by Problem 20,

$$E_K^c = \bigcup_{k=1}^{\infty} I_k,$$

where the intervals $(I_k)_{k=1}^{\infty}$ are open and disjoint. We now show that

$$\sum_{k=1}^{\infty} |I_k| = b - a.$$

Suppose, on the contrary, that $\Sigma_{k=1}^{\infty} |I_k| = \lambda < b - a$. Since E_K has measure zero, we can find $(J_k')_{k=1}^{\infty}$ such that

$$\bigcup_{k=1}^{\infty} J_k' \supset E_K \quad \text{and} \quad \sum_{k=1}^{\infty} |J_k'| < \frac{b - a - \lambda}{2}.$$

The intervals $(J_k')_{k=1}^{\infty}$ and $(I_k)_{k=1}^{\infty}$ together cover [a,b], and hence

$$b - a \le \sum_{k=1}^{\infty} |J_k'| + \sum_{k=1}^{\infty} |I_k| < \frac{b - a - \lambda}{2} + \lambda = \frac{b - a + \lambda}{2} < b - a.$$

This contradiction shows that $\lambda = b - a$. We now see that, for some n,

$$\sum_{k=1}^{n} |I_k| > b - a - \frac{\varepsilon}{2}$$

and we can find closed intervals $J_1, \ldots, J_n$ so that for $j = 1, \ldots, n$,

$$I_j \subset J_j \quad \text{and} \quad \sum_{j=1}^{n} |J_j| > b - a - \varepsilon.$$

At each point t_0 of J_1 we have $\omega(t_0) < 1/K < \varepsilon$, and there is some $\delta(t_0)$ such that, on the interval $[t_0 - \delta(t_0), t_0 + \delta(t_0)]$, the supremum of f(t) minus the infimum of f(t) is less than ε. By the compactness of J_1, finitely many such intervals cover J_1. We can use the endpoints of these intervals to form a partition P_1 of J_1, and similarly, we form partitions $P_2, \ldots, P_n$ of $J_2, \ldots, J_n$, thus obtaining a partition P of [a,b] which is the common refinement of $P_1, \ldots, P_n$.

Calculating U(f,P) - L(f,P), we have

$$U(f,P) - L(f,P) = \sum (M_k - m_k)(t_k - t_{k-1})$$

$$= \left\{ \sum_{P_1} (M_k - m_k)(t_k - t_{k-1}) + \cdots + \sum_{P_n} (M_k - m_k)(t_k - t_{k-1}) \right\}$$

$$+ \sum{}' (M_k - m_k)(t_k - t_{k-1})$$

$$\le \sum_{k=1}^{n} |J_k| + \varepsilon 2M \le \varepsilon(2M + b - a),$$

where $M = \sup\{|f(t)|: t \in [a,b]\}$. It follows that f is Riemann integrable over [a,b].

PROBLEM 28. Using the result in Problem 27, show that there exists a function which is differentiable at every point of an interval and whose derivative is a bounded function, but this derivative is not Riemann integrable.

Solution. The following construction is due to V. Volterra. We remove from the middle of the closed interval [0,1] the open interval of length 1/4. From the middle of the two remaining closed intervals we remove open intervals of length 1/16. In the n-th step we remove from the middle of each of the 2^{n-1} closed intervals which remain after the (n - 1)-th step an open interval of length 4^{-n}. Continuing this process indefinitely we remove from the interval [0,1] a system of open intervals of total length

$$\frac{1}{4} + 2\cdot\frac{1}{4^2} + \cdots + 2^{n-1}\cdot\frac{1}{4^n} + \cdots = \frac{1}{2}.$$

The points which remain form a closed set E; E is clearly not of measure zero. Let d_n denote the length of each of the closed intervals which remain after the n-th step. From the construction it is clear that $d_{n+1} < d_n/2$; hence $d_n \to 0$ as $n \to \infty$. This implies that no subinterval of [0,1], no matter how small, can be contained in the set E.

Now we define the function f. At the points x of the set E we put f(x) = 0. If (α,β) is one of the removed open intervals we define

$$f(x) = (x - \alpha)^2 \sin \frac{1}{x - \alpha}$$

immediately to the right of α, and

$$f(x) = (x - \beta)^2 \sin \frac{1}{\beta - x}$$

immediately to the left of β, until we reach the maximum points $\bar{\alpha}$, $\bar{\beta}$ nearest the middle of (α,β); in the interval $(\bar{\alpha},\bar{\beta})$ we set f(x) equal to this maximum value.

We have thus defined the function f on the entire interval [0,1] and it is a continuous function.

It is clear that f is differentiable in each of the removed intervals (α,β), even at the points $\bar{\alpha}$, $\bar{\beta}$ where it has derivative zero. For x near enough to α $(x > \alpha)$ we have

$$f'(x) = 2(x - \alpha) \sin \frac{1}{x - \alpha} - \cos \frac{1}{x - \alpha};$$

if $x \to \alpha$ the first term of the right-hand side converges to 0 while the second term oscillates between the values +1 and -1. The situation is analogous in a left-hand neighborhood of β.

f'(x) exists at the points α, β and even at every point $x_0 \in E$, and we have

$$f'(x_0) = 0. \tag{E.14}$$

To show this, we assume first that $x > x_0$. If $x \in E$, then

$$f(x) - f(x_0) = 0 - 0 = 0,$$

and if x is contained in some of the removed intervals (α,β) then

$$|f(x) - f(x_0)| = |f(x)| \le (x - \alpha)^2 \le (x - x_0)^2;$$

thus in each case

$$\left|\frac{f(x) - f(x_0)}{x - x_0}\right| \le |x - x_0|.$$

We get the same result for $x < x_0$. Letting $x \to x_0$ this implies equation (E.14).

Thus $f'(x)$ exists everywhere. But at the points of E it is not continuous. Indeed, if $x_0 \in E$, then in every neighborhood of x_0 there is a point, and therefore also an endpoint, of one of the removed intervals, and we know that at such an endpoint the oscillation of the function f is equal to 2. But the set E is not of measure zero and so f' cannot be Riemann integrable according to the result in Problem 27.

PROBLEM 29. Show the following theorem: If f is differentiable in $[a,b]$ then the derivative f' is Riemann integrable over $[a,b]$ if and only if there exists a Riemann integrable function g in $[a,b]$ such that

$$f(x) = f(a) + \int_a^x g(t)\, dt.$$

Solution. We give the verification in three steps.

I. Let g be Riemann integrable and assume that $m \le g(x) \le M$ for all $x \in [a,b]$. Define, for $x \in [a,b]$,

$$G(x) = \int_a^x g(t)\, dt,$$

and assume that G is differentiable on $[a,b]$. Then we have

$$m \le G'(x) \le M$$

for all $x \in [a,b]$.

Proof. Define $K(x) = G(x) - m(x - a)$. Then K is differentiable and $K'(x) = G'(x) - m$. We also have

$$K(x) = \int_a^x g(t)\, dt - m(x - a) = \int_a^x (g(t) - m)dt.$$

Since $g(t) \ge m$ we see that K is monotonically nondecreasing. Hence $K'(x) \ge 0$, or

$$G'(x) \ge m \quad \text{for all } x \in [a,b].$$

In a similar fashion one proves

$$G'(x) \le M \quad \text{for all } x \in [a,b].$$

II. Under the same conditions as in step I we have that G' is Riemann integrable over $[a,b]$.

Proof. Step I is applicable on every closed suinterval of [a,b]. From this it follows that the oscillation of G' is not larger than the oscillation of g. According to the result in Problem 11 we may therefore conclude that G' is Riemann integrable over [a,b].

III. Proof of the theorem: It is clear that the given condition is necessary; take g = f'.

To prove the sufficiency we write

$$f(x) - f(a) = \int_a^x g(t)\ dt.$$

Then it is clear that f' is the derivative of a function of the form

$$\int_a^x g(t)\ dt.$$

According to step II we obtain that f' is Riemann integrable over [a,b], completing the solution.

Remark. Clearly we have

$$\int_a^x f'(t)\ dt = \int_a^x g(t)\ dt \qquad \text{for all } x \in [a,b].$$

However, it is easy to construct examples in which $f' \neq g$.

PROBLEM 30. A function f is said to be *absolutely continuous on* [a,b] if for any $\varepsilon > 0$ there is some $\delta > 0$ such that

$$\sum_{k=1}^{n} |f(b_k) - f(a_k)| < \varepsilon$$

for every finite, pairwise disjoint sequence

$$(a_1,b_1), \quad (a_2,b_2), \quad \ldots, \quad (a_n,b_n)$$

of open intervals of [a,b] for which

$$\sum_{k=1}^{n} (b_k - a_k) < \delta.$$

Show that the Cantor function g defined in Problem 25 is not absolutely continuous on [0,1].

Solution. First we extend the domain of definition of g as follows: we set $g(x) = 0$ for $x < 0$ and $g(x) = 1$ for $x > 1$. Next, we enclose the Cantor set C defined in Problem 23 in a union

$$\bigcup_{k=1}^{\infty} (a_k, b_k)$$

or pairwise disjoint open intervals such that $\Sigma_{k=1}^{\infty} (b_k - a_k)$ is arbitrarily small. It is easily seen that

$$\sum_{k=1}^{\infty} [g(b_k) - g(a_k)] = 1$$

and

$$\sum_{k=1}^{n} [g(b_k) - g(a_k)] \geq \frac{1}{2}$$

for sufficiently large n, while

$$\sum_{k=1}^{n} (b_k - a_k)$$

is arbitrarily small.

PROBLEM 31. Give an example of a differentiable function which takes rational numbers into rational numbers but whose derivative takes rational numbers into irrational numbers.

Solution. Let

$$f(x) = \sum_{n=0}^{\infty} \frac{g(n!x)}{(n!)^2}, \tag{E.14}$$

where $g(y)$ is the periodic function of period 1 defined on $[-1/2,1/2]$ by

$$g(y) = y(1 - 4y^2).$$

The function g vanishes at all integers and has a continuous derivative which is unity at all integers. For any rational x, the series (E.14) has at most finitely many non-zero terms and they are rational.

The formal derivative of (E.14) converges uniformly and absolutely and therefore converges to the derivative of f. For any rational x, the derivative of the series (E.14) is the same as the series for e,

$$\sum_{n=0}^{\infty} \frac{1}{n!},$$

save for at most finitely many terms which are rational. Thus for rational x, the derivative of f is e plus some rational number. But e is irrational (see Remarks to Problem 13 of Chapter 3).

PROBLEM 32. Give an example of an everywhere continuous but nowhere differentiable function.

Solution. For $-\infty < x < \infty$, let

$$T(x) = \sum_{n=1}^{\infty} 2^{-n} \sin^{n^2} x.$$

Putting $q_n = 2^{-n}$, $p_n = 2^{n^2}$, and $h' = 2h$, we obtain

$$T(x + h') - T(x) = \sum_{n=1}^{\infty} 2q_n (\sin p_n h) \cos p_n (x + h).$$

For any $\varepsilon > 0$ there exists some n_0, and for this n_0 an $h_0 > 0$, such that

$$\sum_{n>n_0} 2q_n < \frac{\varepsilon}{2} \quad \text{and, if } |h| < h_0, \quad \sum_{n=1}^{n_0} 2q_n |\sin p_n h| < \frac{\varepsilon}{2}.$$

For $|h'| < 2h_0$ we therefore get $|T(x + h') - T(x)| < \varepsilon$; T is thus uniformly continuous on the entire number line.

For $h' \neq 0$ and holding x fixed, we define

$$Q(h') = \frac{T(x + h') - T(x)}{h'} = \sum_{n=1}^{\infty} q_n h^{-1} (\sin p_n h) \cos p_n (x + h).$$

Considering $Q(h')$ only for the special values

$$h = \frac{\pi}{4} s\, p_m^{-1},$$

where $m > 5$ and $s = \pm 1, \pm 3$, we obtain:

(a) The first $m - 1$ terms of the series for $Q(h')$ in absolute value have a sum smaller than

$$\sum_{n=1}^{m-1} q_n p_n < p_{m-1} \sum_{n=1}^{\infty} q_n = p_{m-1}.$$

(To see this we make use of the inequality $|\sin t| < |t|$ and the result in Problem 58 of Chapter 3, that is, Abel's Inequality.)

(b) All terms of the series for Q(h') with index n > m are zero. Indeed, $p_n h = \pi u_n$, where

$$u_n = \frac{1}{4} s\, p_n\, p_m^{-1} = s\, 2^{n^2-m^2-2},$$

which is an integer because $n^2 - m^2 - 2 > 0$, and hence $\sin p_n h = 0$.

(c) The m-th term of the mentioned series is c_m, where

$$c_m = 4\, p_m\, q_m \frac{\sin \frac{s\pi}{4}}{s\pi} \cos\left(p_m x + \frac{s\pi}{4}\right),$$

whose sign for the four mentioned s-values is determined by the variable of the cosine term. Since we can assign to this variable four consecutive values at distance $\pi/2$ from each other (whatever $p_m x$ may be), it is possible to make the cosine term both $\geq 1/\sqrt{2}$ and $\leq -1/\sqrt{2}$ for suitable values of s. Moreover, for these values of s,

$$|c_m| \geq 4\, p_m\, q_m \frac{1/\sqrt{2}}{3\pi} \frac{1}{\sqrt{2}} = \frac{1}{3\pi} 2^{m^2-m+1} \geq 2^{m^2-m-3},$$

$$|c_m| - p_{m-1} \geq 2^{m^2-m-3} - 2^{m^2-2+1} > 2^m.$$

Hence there are values h', as small as we please in absolute value, for which $Q(h') > 2^m$ and there are values h', arbitrarily small in absolute value, for which $Q(h') < -2^m$. The desired result therefore follows.

Remarks. The foregoing function, due to Lebesgue, provides an example of a function for which at every point the right and the left upper Dini derivatives are $+\infty$ and the right and left lower Dini derivatives are $-\infty$.

PROBLEM 33. A function f defined on [a,b] is said to have the *intermediate value property* provided the closed interval from f(x) to f(y) is contained in the image under f of the closed interval from x to y for each

x and y in $[a,b]$.

One of the more important properties of continuous functions is that they possess the intermediate value property. From the result in Problem 9 we see that if f' exists on $[a,b]$, then f' has the intermediate value property. If f and g are continuous on $[a,b]$, then $f + g$ has the intermediate value property since $f + g$ is continuous. If f' and g' exist on $[a,b]$, then $f' + g'$ possesses the intermediate value property because $f' + g'$ is the derivative of $f + g$.

Give an example of two functions f and g that have the intermediate value property but $f + g$ does not.

Solution. Let

$$F(t) = t^2 \sin \frac{1}{t} \quad \text{if } t \neq 0$$
$$= 0 \quad \text{if } t = 0$$

and

$$G(t) = t^2 \cos \frac{1}{t} \quad \text{if } t \neq 0$$
$$= 0 \quad \text{if } t = 0.$$

Then

$$F'(t) = 2t \sin \frac{1}{t} - \cos \frac{1}{t} \quad \text{if } t \neq 0$$
$$= 0 \quad \text{if } t = 0$$

and

$$G'(t) = 2t \cos \frac{1}{t} + \sin \frac{1}{t} \quad \text{if } t \neq 0$$
$$= 0 \quad \text{if } t = 0.$$

If

$$f(t) = [F'(t)]^2 \quad \text{and} \quad g(t) = [G'(t)]^2$$

then f and g have the intermediate value property because F' and G' do. But

$$(f + g)(t) = 4t^2 + 1 \quad \text{if } t \neq 0,$$
$$= 0 \quad \text{if } t = 0.$$

Hence $f + g$ does not possess the intermediate value property on any interval containing 0.

Remark. Let, for $x \in [0,1]$,

$$h(x) = x \quad \text{if } x \text{ is rational,}$$
$$= 1 - x \quad \text{if } x \text{ is irrational.}$$

This function is seen to be a one-to-one mapping between the interval [0,1] and itself, and hence satisfies the intermediate value property. On the other hand, the function is clearly discontinuous, except at $x = 1/2$.

PROBLEM 34. Let f be a strictly increasing function with a continuous derivative on a compact interval [a,b]. Integration by parts gives

$$\int_a^b f(x)\,dx = bf(b) - af(a) - \int_a^b x\,f'(x)\,dx. \tag{E.15}$$

Let $y = f(x)$, $x = f^{-1}(y)$; then (E.15) can be written

$$\int_a^b f(x)\,dx = bf(b) - af(a) - \int_{f(a)}^{f(b)} f^{-1}(y)\,dy. \tag{E.16}$$

It is easy to see that (E.16) remains valid in case f is only assumed to be strictly increasing and continuous on [a,b]. Indeed, if f is strictly increasing and continuous on [a,b], then it admits an inverse function f^{-1} of the same type. Moreover, we may assume without loss of generality that the graph of f is situated in the first quadrant of the x,y-plane. But (E.16) permits a simple geometrical interpretation in terms of areas of regions represented by the integrals and the quantities bf(b) and af(a) viewed as areas of rectangles. The desired extension is therefore immediate.

Young's Inequality (see Problem 74 of Chapter 2) says that when f is s strictly increasing continuous function with $f(0) = 0$, $f(a) \neq b$, and $a > 0$, $b > 0$, then

$$ab < \int_0^a f(x)\,dx + \int_0^b f^{-1}(y)\,dy. \tag{E.17}$$

Give a proof of (E.17) based on (E.16).

Solution. For $0 < r < a$ it is obvious that

$$(a - r)f(r) < \int_r^a f(x)\,dx.$$

Writing this as

$$af(r) - \int_0^a f(x)\, dx < rf(r) - \int_0^r f(x)\, dx,$$

we apply (E.16) to the integral on the right. We get

$$af(r) - \int_0^a f(x)\, dx < \int_0^{f(r)} f^{-1}(y)\, dy.$$

If $0 < b < f(a)$, we can take $r = f^{-1}(b)$, and (E.17) follows.

PROBLEM 35. A sequence $(f_n)_{n=1}^{\infty}$ of functions defined on an interval $[a,b]$ is said to be *uniformly bounded* if there exists a constant $G > 0$ such that

$$|f_n(x)| < G \quad \text{for all } x \in [a,b] \text{ and all } n.$$

Prove the following theorem, due to Osgood: If $(f_n)_{n=1}^{\infty}$ is a uniformly bounded sequence of continuous functions on a closed bounded interval $[a,b]$ converging (pointwise) to a continuous function f on $[a,b]$, then

$$\lim_{n\to\infty} \int_a^b f_n(x)\, dx = \int_a^b \lim_{n\to\infty} f_n(x)\, dx = \int_a^b f(x)\, dx.$$

Solution. To prove the theorem, we first observe that the general case can be reduced to the special case in which it is assumed that all functions $\{f_n\}_{n=1}^{\infty}$ are nonnegative and $f = 0$. Indeed, since

$$|f_n(x)| < G \quad \text{for all } x \in [a,b] \text{ and all } n,$$

we have that $|f(x)| \le G$ for all $x \in [a,b]$ and so

$$|f_n(x) - f(x)| < 2G \quad \text{for all } x \in [a,b] \text{ and all } n.$$

Since $f_n(x) \to f(x)$, we see that $|f_n(x) - f(x)| \to 0$; since f_n and f are continuous functions on $[a,b]$, so are the functions $|f_n - f|$. If therefore the theorem holds under the assumptions made in the special case, then the integral of $|f_n - f|$ over $[a,b]$ tends to zero as $n \to \infty$ and, using the estimate

$$\left|\int_a^b f_n(x)\, dx - \int_a^b f(x)\, dx\right| = \left|\int_a^b \{f_n(x) - f(x)\}\, dx\right| \le \int_a^b |f_n(x) - f(x)|\, dx,$$

we obtain

$$\lim_{n \to \infty} \int_a^b f_n(x)\ dx = \int_a^b f(x)\ dx.$$

It therefore remains to show that for a uniformly bounded sequence of continuous, nonnegative functions f_n which converge to zero along the interval $[a,b]$ we have

$$\lim_{n \to \infty} \int_a^b f_n(x)\ dx = 0.$$

To begin with we also impose the further restriction that for all n

$$f_{n+1}(x) \leq f_n(x) \quad \text{for all } x \in [a,b],$$

that is, we assume that the sequence is monotonically decreasing. Then, by a well-known theorem of Dini, the convergence is uniform and termwise integration is permitted. Without Dini's theorem we may reason as follows: The integrals form a monotonically decreasing sequence of nonnegative numbers and converge therefore to a limit $I_0 \geq 0$. Dividing the interval $[a,b]$ into two equal parts, the same holds for both subintervals, and for at least one of them the corresponding limit I_1 satisfies $I_1 \geq I_0/2$. Continued bisection of intervals generates an infinite sequence of nested intervals which have a single point x^* in common; for all these intervals the corresponding limits satisfy

$$I_k \geq \frac{I_0}{2^k}.$$

On the other hand, let ε be an arbitrarily small positive number; since $f_n(x) \to 0$, there is an integer n such that

$$|f_n(x^*)| < \frac{\varepsilon}{2};$$

moreover, by the continuity of f_n, there exists a neighborhood $x^* - \delta < x < x^* + \delta$ of x^* such that for all points x of this neighborhood

$$|f_n(x^*) - f_n(x)| < \frac{\varepsilon}{2}$$

holds. Then for all points x of the neighborhood

$$|f_n(x)| < \varepsilon.$$

However, for sufficiently large values of k the intervals we considered in the bisection process further above must be contained in the neighborhood $(x^* - \delta, x^* + \delta)$; thus for these intervals

$$I_k < \varepsilon \frac{b - a}{2^k}.$$

Hence

$$\frac{I_0}{2^k} \le I_k < \varepsilon \frac{b - a}{2^k},$$

that is $I_0 < \varepsilon(b - a)$. Since ε may be chosen arbitrarily small, while I_0 is a fixed nonnegative number, it follows that $I_0 = 0$. Thus the considered special case is settled.

We now suppose that $(f_n)_{n=1}^{\infty}$ is a uniformly bounded sequence of continuous, nonnegative functions which tend to zero along the interval [a,b]. We denote by $f_m^n(x)$ (where $m \le n$) the largest of the values $f_m(x)$, $f_{m+1}(x)$, ..., $f_n(x)$; the function f_m^n is clearly continuous on [a,b]. The integral of f_m^n over [a,b] we denote by I_m^n. Holding m fixed and letting n tend to infinity, the numbers I_m^n form an increasing sequence which is certainly bounded, since if the $f_k(x)$ are below a bound G, then the $f_m^n(x)$ must also be below the bound G and thus $I_m^n < (b - a)G$. Therefore the sequence in question tends to some limit J_m. Since

$$f_m(x) \le f_m^n(x) \qquad \text{for all } x \in [a,b],$$

it is evident that

$$\int_a^b f_m(x)\, dx \le I_m^n \le J_m.$$

If we therefore succeed in showing that $J_m \to 0$ as $m \to \infty$, then we may conclude that

$$\lim_{m \to \infty} \int_a^b f_m(x)\, dx = 0$$

and the proof would then be finished.

Now, let σ be an arbitrarily small positive number and let $n = n_1$ be such that $I_1^n > J_1 - \sigma/2$, for example, the smallest such number, then let $n = n_2$ a number larger than n_1 such that $I_2^n > J_2 - \sigma/4$ and so forth, in general, let $n = n_m$ a number larger than n_{m-1} such that $I_m^n > J_m - \sigma/2^m$. Finally, denote by $g_m(x)$ the smallest of the values

$$f_1^{n_1}(x), \quad f_2^{n_2}(x), \quad \ldots, \quad f_m^{n_m}(x).$$

Evidently, g_m is a continuous function on [a,b]. We assert that

$$\int_a^b g_m(x)\,dx > J_m - \sigma\left\{1 - \frac{1}{2^m}\right\} > J_m - \sigma$$

holds. For m = 1 the assertion is true on the basis of the suppositions made about $I_1^{n_1}$ because $g_1 = f_1^{n_1}$. It is enough therefore to infer m from m - 1. Let us suppose then that

$$\int_a^b g_{m-1}(x)\,dx > J_{m-1} - \sigma\left\{1 - \frac{1}{2^{m-1}}\right\}$$

is true. Since the function g_m at the point x equals the smaller of the two values $g_{m-1}(x)$ and $f_m^{n_m}(x)$, that is, $g_m(x)$ equals the sum of these two values minus the value of the larger amongst the two, while this larger value because of

$$g_{m-1}(x) \le f_{m-1}^{n_{m-1}}(x)$$

does not exceed the larger of the values $f_{m-1}^{n_{m-1}}(x)$ and $f_m^{n_m}(x)$, that is, $f_{m-1}^{n_m}(x)$, we have

$$g_m(x) \ge g_{m-1}(x) + f_m^{n_m}(x) - f_{m-1}^{n_m}(x);$$

the corresponding inequality is also valid for the integrals of these functions and so, since

$$\int_a^b g_{m-1}(x)\,dx > J_{m-1} - \sigma\left\{1 - \frac{1}{2^{m-1}}\right\}, \quad I_m^{n_m} > J_m - \frac{\sigma}{2^m}, \quad I_{m-1}^{n_m} \le J_{m-1},$$

we obtain the desired inequality

$$\int_a^b g_m(x)\,dx > J_{m-1} - \sigma\left\{1 - \frac{1}{2^{m-1}}\right\} + J_m - \frac{\sigma}{2^m} - J_{m-1}$$

$$= J_m - \sigma\left\{1 - \frac{1}{2^m}\right\} > J_m - \ .$$

But the functions g_m form a monotonically decreasing sequence; since $g_m(x) \leq f_m^{n_m}(x) = f_j(x)$, where j for each value of x is a number (in general dependent on x) larger than or equal to m, namely one of the numbers m, m + 1, ..., n_m, it follows from $f_m(x) \to 0$ that $g_m(x) \to 0$. We are therefore back in the special case already settled and the integral of g_m over [a,b] tends to zero as $m \to \infty$. On the basis of the inequality just proved we can see that

$$J_m - \sigma$$

must sink below any positive number, hence specifically below ; thus, for sufficiently large m, we obtain

$$J_m < 2\sigma.$$

But the J_m's are nonnegative and may be picked as small as we please; hence

$$J_m \to 0 \quad \text{as } m \to \infty$$

and the proof of the theorem of Osgood is complete.

Remark. The foregoing proof of Osgood's theorem is due to F. Riesz.

PROBLEM 36. Let f be a continuous and increasing function on the interval [a,b]. Show that f has a finite derivative almost everywhere on [a,b], that is, at every point of [a,b] with the possible exception of the points of a set having measure zero.

Solution. The proof will be based on a lemma.

LEMMA. Let g be a continuous function on an interval [a,b], and let E be the set of points x interior to this interval such that there exists a point t lying to the right of x and satisfying g(t) > g(x). Then the set E is either empty or it decomposes into countably many disjoint open intervals (a_k,b_k) and

$$g(a_k) \leq g(b_k) \tag{E.18}$$

for each of these intervals.

Proof. We first note that the set E is open, since if $t > x_0$ and $g(t) > g(x_0)$, then by virtue of the continuity, the relations t > x, g(t) > g(x) remain valid when x varies in a neighborhood of the point x_0. Thus E, if not empty, decomposes into countably many disjoint open intervals (a_k,b_k) (see Problem 20); the points a_k, b_k do not belong to the set E. Let x_0 be a point between a_k and b_k; we shall show that

$$g(x_0) \leq g(b_k); \tag{E.19}$$

(E.18) will follow by letting x_0 tend to a_k. To verify (E.19), let x_1 be a point at which the function g assumes its largest value on the interval $[x_0,b]$. The point x_1 cannot belong to E, for there exists no t such that $x_1 < t \leq b$ and $g(t) > g(x_1)$. Since the part of $[x_0,b]$ to the left of b_k belongs entirely to E, we necessarily have $b_k \leq x_1 \leq b$. We cannot have $g(b_k) < g(x_1)$ since b_k does not belong to E. Thus $g(b_k) = g(x_1)$ and so $g(x_0) \leq g(b_k)$. This completes the proof of the *LEMMA*.

Now let f be a continuous and increasing function on the interval $[a,b]$. To examine the differentiability of f, we shall compare its Dini derivatives at a point x:

$$D^+, \quad D_+, \quad D^-, \quad D_-,$$

that is, the right-hand limit superior and inferior and the left-hand limit superior and inferior, respectively, of the ratio

$$\frac{f(t) - f(x)}{t - x}$$

as a function of t, at the point $t = x$. The values $\pm\infty$ are admitted. Then function f is differentiable at x if all the four Dini derivatives have the same finite value there, and then

$$f'(x) = D^+ = D_+ = D^- = D_-.$$

As an immediate consequence of the definition we have

$$-\infty \leq D_- \leq D^- \leq \infty, \qquad -\infty \leq D_+ \leq D^+ \leq \infty;$$

in the case of an increasing function the Dini derivatives are of course non-negative.

Our problem consists in showing that for the increasing function f we have almost everywhere

$$\text{(i)} \quad D^+ < \infty; \qquad \text{(ii)} \quad D^+ < D_-.$$

In fact, applying (ii) to the function $-f(-x)$, it follows that we have almost everywhere

$$D^- \le D_+$$

and combining this with (i) and (ii) we obtain

$$0 \le D^+ \le D_- \le D^- \le D_+ \le D^+ \le \infty \quad \text{almost everywhere on } [a,b];$$

hence the equality signs must hold, which was to be proved.

Assertion (i) means that the set E_∞ of points x for which $D^+ = \infty$, is of measure zero. This set E is included in the set E_C of points x for which we have $D^+ > C$, where C denotes a number chosen as large as we please. But the relation $D^+ > C$ implies the existence of some $t > x$ such that

$$\frac{f(t) - f(x)}{t - x} > C,$$

that is,

$$g(t) > g(x),$$

where $g(x) = f(x) - Cx$. Hence the set E_C is included in the set E attached to the function g by the *LEMMA* and so it can be covered by a sequence of disjoint open intervals (a_k,b_k) for which

$$g(a_k) \le g(b_k), \quad \text{that is,} \quad f(b_k) - Cb_k \ge f(a_k) - Ca_k;$$

hence

$$C(b_k - a_k) \le f(b_k) - f(a_k).$$

This yields, by addition,

$$C \sum (b_k - a_k) \le \sum [f(b_k) - f(a_k)] \le f(b) - f(a); \tag{E.20}$$

here we used the fact that the total increase of an increasing function on an interval cannot be less than the sum of the increases on disjoint subintervals. The inequality (E.20) shows that, for sufficiently large C, the total length of the intervals (a_k,b_k) will be as small as we please. That is, the set E_∞ is of measure zero.

The statement (ii) is proved by analogous reasoning which is repeated alternately in two different forms. Let c and C be two positive numbers, $c < C$. We first show that the set E_{cC} of points x for which $D^+ > C$ and $D_- < c$, is of measure zero.

Consider first the condition $D_- < c$. Applying the *LEMMA* to the function $g(x) = f(-x) + cx$ on the interval $[-b,-a]$, we obtain, for reasons similar to those just used, that the set E_{cC} can be covered by a sequence of disjoint open intervals (a_k,b_k), such that

$$f(b_k) - f(a_k) \leq c(b_k - a_k). \tag{E.21}$$

Next we consider, inside each of the intervals (a_k,b_k), the points where $D^+ > C$; applying the *LEMMA* to the function $g(x) = f(x) - Cx$ on the interval (a_k,b_k), we see that these points can be covered by a sequence of disjoint intervals (a_{km},b_{km}) such that

$$C(b_{km} - a_{km}) \leq f(b_{km}) - f(a_{km});$$

hence

$$C \sum_m (b_{km} - a_{km}) \leq \sum_m \{f(b_{km}) - f(a_{km})\} \leq f(b_k) - f(a_k).$$

Taking account of (E.21), it follows that

$$C \sum_{k,m} (b_{km} - a_{km}) \leq c \sum_k (b_k - a_k).$$

If $|S_1|$ and $|S_2|$ denote the total length of the systems

$$S_1 = \{(a_k,b_k)\} \quad \text{and} \quad S_2 = \{(a_{km},b_{km})\},$$

respectively, it follows that

$$|S_2| \leq \frac{c}{C}|S_1|.$$

Repeating the two steps alternately, we obtain a sequence S_1, S_2, ... of systems of intervals, each imbedded in the preceding, and we have

$$|S_{2n}| \leq \frac{c}{C}|S_{2n-1}| \leq \frac{c}{C}|S_{2n-2}| \quad \text{for } n = 1,2,\ldots$$

It follows that

$$|S_{2n}| \leq \left(\frac{c}{C}\right)^n |S_1| \to 0 \quad \text{as } n \to \infty.$$

Thus the set E_{cC} can be covered by a system of intervals of total length as small as we please and E_{cC} is seen to have measure zero.

Now we form the union E^* of all the sets E_{cC} corresponding to pairs c, C of positive rational numbers ($c < C$). As a union of countably many sets of measure zero (see Problem 22), E* itself is of measure zero. If at a point x we have $D_- < D^+$, we can interpolate between D_- and D^+ two rational numbers,

$$D_- < c < C < D^+;$$

then x is a point of the set E_{cC} and consequently of E*. Thus the points x where (ii) does not hold, form a set of measure zero. This completes the solution.

Remark. The foregoing proof is due to F. Riesz.

PROBLEM 37. We assume that the upper and lower integrals

$$\overline{\int_a^b} f(t)\, dt \quad \text{and} \quad \underline{\int_a^b} f(t)\, dt$$

of a bounded function on a bounded closed interval [a,b] have been defined (see Problem 11). We also assume the following elementary properties of upper and lower integrals:

(a) $\overline{\int_a^b} f(t)\, dt \geq \underline{\int_a^b} f(t)\, dt$ for each bounded function f.

(b) For $a < c < b$,

$$\overline{\int_a^b} f(t)\, dt = \overline{\int_a^c} f(t)\, dt + \overline{\int_c^b} f(t)\, dt$$

and

$$\underline{\int_a^b} f(t)\, dt = \underline{\int_a^c} f(t)\, dt + \underline{\int_c^b} f(t)\, dt.$$

(c) The upper and lower integrals are unchanged if f is replaced by a new function which differs from f at one point only.

Show that if f is a bounded function on [a,b] and has a right-hand limit at each point of [a,b), then f is Riemann integrable over [a,b].

Solution. For any $\delta > 0$, define

$$h_\delta(x) = \overline{\int_a^x} f(t)\, dt - \underline{\int_a^x} f(t)\, dt - \delta(x - a) \quad \text{for } a < x \leq b$$

$$= 0 \quad \text{for } x = a.$$

It is enough to show that $h_\delta(b) < 0$ for all $\delta \geq 0$, since this will imply

$$\overline{\int_a^b} f(t)\ dt \le \underline{\int_a^b} f(t)\ dt.$$

Suppose this is not true. Then there exists $\alpha > 0$ such that $h_\alpha(b) > 0$. Let

$$x = \inf\{t \in [a,b]: h_\alpha(t) > 0\}.$$

We shall show that the assumptions $h_\alpha(x) > 0$ and $h_\alpha(x) \le 0$ both lead to contradictions.

If $h_\alpha(x) > 0$, let $M = \sup\{|f(t)|: a \le t \le b\}$ and choose $0 < \gamma < x - a$ and $2\gamma M < h_\alpha(x)$. Then

$$\overline{\int_a^{x-\gamma}} f(t)\ dt = \overline{\int_a^x} f(t)\ dt - \overline{\int_{x-\gamma}^x} f(t)\ dt \ge \overline{\int_a^x} f(t)\ dt - M\gamma,$$

$$\underline{\int_a^{x-\gamma}} f(t)\ dt = \underline{\int_a^x} f(t)\ dt - \underline{\int_{x-\gamma}^x} f(t)\ dt \le \underline{\int_a^x} f(t)\ dt + M\gamma,$$

so

$$h_\alpha(x - \gamma) = \overline{\int_a^{x-\gamma}} f(t)\ dt - \underline{\int_a^{x-\gamma}} dt - \alpha(x - \gamma - a)$$

$$\ge h_\alpha(x) - (2M - \alpha)\gamma > 0.$$

This contradicts the fact that x is a lower bound.

If $h_\alpha(x) \le 0$, then, by assumption, since

$$f(x + 0) = \lim_{h \to \infty} f(x + h) \quad \text{with } h > 0$$

exists, we can choose γ such that $0 < \gamma < b - x$ and

$$|f(x + 0) - f(t)| < \frac{\alpha}{2} \quad \text{for } 0 < t - x < \gamma.$$

By property (c) we can assume that $f(x) = f(x + 0)$ for purposes of evaluating h_α. But then for $x < t < x + \gamma$ we have

$$\overline{\int_a^t} f(y)\ dy = \overline{\int_a^x} f(y)\ dy + \overline{\int_x^t} f(y)\ dy$$

$$\le \overline{\int_a^x} f(y)\ dy + [f(x + 0) + \alpha/2](t - x),$$

$$\underline{\int_a^t} f(y)\ dy = \underline{\int_a^x} f(y)\ dy + \underline{\int_x^t} f(y)\ dy$$

$$\ge \underline{\int_a^x} f(y)\ dy + [f(x + 0) - \alpha/2](t - x).$$

Therefore

$$h_\alpha(t) \le \bar{\int_a^x} f(y)\,dy + [f(x+0) + \alpha/2](t - x)$$

$$- \underline{\int_a^x} f(y)\,dy - [f(x+0) - \alpha/2](t - x) - \alpha(t - x)$$

$$= h_\alpha(x) \le 0.$$

This contradicts the fact that x is the greatest lower bound.

Remarks. Of course the dual result on left-hand limits is an immediate consequence by consideration of $f(-x)$. Clearly the result includes the propositions that continuous functions are Riemann integrable and that monotone functions are Riemann integrable. Given the "continuous almost everywhere" characterization of Riemann integrability (see Problem 27), we have that bounded functions with right-hand limits at every point of an interval are continuous almost everywhere on the interval.

PROBLEM 38. Let $(q_n)_{n=0}^\infty$ be the sequence of polynomials defined by

$$q_0(t) = 0, \quad q_{n+1}(t) = q_n(t) + \tfrac{1}{2}\{t - q_n^2(t)\}, \quad n \ge 0. \tag{E.22}$$

Show that in the interval [0,1] the sequence $(q_n)_{n=0}^\infty$ is increasing and that for any $\varepsilon > 0$, there exists a positive integer n_0 (depending only on ε) such that $n > n_0$ implies

$$|\sqrt{t} - q_n(t)| < \varepsilon \quad \text{for every } t \text{ in } [0,1],$$

that is, $(q_n)_{n=0}^\infty$ converges uniformly to $\sqrt{t}$ on [0,1].

Solution. To prove the claim it is enough to show that, for all t in [0,1], we have

$$0 \le \sqrt{t} - q_n(t) \le \frac{2\sqrt{t}}{2 + n\sqrt{t}}, \tag{E.23}$$

for (E.23) implies that $0 \le \sqrt{t} - q_n(t) \le 2/n$.

We prove (E.23) by induction on n. It is true for $n = 0$. If $n \ge 0$ it follows from the inductive assumption (E.23) that

$$0 \le \sqrt{t} - q_n(t) \le \sqrt{t},$$

and hence $0 \le q_n(t) \le \sqrt{t}$, and therefore from (E.22) we have

$$\sqrt{t} - q_{n+1}(t) = [\sqrt{t} - q_n(t)]\{1 - \tfrac{1}{2}[\sqrt{t} + q_n(t)]\},$$

so that $\sqrt{t} - q_{n+1}(t) \ge 0$, and from (E.23)

$$\sqrt{t} - q_{n+1}(t) \le \frac{2\sqrt{t}}{2 + n\sqrt{t}}\left\{1 - \frac{\sqrt{t}}{2}\right\}$$

$$\le \frac{2\sqrt{t}}{2 + n\sqrt{t}}\left\{1 - \frac{\sqrt{t}}{2 + (n + 1)\sqrt{t}}\right\} = \frac{2\sqrt{t}}{2 + (n + 1)\sqrt{t}}.$$

To verify that $(q_n)_{n=0}^{\infty}$ is an increasing sequence, we note that

$$q_{n+1}(t) - q_n(t) = \tfrac{1}{2}\{t - q_n^2(t)\} = \tfrac{1}{2}\{\sqrt{t} - q_n(t)\}\{\sqrt{t} + q_n(t)\};$$

but we know already that $0 \le \sqrt{t} - q_n(t) \le \sqrt{t}$ and $0 \le q_n(t) \le \sqrt{t}$.

PROBLEM 39. Show that for any $\varepsilon > 0$ there exists a polynomial p such that

$$|p(x) - |x|| < \varepsilon \quad \text{for all } x \text{ in } [-1,1].$$

Solution. By Problem 38, for any $\varepsilon > 0$ there exists a polynomial q such that

$$|q(t) - \sqrt{t}| < \varepsilon \quad \text{for all } t \text{ in } [0,1].$$

Replacing t by x^2 and noting that

$$x = \sqrt{x^2},$$

the desired result follows.

Remarks. Consider the function $|x - c|$ on an arbitrary closed bounded interval $[a,b]$. We choose a number d such that the interval $[c - d, c + d]$ includes the interval $[a,b]$. By the substitution

$$s = \frac{x - c}{d}$$

the interval $c - d \le x \le c + d$ goes over into the interval $-1 \le s \le 1$. From Problem 39 we know that for any $\varepsilon > 0$ there exists a polynomial $p(s)$ such that

$$|p(s) - |s|| \le \frac{\varepsilon}{d} \quad \text{for } -1 \le s \le 1;$$

this implies that

$$\left|d\cdot p\left(\frac{x - c}{d}\right) - |x - c|\right| = d|p(s) - |s|| \le d\cdot\frac{\varepsilon}{d} = \varepsilon$$

on the interval $[c - d, c + d]$ and, a fortiori, on the interval $[a,b]$. Thus the function

$$P(x) = d\cdot p\left(\frac{x - c}{d}\right),$$

which is a polynomial in the variable x, approximates the function $|x - c|$ on $[a,b]$ with accuracy ε. Hence the polynomial

$$Q(x) = \frac{1}{2}[P(x) + x - c]$$

approximates the function

$$L_c(x) = \frac{1}{2}\{(x - c) + |x - c|\}$$

on the interval $[a,b]$ with accuracy $\varepsilon/2$:

$$|Q(x) - L_c(x)| = \frac{1}{2}\left|P(x) - |x - c|\right| \le \frac{\varepsilon}{2}.$$

Note that

$$L_c(x) = 0 \qquad \text{for } x \le c,$$

$$\quad = x - c \qquad \text{for } x \ge c;$$

L_c is a polygonal function whose graph has an angle at the basic point $x = c$.

Suppose now that g is any polygonal function on $[a,b]$ whose graph has angles at the basic points $a = a_0 < a_1 < \cdots < a_n = b$. Then g is a linear combination of the L_c. Indeed, let

$$g_0(x) = g(a) + C_0 L_{a_0}(x) + C_1 L_{a_1}(x) + \cdots + C_{n-1} L_{a_{n-1}}(x),$$

and define the constants C_j by the equations

$$g_0(a_j) = g(a_j), \quad j = 0,1,\ldots,n.$$

The first of these equations is an identity, the second is

$$g(a) + C_0 L_{a_0}(a_1) = g_0(a_1)$$

and defines C_0, the third defines C_1, and so on. The two polygonal functions g and g_0 coincide at all basic points and are therefore identical. This shows that g admits approximation by polynomials.

PROBLEM 40. Prove the following *Approximation Theorem of Weierstrass:* Let [a,b] be a closed bounded interval and f a continuous function on [a,b]. Then, for any $\varepsilon > 0$, there exists a polynomial P such that

$$|f(x) - P(x)| < \varepsilon \quad \text{for all } x \in [a,b]$$

and we say that f admits uniform approximation by P.

Solution. If f is continuous on a closed bounded interval [a,b], then f is uniformly continuous on it and hence admits uniform approximation with an arbitrarily small error by a polygonal function on [a,b]. In fact, f is uniformly continuous on a closed bounded interval if and only if f is such that

$$|f(x) - g(x)| < \varepsilon \quad \text{for all } x \in [a,b]$$

with g a polygonal function. But we know from the Remarks following Problem 39 that polygonal functions on a closed bounded interval can be approximated uniformly with an arbitrarily small error by polynomials.

Remarks. In the foregoing, the proof of the Approximation Theorem of Weierstrass was reduced to the problem of approximating the function $|x|$ by polynomials, a procedure due to H. Lebesgue.

For emphasis we add that f is, of course, assumed to be a real-valued function.

PROBLEM 41. Let f be a continuous function on the set of real numbers R^1. Show that if f can be approximated uniformly throughout R^1 by polynomials, then f is itself a polynomial.

Solution. If polynomials P_n approach f uniformly, then for some n we have

$$|P_i(x) - f(x)| < 1 \quad \text{for all } i \geq n \text{ and all } x \text{ in } R^1.$$

Hence for all $i \geq n$, $|P_i(x) - P_n(x)| < 2$ for all real x, so that $P_i - P_n$ is a bounded polynomial, and hence is constant. Taking limits as $i \to \infty$, $f - P_n$ is constant, and hence $f = P_n + C$ is a polynomial.

PROBLEM 42. Let f be a continuous function on a closed bounded interval [a,b] and suppose that

$$\int_a^b x^n f(x) = 0 \quad \text{for } n = 0,1,2,\ldots$$

Show that $f = 0$.

Solution. By Problem 41, f admits uniform approximation by a polynomial P; thus, for all $x \in [a,b]$,

$$f(x) = P(x) + \varepsilon h(x),$$

where ε is an arbitrary positive real number and $|h(x)| < 1$ on [a,b]. We see therefore that

$$\int_a^b f^2(x)\, dx = \int_a^b f(x)\{P(x) + \varepsilon h(x)\}\, dx.$$

But

$$\int_a^b f(x)\, P(x)\, dx = 0$$

by hypothesis. Noting that

$$\int_a^b f^2(x)\, dx < \varepsilon \int_a^b |f(x)|\, dx,$$

we see that we have reached a contradiction, for, if $f \neq 0$, then ε would be bounded below by

$$\frac{\int_a^b f^2(x)\, dx}{\int_a^b |f(x)|\, dx}.$$

Remarks. If $f \neq 0$, then $|f(x_1)| = d > 0$ for some $x_1 \in (a,b)$ and there is some $\delta > 0$ such that $|f(x)| > d/2$ for all x in the interval $x_1 - \delta \leq x \leq x_1 + \delta$, by the continuity of f on [a,b]. Thus

$$\int_a^b |f(x)|\, dx > \frac{d}{2}\cdot 2\delta = d\delta > 0.$$

In the same way we can see that if $f \neq 0$, then

$$\int_a^b f^2(x)\, dx > 0.$$

Thus

$$\frac{\int_a^b f^2(x)\, dx}{\int_a^b |f(x)|\, dx} > 0.$$

PROBLEM 43. Let f be a continuous function on $[-\pi,\pi]$ and suppose that

$$\int_{-\pi}^{\pi} f(t) \begin{matrix}\cos nt \\ \sin nt\end{matrix}\, dt = 0 \qquad \text{for } n = 0,1,2,\ldots$$

Show that $f = 0$.

Solution. Suppose not, that is, suppose $|f(t_0)| > 0$, say $f(t_0) = \sigma > 0$. Then by continuity there are two positive numbers ε and δ such that $f(t) > \varepsilon$ for all t in the interval I, where

$$I = [t_0 - \delta, t_0 + \delta].$$

It will be enough to show that there is a sequence (T_n) of trigonometric polynomials such that

(i) $T_n(t) \geq 0$ for $t \in I$,

(ii) $T_n(t)$ tends uniformly to $+\infty$ in every interval J inside I,

(iii) the T_n are uniformly bounded outside I.

For then the integral

$$\int_{-\pi}^{\pi} f(t)\ T_n(t)\ dt$$

may be split into two, extended respectively over I and over the rest of $(-\pi,\pi)$. By (i), the first integral exceeds

$$|J| \min_{t \in J} T_n(t),$$

and so, by (ii), tends to $+\infty$ with n. The second integral is bounded, in view of (iii). Thus $\int_{-\pi}^{\pi} f(t)\ T_n(t)\ dt = 0$ is not possible for T_n with large n and we have reached a contradiction.

If we set $T_n(t) = [x(t)]^n$, $x(t) = 1 + \cos(t - t_0) - \cos\delta$, then $x(t) \geq 1$ in I, $x(t) > 1$ in J, $|x(t)| \leq 1$ outside I. Moreover, conditions (i), (ii), and (iii) are satisfied.

PROBLEM 44. Let f be a continuous function on a closed bounded interval $[a,b]$. Show that there exists a monotone increasing sequence of polynomials $(p_n)_{n=1}^{\infty}$ converging uniformly to f on $[a,b]$.

Solution. Let $(e_n)_{n=1}^{\infty}$ be any sequence of positive numbers such that $\Sigma_{n=1}^{\infty} e_n = 1$. By the Weierstrass approximation theorem (see Problem 40) we can find polynomials $p_1, p_2, \ldots, p_n, \ldots$ such that for all $x \in [a,b]$

$$|f(x) + (1 - \tfrac{1}{2}e_1) - p_1(x)| < \tfrac{1}{2}e_1,$$

$$|f(x) + (1 - e_1 - \tfrac{1}{2}e_2) - p_2(x)| < \tfrac{1}{2}e_2,$$

$$\cdots$$

$$|f(x) + (1 - e_1 - \cdots - e_{n-1} - \tfrac{1}{2}e_n) - p_n(x)| < \tfrac{1}{2}e_n.$$

Letting $f_0(x) = f(x) + 1$, and

$$f_n(x) = f(x) + \left\{1 - \sum_{k=1}^{n} e_k\right\} \quad \text{for } n \geq 1,$$

we have

$$f_0(x) \geq p_1(x) \geq f_1(x) \geq p_2(x) \geq \cdots \geq f_n(x) \geq p_n(x) \geq \cdots.$$

But $f_n \to f$ uniformly on $[a,b]$. (Clearly, a similar result holds for the monotone decreasing case.)

PROBLEM 45. Let f be a continuous finction on a closed bounded interval $[a,b]$. Show that f admits uniform approximation by a polynomial with rational coefficients.

Solution. Let $\varepsilon > 0$ be given. By Problem 40 there exists a polynomial P such that

$$\sup\{|f(t) - P(t)|: t \in [a,b]\} < \varepsilon/2,$$

where

$$P(t) = a_0 + a_1 + \cdots + a_n t^n$$

and $a_0, a_1, \ldots, a_n$ are real numbers. Let $c = \max\{|a|,|b|\}$; for each j, $0 \le j \le n$, choose a rational number b_j such that

$$|b_j - a_j| < \frac{\varepsilon}{2(n + 1)c^j}$$

and let

$$Q(t) = b_0 + b_1 + \cdots + b_n t^n.$$

Then

$$|P(t) - Q(t)| = \left|\sum_{j=0}^{n} (b_j - a_j)t^j\right| < \varepsilon/2$$

and so

$$\sup\{|P(t) - Q(t)|: t \in [a,b]\} < \varepsilon/2.$$

Hence $\sup\{|f(t) - Q(t)|: t \in [a,b]\} < \varepsilon$.

PROBLEM 46. Let the function $h(x)$ be continuous on the interval $0 \le x \le 1$, and let $f(x)$ be an arbitrary continuous function on $0 \le x \le 1$. Show that a necessary and sufficient condition that every such $f(x)$ can be uniformly approximated on $[0,1]$ as closely as desired by a polynomial in $h(x)$ is that $h(x)$ be strictly monotonic in $[0,1]$.

Solution. We say that h is strictly monotonic on $[0,1]$ provided that, for x_1 and x_2 in $[0,1]$, the inequality $x_1 < x_2$ always implies $h(x_1) < h(x_2)$, or always implies $h(x_1) > h(x_2)$.

If $h(x)$ is continuous and strictly monotonic on $[0,1]$, the transformation $z = h(x)$, $x = h^{-1}(z)$, sets up a one-to-one continuous correspondence between the points of $0 \le x \le 1$ and the points of some finite interval $a \le z \le b$, with the endpoints of the one interval corresponding to the endpoints of the other interval. The given function $f(x) = f[h^{-1}(z)]$ is continuous on the latter interval, so by the Weierstrass approximation theorem (see Problem 40) it can be uniformly approximated on $a \le z \le b$ by a polynomial in z. If $\varepsilon > 0$ is given, there exists

$$a_0 + a_1 z + \cdots + a_n z^n$$

such that

$$\left| f[h^{-1}(z)] - \sum_{k=0}^{n} a_k z^k \right| < \varepsilon, \qquad a \le z \le b,$$

and this is equivalent to

$$\left| f(x) - \sum_{k=0}^{n} a_k [h(x)]^k \right| < \varepsilon, \qquad 0 \le x \le 1;$$

here we define $[h(x)]^0 = 1$ even if $h(x_0) = 0$ for some x_0 in $[0,1]$.

Conversely, suppose $h(x)$ is not strictly monotonic on $[0,1]$, and hence we have for two numbers x_1 and x_2 on $[0,1]$ the relations $x_1 \ne x_2$ and $h(x_1) = h(x_2)$. If an arbitrary function $f(x)$ can be uniformly approximated within $\varepsilon > 0$ by a polynomial in $h(x)$, we so approximate the function $f(x) = x$, where $2\varepsilon < |x_1 - x_2|$:

$$\left| x - \sum_{k=0}^{n} a_k [h(x)]^k \right| < \varepsilon; \qquad a \le x \le b.$$

There follow the inequalities

$$\left| x_1 - \sum_{k=0}^{n} a_k [h(x_1)]^k \right| < \varepsilon,$$

$$\left| x_2 - \sum_{k=0}^{n} a_k [h(x_2)]^k \right| < \varepsilon,$$

$$|x_1 - x_2| < 2\varepsilon.$$

This contradiction completes the proof.

Remarks. It is clear that the result in Problem 45 remains valid if we replace the interval $[0,1]$ by any closed bounded interval $[\alpha,\beta]$.

It can be seen therefore that if f is real-valued and continuous on the interval $[-\pi/2,\pi/2]$, then it is possible to approximate f uniformly on this interval by linear combinations of $(\sin x)^k$ with $k = 0,1,2,\ldots$; however, f cannot be approximated uniformly on this interval by linear combinations of $(\cos x)^k$ with $k = 0,1,2,\ldots$

PROBLEM 47. Let f be a function with at least k derivatives. Given that for some real number r,

$$\lim_{x\to\infty} x^r f(x) = 0 \quad \text{and} \quad \lim_{x\to\infty} x^r f^{(k)}(x) = 0,$$

show that

$$\lim_{x\to\infty} x^r f^{(j)}(x) = 0, \quad 0 \le j \le k.$$

Solution. For each integer j, $1 < j < k$, expand $f(x + j)$ in a Taylor polynomial about the point x:

$$f(x + j) = f(x) + j\, f'(x) + \frac{j^2}{2!} f''(x) + \cdots + \frac{j^{k-1}}{(k - 1)!} f^{(k-1)}(x)$$

$$+ \frac{j^k}{k!} f^{(k)}(\theta_j),$$

where $x < \theta_j < x + j$. This may be considered as a system of linear equations in the unknowns

$$f(x), \quad f'(x), \quad \ldots, \quad f^{(k-1)}(x).$$

The matrix of coefficients has for its i-th row

$$1, \ \frac{i}{1!}, \ \frac{i^2}{2!}, \ \ldots, \ \frac{i^{k-1}}{(k - 1)!}.$$

From the corresponding determinant we may factor out the denominators common to the elements of each column and will have

$$\frac{1}{1!\ 2! \cdots (k - 1)!}$$

times the familiar Vandermond determinant. Hence the determinant of coefficients equals 1. The system of equations has therefore a solution and this solution gives $f^{(j)}(x)$ as a linear combination of

$$f(x+1), \quad f(x+2), \quad \dots, \quad f(x+k), \quad f^{(k)}(\theta_1), \quad \dots, \quad f^{(k)}(\theta_k).$$

Now if $x < t < x + k$, then

$$\lim_{x\to\infty} x^r f(t) = \lim_{x\to\infty} \left(\frac{x}{t}\right)^r t^r f(t) = \lim_{x\to\infty} \left(\frac{x}{t}\right)^r \lim_{x\to\infty} t^r f(t) = 1\cdot 0 = 0,$$

and similarly,

$$\lim_{x\to\infty} x^r f^{(k)}(t) = 0.$$

Hence $x^r f^{(j)}(x)$ is a linear combination of terms of the form

$$x^r f(t) \quad \text{or} \quad x^r f^{(k)}(t), \quad \text{where } x < t < x + k.$$

Since these all go to 0 as $x \to \infty$, the result follows.

PROBLEM 48. Let f be Riemann integrable over every bounded interval and

$$f(x+y) = f(x) + f(y) \quad \text{for any real numbers } x \text{ and } y. \tag{E.24}$$

Show that $f(x) = cx$, where $c = f(1)$.

Solution. Integrating $f(u + y) = f(u) + f(y)$ with respect to u over the interval $[0,x]$, we easily see that

$$yf(x) = \int_0^{x+y} f(u)\,du - \int_0^x f(u)\,du - \int_0^y f(u)\,du \tag{E.25}$$

holds. Since the right-hand side of (E.25) is invariant under the interchange of x and y, it follows that $xf(y) = yf(x)$. Thus, for $x \neq 0$, $f(x)x^{-1} = c$, a constant; hence $f(x) = cx$. Since (E.24) implies $f(0) = 0$, $f(x) = cx$ also holds for $x = 0$. Taking $x = 1$ in $f(x) = cx$, we obtain $c = f(1)$.

Remarks. If f satisfies (E.24), then f being continuous at a single point implies that f is continuous everywhere. Indeed,

$$|f(x+h) - f(x)| = |f(h)| = |f(y+h) - f(y)|.$$

Suppose that f satisfies (E.24) but f(x) is not of the form cx, where c is a constant. Then the graph of f is dense in the plane.

Indeed, let $c = f(1)$ and choose x so that $f(x) \neq cx$. The graph of f contains all points of the form $[u + vx, uc + vf(x)]$, u, v rational. Let A be the matrix

$$\begin{pmatrix} 1 & c \\ x & f(x) \end{pmatrix}.$$

Then A is non-singular, hence a homeomorphism of the plane onto itself. In particular, A preserves dense sets; one dense set is $\{(u,v): u \text{ and } v \text{ rational}\}$ and A maps this set [via multiplication, XA, where $X = (u,v)$], onto the subset of the graph of f mentioned above. Thus the graph of f is dense.

PROBLEM 49. Show that

$$\sum_{j=0}^{m-1} \frac{m^j}{j!} \geq e^{m-1}$$

for all positive integers m.

Solution. By Taylor's theorem

$$\sum_{j=0}^{m-1} \frac{m^j}{j!} = e^m - \frac{e^m}{(m-1)!}\int_0^m e^{-t}\, t^{m-1}\, dt.$$

Hence it will suffice to establish that

$$\int_0^m e^{-t}\, t^{m-1}\, dt \leq (m-1)!\left(1 - \frac{1}{e}\right) \tag{E.26}$$

for all positive integers m. This will be done by induction.

The relation (E.26) holds, with equality, if $m = 1$. Suppose that the relation (E.26) holds for $m = k$. Integrating the left member by parts and multiplying both members of the last mentioned inequality by m shows that

$$k^k\, e^{-k} + \int_0^k e^{-t}\, t^k\, dt \leq k!\left(1 - \frac{1}{e}\right).$$

But

$$\int_k^{k+1} e^{-t}\, t^k\, dt \leq \max_{k \leq t \leq k+1} e^{-t}\, t^k = e^{-k}\, k^k.$$

Therefore

$$\int_0^{k+1} e^{-t}\, t^k\, dt = \int_k^{k+1} e^{-t}\, t^k\, dt + \int_0^k e^{-t}\, t^k\, dt$$

$$\le k^k\, e^{-k} + \int_0^k e^{-t}\, t^k\, dt \le k!\left(1 - \frac{1}{e}\right).$$

PROBLEM 50. Show that the following relation is valid for all real $n \ge 1$:

$$\int_0^n \frac{n^{[x]}}{[x]!}\, dx \ge e^{n-1}.$$

Solution. Let

$$I(t) = \int_0^y \frac{t^{[x]}}{[x]!}\, dx;$$

if m is a positive integer and $m < t < m + 1$, then

$$I(t) = \sum_{k=0}^{m-1} \frac{t^k}{k!} + \frac{t^m(t - m)}{m!}.$$

Direct calculation yields

$$I'(t) - I(t) = \frac{t^{m-1}}{m!}(t - m)(m + 1 - t),$$

which is positive on $(m,m + 1)$. If $G(t) = \log I(t)$, we have, therefore, $G'(t) > 1$ on $(m,m + 1)$, and G is continuous at m, whence by the mean-value theorem, $G(t) > G(m) + (t - m)$. By Problem 49, $I(m) \ge e^{m-1}$, so $G(m) \ge m - 1$. Thus $G(t) > t - 1$, and therefore $I(t) > e^{t-1}$ on $(m,m + 1)$.

PROBLEM 51. Let f be a continuous function on the interval $[a,b]$ and $x_1, x_2, \ldots, x_n$ be points of $[a,b]$. Show that there is a point t in $[a,b]$ such that

$$f(t) = \frac{1}{n}[f(x_1) + f(x_2) + \cdots + f(x_n)].$$

Solution. By Problem 28 of Chapter 2,

$$\min\{f(x_1),f(x_2),\ldots,f(x_n)\} \le \frac{1}{n}[f(x_1) + f(x_2) + \cdots + f(x_n)]$$

$$\le \max\{f(x_1),f(x_2),\ldots,f(x_n)\}.$$

The rest follows by the intermediate value property (see Problem 33).

PROBLEM 52. Give a proof of the relation

$$\int_0^\infty \frac{\sin x}{x}\, dx = \frac{\pi}{2}. \tag{E.27}$$

Solution. Our proof will be based on the result in Problem 13: If f is Riemann integrable in the interval $[a,b]$, then

$$\lim_{p\to\infty} \int_a^b f(x) \sin px\, dx = 0. \tag{E.28}$$

First we note that if $x \neq 2k\pi$ $(k = 0,1,2,\ldots)$, then

$$\frac{1}{2} + \cos x + \cos 2x + \cdots + \cos nx = \frac{\sin(n + \frac{1}{2})x}{2 \sin \frac{x}{2}}. \tag{E.29}$$

Indeed, since

$$\sin \frac{2n + 1}{2} x = \sin \frac{x}{2} + \left(\sin \frac{3}{2} x - \sin \frac{x}{2}\right) + \left(\sin \frac{5}{2} x - \sin \frac{3}{2} x\right)$$
$$+ \cdots + \left(\sin \frac{2n + 1}{2} x - \sin \frac{2n - 1}{2}\right),$$

we get

$$\sin \frac{2n + 1}{2} x = \left(\frac{1}{2} + \cos x + \cos 2x + \cdots + \cos nx\right) 2 \sin \frac{x}{2}$$

by using the identity $\sin A - \sin B = 2 \sin \frac{A - B}{2} \cos \frac{A + B}{2}$.

Integrating both sides of the identity (E.29), we obtain

$$\int_0^\pi \frac{\sin(n + \frac{1}{2})x}{2 \sin x/2}\, dx = \frac{\pi}{2}, \qquad (n = 0,1,2,\ldots). \tag{E.30}$$

Let

$$g(x) = \frac{1}{x} - \frac{1}{2 \sin x/2} = \frac{2 \sin x/2 - x}{2x \sin x/2}, \qquad 0 < x \le \pi. \tag{E.31}$$

The function g is continuous for $0 < x \le \pi$. By applying twice L'Hospital's rule to (E.31), we see that $\lim_{x\to 0} g(x) = 0$. Consequently, if we put $g(0) = 0$, g will be continuous for $0 \le x \le \pi$. Therefore g certainly satisfies the condition of the proposition in Problem 13; then the relation (E.28), with $p = n + \frac{1}{2}$, gives

$$\lim_{n\to\infty} \int_0^{\pi} \left(\frac{1}{x} - \frac{1}{2 \sin x/2}\right) \sin(n + \tfrac{1}{2})x \, dx = 0.$$

Taking into account (E.30), we see that

$$\lim_{n\to\infty} \int_0^{\pi} \frac{\sin(n + \frac{1}{2})x}{x} \, dx = \frac{\pi}{2},$$

or, making the substitution $u = (n + \frac{1}{2})x$,

$$\lim_{n\to\infty} \int_0^{(n+\frac{1}{2})\pi} \frac{\sin u}{u} \, du = \frac{\pi}{2}. \tag{E.32}$$

If we can show that

$$\int_0^{\infty} \frac{\sin u}{u} \, du$$

is convergent, then (E.32) yields the relation (E.27) and we are done. But from the result in Problem 15 we get, for $0 < a \le t \le b$,

$$\int_a^b \frac{\sin u}{u} \, du = \frac{1}{a} \int_a^t \sin u \, du + \frac{1}{b} \int_t^b \sin u \, du.$$

Since $|\int_{\alpha}^{\beta} \sin u \, du| \le 2$ for any α and β, we therefore get

$$\left|\int_a^b \frac{\sin u}{u} \, du\right| \le 2\left(\frac{1}{a} + \frac{1}{b}\right); \tag{E.33}$$

hence the quantity on the left-hand side of (E.33) is less than ε when $b > a \ge A$, provided $A > 4/\varepsilon$.

Remarks. Another interesting method for the evaluation of the integral

$$I = \int_0^{\infty} \frac{\sin x}{x} \, dx$$

depends on the partial fraction decomposition of the function $1/(\sin t)$, that is, on the relation

$$\frac{1}{\sin t} = \frac{1}{t} + \sum_{n=1}^{\infty} (-1)^n \left(\frac{1}{t - n\pi} + \frac{1}{t + n\pi}\right),$$

where t is arbitrary, but not a multiple of π. To verify this relation, we consider the function

$$f(x) = \cos ax,$$

where a is not an integer and $-\pi \le x \le \pi$. Since

$$\frac{1}{2} a_0 = \frac{1}{\pi} \int_0^{\pi} \cos ax\, dx = \frac{\sin a\pi}{a\pi}$$

and, for $n > 0$,

$$a_n = \frac{2}{\pi} \int_0^{\pi} \cos ax \cos nx\, dx = \frac{1}{\pi} \int_0^{\pi} [\cos(a + n)x + \cos(a - n)x]\, dx$$

$$= (-1)^n \frac{2a}{a^2 - n^2} \cdot \frac{\sin a\pi}{\pi},$$

we see that the Fourier cosine series expansion of the function f yields

$$\frac{\pi}{2} \frac{\cos ax}{\sin a\pi} = \frac{1}{2a} + \sum_{n=1}^{\infty} (-1)^n \frac{a \sin nx}{a^2 - n^2} \qquad (-\pi \le x \le \pi).$$

Setting $x = 0$ we get

$$\frac{1}{\sin a\pi} = \frac{1}{a\pi} + 2 \sum_{n=1}^{\infty} \frac{(-1)^n a\pi}{(a\pi)^2 - (n\pi)^2}$$

and putting $a\pi = t$ yields

$$\frac{1}{\sin t} = \frac{1}{t} + \sum_{n=1}^{\infty} (-1)^n \frac{2t}{t^2 - (n\pi)^2} = \frac{1}{t} + \sum_{n=1}^{\infty} (-1)^n \left(\frac{1}{t - n\pi} + \frac{1}{t + n\pi}\right);$$

here t is an arbitrary real number, but not an integral multiple of π.

We now write

$$I = \int_0^{\infty} \frac{\sin x}{x}\, dx = \sum_{k=0}^{\infty} \int_{k\pi/2}^{(k+1)\pi/2} \frac{\sin x}{x}\, dx.$$

For $k = 2m$ we consider the substitution $x = m\pi + t$ and for $k = 2m - 1$ we consider the substitution $x = m\pi - t$. This leads to

$$\int_{2m\pi/2}^{(2m+1)\pi/2} \frac{\sin x}{x}\, dx = (-1)^m \int_0^{\pi/2} \frac{\sin t}{m\pi + t}\, dt$$

and

$$\int_{(2m-1)\pi/2}^{2m\pi/2} \frac{\sin x}{x}\, dx = (-1)^{m-1} \int_0^{\pi/2} \frac{\sin t}{m\pi - t}\, dt.$$

It follows that

$$I = \int_0^{\pi/2} \frac{\sin t}{t}\, dt + \sum_{m=1}^{\infty} \int_0^{\pi/2} (-1)^m \left(\frac{1}{t - m\pi} + \frac{1}{t + m\pi}\right) \sin t\, dt.$$

But the series

$$\sum_{m=1}^{\infty} (-1)^m \left(\frac{1}{t - m\pi} + \frac{1}{t + m\pi}\right) \sin t$$

converges uniformly in the interval $0 \le t \le \pi/2$ because it is majorized by the convergent series

$$\frac{1}{\pi} \sum_{m=1}^{\infty} \frac{1}{m^2 - \frac{1}{4}}$$

and so can be integrated term-wise. We therefore have

$$I = \int_0^{\pi/2} \sin t \left\{ \frac{1}{t} + \sum_{m=1}^{\infty} (-1)^m \left(\frac{1}{t - m\pi} + \frac{1}{t + m\pi}\right) \right\} dt.$$

But we already know that

$$\frac{1}{\sin t} = \frac{1}{t} + \sum_{m=1}^{\infty} \left(\frac{1}{t - m\pi} + \frac{1}{t + m\pi}\right).$$

where t is arbitrary, but not a multiple of π; we may therefore conclude that

$$I = \int_0^{\pi/2} \sin t \frac{1}{\sin t}\, dt = \int_0^{\pi/2} dt = \frac{\pi}{2}.$$

(This elegant calculation of the integral I is due to N. I. Lobatshewski.)

PROBLEM 53. Show that, if x is a positive integer,

$$\frac{1}{2} e^x = 1 + \frac{x}{1!} + \frac{x^2}{2!} + \cdots + \frac{x^x}{x!} \theta(x), \tag{E.34}$$

where $\theta(x)$ lies between 1/2 and 1/3 and is decreasing as x increases from 0 to ∞.

Solution. From (E.34) we get

$$\theta(x) = 1 + \frac{1}{2}\frac{e^x\,x!}{x^x} - \frac{x!}{x^x}\left\{1 + \frac{x}{1!} + \cdots + \frac{x^x}{x!}\right\}.$$

Using integration by parts, we note that

$$1 + \frac{x}{1!} + \cdots + \frac{x^x}{x!} = e^x\,\frac{x^{x+1}}{x!}\int_1^\infty (te^{-t})^x\,dt$$

and

$$\frac{x!}{x^{x+1}} = \int_0^\infty (te^{-t})^x\,dt,$$

we obtain

$$\theta(x) = 1 + \frac{xe^x}{2}\left\{\int_0^1 (te^{-t})^x\,dt - \int_1^\infty (te^{-t})^x\,dt\right\}.$$

Let w be defined by

$$we^{-w} = te^{-t}$$

for all $t > 0$ with $w(t) \geq 1$ when $0 < t \leq 1$ and $w(t) \leq 1$ when $1 \leq t$. The function w is given explicitly by

$$w(t) = ts(t),$$

where s is defined by

$$\frac{\log s}{s - 1} = t \qquad \text{for } t > 0;$$

in other words, s is the inverse function of $(\log t)/(t - 1)$. Hence

$$\int_1^\infty (te^{-t})^x\,dt = \int_1^\infty (we^{-w})^x\,dt$$

and substituting

$$t = w(u), \quad u = w(t)$$

we have

$$\int_1^\infty (te^{-t})^x\,dt = -\int_0^1 (ue^{-u})^x\,w'(u)\,du.$$

Thus

$$\theta(x) = 1 + \frac{xe^x}{2}\int_0^1 (te^{-t})^x (1 + w')\, dt.$$

Using integration by parts, after having multiplied and divided by $(1 - t)/t$ and set

$$w_1(t) = \frac{t}{1 - t}\{1 + w'(t)\},$$

we get

$$\theta(x) = 1 + \frac{1}{2} w_1(1) - \frac{e^x}{2}\int_0^1 (te^{-t})^x w_1'(t)\, dt. \tag{E.35}$$

From the definition of the function w we get

$$w'(t) = \frac{1 - t}{t}\frac{w}{1 - w},$$

$$w_1(t) = \frac{t}{1 - t}\frac{w}{1 - w},$$

$$w_1'(t) = \frac{1 - t}{t}\left(\frac{t}{(1 - t)^3} + \frac{w}{(1 - w)^3}\right);$$

we obtain $w_1(0+) = -1$, and since

$$\frac{\log s}{s - 1} = t,$$

we see that $w_1(1) = -4/3$. Substituting this later value into (E.35), we obtain

$$\theta(x) = \frac{1}{3} - \frac{1}{2}\int_0^1 (te^{1-t})^x w_1'(t)\, dt.$$

To see that $\theta(x)$ is decreasing from

$$\theta(0+) = \frac{1}{3} + \frac{1}{2}\{w_1(0+) - w_1(1)\} = \frac{1}{2}$$

to

$$\theta(\infty) = \frac{1}{3},$$

it is sufficient to show that $w_1'(t) \le 0$, that is,

$$\frac{t}{(1 - t)^3} \le \frac{w}{(w - 1)^3}. \tag{E.36}$$

But

$$w = ts \quad \text{and} \quad t = \frac{\log s}{s - 1}$$

and so (E.36) is seen to be equivalent to

$$t = \frac{\log s}{s - 1} \le \frac{1 + s^{1/3}}{s + s^{1/3}} \quad \text{for } s > 0.$$

However, the latter inequality is true (see Problem 95 of Chapter 2) and we have the desired result.

Remarks. The result in Problem 53 was posed as a question by Ramanujan. The Solution given above is due to Karamata.

We draw attention to the fact that the result in Problem 49 can easily be deduced from the result in Problem 53. Indeed, since (by Problem 53)

$$\frac{e^n}{2} = \sum_{k=0}^{n-1} \frac{n^k}{k!} + (n^n/n!)\theta_n, \qquad 1/3 < \theta_n < 1/2,$$

we have

$$\sum_{x=0}^{m-1} m^x/x! - e^{m-1} = e^m/2 - (m^m/m!)\theta_m$$

$$\ge e^m[1/2 - 1/e - (m^m/2m!)e^{-m}]$$

$$= e^m f(m), \quad \text{say.}$$

Since $(m^m/m!)e^{-m}$ is a decreasing sequence $(m = 1,2,\ldots)$, $f(m)$ is increasing, and since $f(m) \ge 0$ for $m = 3$, we get

$$\sum_{x=0}^{m-1} \frac{m^x}{x!} \ge e^{m-1}$$

for $m = 3,4,\ldots$ But this gives us the desired result, since the inequality is obviously true for $m = 1$ and $m = 2$.

We also note that

$$\frac{e^n}{2} = \sum_{k=0}^{n-1} \frac{n^k}{k!} + \frac{n^n}{n!}\theta_n, \qquad 1/3 < \theta_n < 1/2,$$

may be rewritten in the form

$$\frac{n!}{n^n}\left\{2\sum_{k=0}^{n}\frac{n^k}{k!} - e^n\right\} = 2(1 - \theta_n),$$

from which it is evident that $(n!/n^n)(2\ \Sigma_{k=0}^n\ n^k/k! - e^n) \to 4/3$ as $n \to \infty$ because $\theta_n \to 1/3$ as $n \to \infty$ (see Solution of Problem 53).

PROBLEM 54. Suppose that f is Riemann integrable over [a,b], $m \le f(x) \le M$ for all x in [a,b], g is continuous on [m,M], and $h(x) = g[f(x)]$. Show that h is Riemann integrable over [a,b].

Solution. Let $\varepsilon > 0$ be given. By the uniform continuity of g on [m,M] we can find some $\delta_1 > 0$ such that

$$|g(s) - g(t)| < \varepsilon$$

if $|s - t| < \delta_1$ and $s, t \in [m,M]$. Let $\delta = \min\{\delta_1,\varepsilon\}$. Corresponding to δ^2, choose a partition $P = \{x_0, x_1, \ldots, x_n\}$ of [a,b] such that

$$U(P,f) - L(P,f) < \delta^2,$$

which is possible by Problem 11. As usual, let

$$m_k = \inf\{f(x)\colon x \in [x_{k-1},x_k]\} \quad \text{and} \quad M_k = \sup\{f(x)\colon x \in [x_{k-1},x_k]\}$$

and let m_k^* and M_k^* be the analogous numbers for the function h. Divide the numbers 1,2,...,n into two classes: $k \in A$ if $M_k - m_k < \delta$ and $k \in B$ if $M_k - m_k \ge \delta$.

For $k \in A$, our choice shows that $M_k^* - m_k^* \le \varepsilon$.

For $k \in B$, $M_k^* - m_k^* \le 2K$, where $K = \sup\{|g(t)|: m \le t \le M$. But $U(P,f) - L(P,f) < \delta^2$ and so, letting $\Delta x_k = x_k - x_{k-1}$,

$$\delta \sum_{k\in B} \Delta x_k \le \sum_{k\in B} (M_k - m_k)\ \Delta x_k < \delta^2 \quad \text{or} \quad \sum_{k\in B} \Delta x_k < \delta.$$

It follows that

$$U(P,h) - L(P,h) = \sum_{k\in A} (M_k^* - m_k^*)\ \Delta x_k + \sum_{k\in B} (M_k^* - m_k^*)\ \Delta x_k$$

$$\le \varepsilon(b - a) + 2K\delta < \varepsilon(b - a + 2K).$$

Since $\varepsilon > 0$ was arbitrary, the result in Problem 11 implies that h is Riemann integrable on [a,b].

Remarks. Letting $g(t) = t^2$, we see that if f is Riemann integrable on [a,b], then so is the function f^2. In view of the identity

$$4uv = (u + v)^2 - (u - v)^2,$$

we can easily see that if u and v are Riemann integrable functions on [a,b], then so is the product uv.

Letting $g(t) = |t|$, we see that if f is Riemann integrable on [a,b], then so is the function $|f|$. Moreover, choosing c as either -1 or 1 to make

$$c \int_a^b f(x)\, dx \geq 0,$$

it follows that

$$\left| \int_a^b f(x)\, dx \right| = c \int_a^b f(x)\, dx = \int_a^b c\, f(x)\, dx \leq \int_a^b |f(x)|\, dx$$

because $c\, f(x) \leq |f(x)|$ for all $x \in [a,b]$.

PROBLEM 55. Let f and g be Riemann integrable functions on [a,b]. Show the following Cauchy-Schwarz Inequality for integrals:

$$\left(\int_a^b f(x)\, g(x)\, dx\right)^2 \leq \left\{\int_a^b (f(x))^2\, dx\right\}\left\{\int_a^b (g(x))^2\, dx\right\}.$$

Solution. For any real number t,

$$\int_a^b (tf(x) + g(x))^2\, dx \geq 0,$$

that is,

$$t^2 \int_a^b \{f(x)\}^2\, dx + 2t \int_a^b f(x)\, g(x)\, dx + \int_a^b \{g(x)\}^2\, dx \geq 0,$$

which directly implies the desired inequality.

PROBLEM 56. Let f and g be two Riemann integrable functions on [a,b]. Let

$$u \leq f(x) \leq U, \quad v \leq g(x) \leq V$$

for all $x \in [a,b]$, where u, U, v, V are fixed real constants. Show that

$$\left|\frac{1}{b-a}\int_a^b f(x)\ g(x)\ dx - \frac{1}{(b-a)^2}\int_a^b f(x)\ dx\cdot\int_a^b g(x)\ dx\right| \le \frac{1}{4}(U-u)(V-v). \tag{E.37}$$

Solution. By making the substitution $x = (t - a)/(b - a)$ the problem is reduced to the special case $a = 0$, $b = 1$. In that case we write

$$F = \int_0^1 f(x)\ dx, \qquad G = \int_0^1 g(x)\ dx,$$

and

$$D(f,g) = \int_0^1 f(x)\ g(x)\ dx - FG.$$

Then (E.37) reads

$$|D(f,g)| \le \frac{1}{4}(U-u)(V-v). \tag{E.38}$$

Note that

$$D(f,f) = \int_0^1 (f(x))^2\ dx - \left\{\int_0^1 f(x)\ dx\right\}^2 \ge 0 \tag{E.39}$$

holds by the Cauchy-Schwarz Inequality (see Problem 55). On the other hand,

$$D(f,f) = (U-F)(F-u) - \int_0^1 \{U - f(x)\}\{f(x) - u\}\ dx,$$

which implies that

$$D(f,f) \le (U-F)(F-u). \tag{E.40}$$

One can easily verify that

$$D(f,f) \le \int_0^1 \{f(x) - F\}\{g(x) - G\}\ dx.$$

Using the Cauchy-Schwarz Inequality, we get

$$(D(f,g))^2 \le \int_0^1 \{f(x) - F\}^2\ dx\cdot\int_0^1 \{g(x) - G\}^2\ dx = D(f,f)D(g,g).$$

According to (E.39) and (E.40), we infer that

$$(D(f,g))^2 \le (U - F)(F - u)(V - G)(G - v). \tag{E.41}$$

Since

$$4(U - F)(F - u) \le (U - u)^2,$$

$$4(V - G)(G - v) \le (V - v)^2,$$

we conclude that (E.41) implies (E.38).

Remark. Taking

$$f(x) = -1 \quad \text{for } 0 \le x \le 1/2$$
$$= 1 \quad \text{for } 1/2 < x \le 1$$

and putting $g(x) = f(x)$, we see that the constant 1/4 in (E.38) is the best possible.

PROBLEM 57. Let f be a function on an interval whose length is not less than 2 and suppose that

$$|f(x)| \le 1 \quad \text{and} \quad |f''(x)| \le 1 \quad \text{for all } x \text{ of this interval.}$$

Show that

$$|f'(x)| \le 2 \quad \text{for all } x \text{ of this interval,} \tag{E.42}$$

where the constant 2 is the best possible.

Solution. Without loss of generality we can suppose that $0 \le x \le 2$. Then, by Taylor's theorem,

$$f(x) - f(0) = xf'(x) - \frac{1}{2}x^2 f''(t_1) \quad \text{for } 0 \le t_1 \le x \le 2,$$

$$f(2) - f(x) = (2 - x)f'(x) + \frac{1}{2}(2 - x)^2 f''(t_2) \quad \text{for } 0 \le x \le t_2 \le 2,$$

and, therefore,

$$f(2) - f(0) = 2f'(x) - \frac{1}{2}x^2 f''(t_1) + \frac{1}{2}(2 - x)^2 f''(t_2),$$

that is,

$$2|f'(x)| \le 1 + 1 + \frac{1}{2}x^2 + \frac{1}{2}(2 - x)^2 = 4 - x(2 - x) \le 4,$$

or

$$|f'(x)| \le 2.$$

The function f defined by

$$f(x) = \frac{1}{2}x^2 - 1$$

shows that the sign of equality can actually hold in (E.42).

PROBLEM 58. In this problem we consider one of the four proofs which Gauss gave of the Fundamental Theorem of Algebra:

If $f(x) = x^n + a_1x^{n-1} + \cdots + a_n$ $(n > 0)$, where $a_1, \ldots, a_n$ are real or complex numbers, then f has at least one real or complex root.

Put $x = r(\cos\theta + i\sin\theta)$; then $x^k = r^k(\cos k\theta + i\sin k\theta)$, hence $f(x) = P + iQ$, where

$$P = r^n \cos n\theta + \cdots, \qquad Q = r^n \sin n\theta + \cdots$$

and all other terms of P and Q contain only smaller powers of r and terms not containing r will be constant.

The Fundamental Theorem of Algebra will be proved if we can show that $P^2 + Q^2$ is zero for a certain pair of values r and θ. We introduce the function

$$U = \arc\tan\frac{P}{Q}.$$

Then

$$\frac{\partial U}{\partial r} = \frac{(\partial P/\partial r)Q - P(\partial Q/\partial r)}{P^2 + Q^2}, \qquad \frac{\partial U}{\partial \theta} = \frac{(\partial P/\partial \theta)Q - P(\partial Q/\partial \theta)}{P^2 + Q^2};$$

hence

$$\frac{\partial^2 U}{\partial r\partial\theta} = \frac{H(r,\theta)}{(P^2 + Q^2)^2}.$$

Here $H(r,\theta)$ is a continuous function of its variables whose exact form is of no interest to us.

Finally, we need the double integrals

$$I_1 = \int_0^R \int_0^{2\pi} \frac{\partial^2 U}{\partial r\partial\theta}\, d\theta\, dr \quad \text{and} \quad I_2 = \int_0^{2\pi} \int_0^R \frac{\partial^2 U}{\partial r\partial\theta}\, dr\, d\theta;$$

R is a positive constant whose value we shall determine later on.

If the function $P^2 + Q^2$ would not become zero anywhere, then the integrand would be continuous and this would, of course, imply that $I_1 = I_2$. But we shall see that $I_1 \neq I_2$ when R becomes large. This means, however, that the function $P^2 + Q^2$ must become zero at some point in the interior of the circle $x^2 + y^2 = R^2$, proving the Fundamental Theorem of Algebra.

Show that the equality $I_1 = I_2$ cannot be fulfilled.

Solution. We compute the interior integral in I_1 and obtain

$$\int_0^{2\pi} \frac{\partial^2 U}{\partial r \partial \theta} \, d\theta = \left. \frac{\partial U}{\partial r} \right|_0^{2\pi} = 0$$

since clearly $\partial U/\partial r$ is a function dependent on θ and having period 2π. From this it follows that $I_1 = 0$. We next consider the integral I_2. Here

$$\int_0^{R} \frac{\partial^2 U}{\partial r \partial \theta} \, dr = \left. \frac{\partial U}{\partial \theta} \right|_0^{R}.$$

For what follows it is important to consider the highest power of r in the numerator and the denominator of $\partial U/\partial \theta$. Since

$$\frac{\partial P}{\partial \theta} = - n\, r^n \sin n\theta + \cdots$$

$$\frac{\partial Q}{\partial \theta} = n\, r^n \cos n\theta + \cdots$$

we get

$$\frac{\partial P}{\partial \theta} Q - P \frac{\partial Q}{\partial \theta} = - n\, r^{2n} + \cdots$$

But

$$R^2 + Q^2 = r^{2n} + \cdots,$$

and so, finally,

$$\frac{\partial U}{\partial \theta} = \frac{- n\, r^{2n} + \cdots}{r^{2n} + \cdots}.$$

Since the remaining terms in the numerator and the denominator are made up of smaller powers of r whose coefficients are bounded functions of θ, we have not only that

$$\lim_{r \to \infty} \frac{\partial U}{\partial \theta} = - n,$$

but even that this convergence to $-n$ is uniform in θ, that is, given any $\varepsilon > 0$, there is a real number $M = M(\varepsilon) > 0$ not dependent on θ such that $|\partial U/\partial\theta + n| < \varepsilon$ for all θ whenever $r > M$. Since $\partial U/\partial\theta = 0$ for $r = 0$ (note that in this case we have $\partial P/\partial\theta = \partial Q/\partial\theta = 0$), the interior integral of I_2 leads to the value $\partial U/\partial\theta$ for $r = R$. In case $R \to \infty$, this value tends to $-n$ uniformly in θ. Hence we obtain

$$\lim_{R\to\infty} I_2 = -2\pi n.$$

We see therefore that the integral I_2 is negative for sufficiently large R. Thus, the equality $I_1 = I_2$ cannot be fulfilled and this completes the proof.

PROBLEM 59. A polygon is said to be convex if it contains the line segments connecting any two of its points. Prove the following proposition: Any convex polygon K (in the complex plane) which contains all the zeros of a polynomial $P(z)$ also contains all the zeros of the derivative $P'(z)$, regardless of whether z is real or complex.

Solution. Let P be a polynomial of degree n and have the zeros $z_1, z_2, \ldots, z_n$. Then

$$P(z) = a_0(z - z_1)(z - z_2) \cdots (z - z_n).$$

Thus

$$\frac{P'(z)}{P(z)} = \sum_{k=1}^{n} \frac{1}{z - z_k}.$$

Let K be the smallest convex polygon containing $z_1, z_2, \ldots, z_n$. If z is a zero of P' and coincides with one of the z_k, then there is nothing to prove; if z is a zero of P' but different from all z_k, then

$$\sum_{k=1}^{n} \frac{1}{z - z_k} = 0$$

holds and it is sufficient to show that the foregoing equation cannot be satisfied for any point z outside of K. We do this now.

Let $z_1, z_2, \ldots, z_n$ be arbitrary points of the complex plane, $m_1 > 0$, $m_2 > 0, \ldots, m_n > 0$, $m_1 + m_2 + \cdots + m_n = 1$ and

$$z = m_1 z_1 + m_2 z_2 + \cdots + m_n z_n.$$

Interpreting the numbers $m_1, m_2, \ldots, m_n$ as masses fixed at the points z_1, $z_2, \ldots, z_n$, the point z defined by $z = m_1 z_1 + m_2 z_2 + \cdots + m_n z_n$ is the center of gravity of this mass distribution. If we consider all such mass distributions at the points $z_1, z_2, \ldots, z_n$ the corresponding centers of gravity cover the interior of a convex polygon, the smallest one containing the points $z_1, z_2, \ldots, z_n$.

The equation

$$\sum_{k=1}^{n} \frac{1}{z - z_k} = 0$$

implies

$$\frac{\overline{z} - \overline{z}_1}{|z - z_1|^2} + \frac{\overline{z} - \overline{z}_2}{|z - z_2|^2} + \cdots + \frac{\overline{z} - \overline{z}_n}{|z - z_n|^2} = 0.$$

Thus

$$z = m_1 z_1 + m_2 z_2 + \cdots + m_n z_n, \qquad m_1 + m_2 + \cdots + m_n = 1,$$

where the k-th "mass" m_k is proportional to

$$\frac{1}{|z - z_k|^2}, \qquad k = 1,2,\ldots,n.$$

Hence if z were outside the smallest convex polygon K that contains the z_k's, there could be no equilibrium.

Remarks. From Rolle's Theorem (see Solution to Problem 5) we know that any interval on the real line which contains all the zeros of a real-valued polynomial P also contains all the zeros of the derivative P'; the result in Problem 59 generalizes this fact.

Any convex polygon which contains all the zeros of a polynomial P also contains all the zeros of its derivatives.

PROBLEM 60. Let $f(x)$ be a complex valued function for $-\infty < x < \infty$ satisfying

$$f(x + y) = f(x)\, f(y) \tag{E.43}$$

and

$$|f(x)| = 1 \tag{E.44}$$

for all x and y of the number line $(-\infty,\infty)$.

Show that if f is continuous on $(-\infty, \infty)$, then

$$f(x) = e^{i\lambda x}, \quad -\infty < x < \infty,$$

where λ is real.

Solution. Suppose first that f is everywhere differentiable. Then differentiation of (E.43) with respect to x gives

$$f'(x + y) = f'(x)\ f(y).$$

Putting $x = 0$ and $f'(0) = i\lambda$, where λ is a complex number, we get

$$f'(y) = i\lambda\ f(y)$$

and integration gives (by changing back to the variable x)

$$f(x) = e^{i\lambda x}.$$

Condition (E.44) tells us that $|f(x)| = 1$ for all real x, hence λ necessarily has to be real. Thus, under the additional assumption that f is differentiable, the claim is established.

To complete the solution, we show that any continuous solution of the functional equation

$$f(x + y) = f(x)\ f(y)$$

is automatically differentiable. We put

$$F_\varepsilon(a) = \int_a^{a+\varepsilon} f(x)\ dx.$$

For fixed ε this function is differentiable with respect to a. We have

$$F_\varepsilon(a)\ f(y) = \int_a^{a+\varepsilon} f(x)\ f(y)\ dx = \int_a^{a+\varepsilon} f(x + y)\ dx$$

$$= \int_{a+y}^{a+y+\varepsilon} f(x)\ dx = F_\varepsilon(a + y).$$

Assuming that $F_\varepsilon(a) \neq 0$, we obtain

$$f(y) = \frac{F_\varepsilon(a + y)}{F_\varepsilon(a)}.$$

Here $F_\varepsilon(a + y)$ is a differentiable function with respect to y, hence f(y) is seen to be differentiable. It only remains to verify that there exist an a and an ε such that $F_\varepsilon(a) \neq 0$. Keeping a fixed and differentiating F (a) with respect to ε, we obtain

$$\frac{\partial}{\partial\varepsilon} F_\varepsilon(a) = \frac{\partial}{\partial\varepsilon}\int_a^{a+\varepsilon} f(x)\, dx = f(a + \varepsilon).$$

If for some a we had $F_\varepsilon(a) \equiv 0$ for all ε, then

$$\frac{\partial}{\partial\varepsilon} F_\varepsilon(a) \equiv 0$$

would follow and thus $f(a + \varepsilon) \equiv 0$. This (trivial) case we can exclude. Then, however, necessarily $F_\varepsilon(a) \not\equiv 0$ must hold and thus there has to be some ε such that $F_\varepsilon(a) \neq 0$ and the proof is complete.

PROBLEM 61. A real-valued function f defined on a set E of real numbers is said to be *uniformly continuous on* E if, given any $\varepsilon > 0$, there exists a $\delta > 0$ such that for all x and y in E with $|x - y| < \delta$ we have $|f(x) - f(y)| < \varepsilon$.

A sequence $(x_n)_{n=1}^\infty$ of real numbers is said to be a *Cauchy sequence* if for any $\varepsilon > 0$ there is an integer n_0 such that $|x_n - x_{n'}| < \varepsilon$ if $n \geq n_0$ and $n' \geq n_0$.

Show: If a real valued function f is uniformly continuous on a set E of real numbers and if $(x_n)_{n=1}^\infty$ is any Cauchy sequence of elements in E, then $\{f(x_n)\}_{n=1}^\infty$ is also a Cauchy sequence. Conversely, if a real-valued function f, defined on a bounded set E of real numbers transforms Cauchy sequences of elements of E into Cauchy sequences, then f is uniformly continuous on E.

Solution. Let f be uniformly continuous on E and let $(x_n)_{n=1}^\infty$ be a Cauchy sequence of elements in E. Given $\varepsilon > 0$, there exists $\delta > 0$ such that if x' and x'' are in E and $|x' - x''| < \delta$ then

$$|f(x') - f(x'')| < \varepsilon.$$

But, associated with the number $\delta > 0$, there exists an index n_0 such that if $m,n \geq n_0$ then $|x_m - x_n| < \delta$; therefore,

$$|f(x_n) - f(x_m)| < \varepsilon,$$

which shows that $\{f(x_n)\}_{n=1}^\infty$ is a Cauchy sequence.

On the other hand, assume that the function f is not uniformly continuous on a bounded set E. The negation of uniform continuity may be thus expressed: there exists an $\varepsilon > 0$ such that for any $\delta > 0$ there exist points x_n', x_n'' in E such that $|x_n' - x_n''| < \delta$, however, $|f(x_n') - f(x_n'')| \geq \varepsilon$.

Since E is bounded, it is possible to extract a convergent subsequence of $(x_n')_{n=1}^{\infty}$, namely $(x_{n_k}')_{k=1}^{\infty}$; it is not implied, however, that the limit x_0 of $(x_{n_k}')_{k=1}^{\infty}$ belongs to E; at any rate, $(x_{n_k}')_{k=1}^{\infty}$ is a Cauchy sequence. In the same way, there exists a Cauchy sequence $(x_{n_k}'')_{k=1}^{\infty}$ having the same limit x_0, as it follows from the inequality

$$|x_{n_k}'' - x_0| \leq |x''_{n_k} - x_{n_k}'| + |x_{n_k}' - x_0|$$

since both $|x_{n_k}'' - x_{n_k}'|$ and $|x_{n_k}' - x_0|$ may be arbitrarily small.

Therefore, the sequence $(x_k)_{k=1}^{\infty}$, where $x_1 = x_{n_1}'$, $x_2 = x_{n_1}''$, ..., $x_{2k-1} = x_{n_k}'$, $x_{2k} = x_{n_k}''$, ..., obtained by "mixing" the two convergent subsequences $(x_{n_k}')_{k=1}^{\infty}$ and $(x_{n_k}'')_{k=1}^{\infty}$, is again a Cauchy sequence.

By hypothesis, we deduce that $\{f(x_k)\}_{k=1}^{\infty}$ is also a Cauchy sequence; thus, given any $\varepsilon > 0$, there exists an index k_0 such that if $k,k' \geq k_0$, then $|f(x_k) - f(x_{k'})| < \varepsilon$. However, this is not the case, since for any k_0 we have

$$|f(x_{2k_0-1}) - f(x_{2k_0})| = |f(x_{n_{k_0}}') - f(x_{n_{k_0}}'')| \geq \varepsilon.$$

This shows that f must be uniformly continuous on the bounded set E.

PROBLEM 62. Let E be a set of real numbers and A be a subset of E. We say that A is *dense in* E if the closure of A is E.

Prove the following proposition, called the *Principle of Extension by Continuity:* Let E be a set of real numbers and A be a subset of E which is dense in E. Moreover, let f be a real-valued function defined and uniformly continuous on A. Then there exists a unique continuous function g on E such that g(x) = f(x) for every x in A. In addition, if E is bounded, then g is uniformly continuous on E.

Solution. First, we observe that if there is at all a continuous function g on E, with the properties stated, then, for every x in E and for every

x in E and for every sequence $(x_n)_{n=1}^{\infty}$ of elements of A such that $\lim_{n\to\infty} x_n = x$ (which exists because x is a limit point of A), we must have

$$g(x) = \lim_{n\to\infty} g(x_n) = \lim_{n\to\infty} f(x_n)$$

since $x_n \in A$. Thus, $g(x)$ must be the unique limit of the convergent sequence $\{f(x_n)\}_{n=1}^{\infty}$ and, consequently, only one continuous function g can exist with the required property.

We have also found out how we should define the value of g at every point $x \in E$. In fact, if $x \in E$, it is a limit point of A, hence there exists a sequence $(x_n)_{n=1}^{\infty}$ of elements of A such that $\lim_{n\to\infty} x_n = x$; thus $(x_n)_{n=1}^{\infty}$ is a Cauchy sequence because any convergent sequence is a Cauchy sequence. But f is uniformly continuous on A and so, by Problem 57, $\{f(x_n)\}_{n=1}^{\infty}$ is a Cauchy sequence. However, any Cauchy sequence of real numbers is a convergent sequence, and so there exists a real number

$$y = \lim_{n\to\infty} f(x_n).$$

We define $g(x) = y$.

Now, we justify this definition by showing that if $(z_n)_{n=1}^{\infty}$ is another Cauchy sequence in A such that $\lim_{n\to\infty} z_n = x$, then we have $\lim_{n\to\infty} f(z_n) = \lim_{n\to\infty} f(x_n)$. This, however, follows from the uniform continuity of f on A. Given any $\varepsilon > 0$, there exists a $\delta > 0$ such that if $x,z \in A$ and $|x - z| < \delta$ then $|f(x) - f(z)| < \varepsilon$. Associated with this number δ, there exists an index n_0 such that both $|x_n - x| < \delta/2$ and $|z_n - x| < \delta/2$ for every index $n \geq n_0$; hence, for any index $n \geq n_0$,

$$|x_n - z_n| \leq |x_n - x| + |x - z_n| < \delta.$$

Thus, for $n \geq n_0$,

$$|f(x_n) - f(z_n)| < \varepsilon,$$

which shows that $\{f(x_n)\}_{n=1}^{\infty}$ and $\{f(z_n)\}_{n=1}^{\infty}$ are Cauchy sequences converging to the same limit.

Finally, we must show that g is continuous on E. Since every point x of E is in some bounded subset T of the set E, it is enough to prove that if T is any bounded subset of E, then g is uniformly continuous on T. This implies, indeed, that g is continuous on T, and, therefore, at the arbitrary point x of E. At the same time, since we have not excluded that $T = E$ whenever E is assumed to be bounded, this will also establish that g is uniformly continuous on E when E is bounded.

We therefore assume that T is a bounded subset of E. By Problem 61, it is sufficient to show that if $(x_n)_{n=1}^{\infty}$ is any Cauchy sequence of elements of T, then $\{g(x_n)\}_{n=1}^{\infty}$ is also a Cauchy sequence.

Given any $\varepsilon > 0$, since f is uniformly continuous on A, there exists $\delta > 0$ such that $x', z' \in A$, $|x' - z'| < \delta$, implies

$$|f(x') - f(z')| < \frac{\varepsilon}{3}.$$

Since $(x_n)_{n=1}^{\infty}$ is a Cauchy sequence, associated with this δ, there exists an index n_0 such that if $m,n \geq n_0$ then

$$|x_m - x_n| < \frac{\delta}{3}.$$

Consider any two indices $m,n \geq n_0$. Since

$$x_m = \lim_{k\to\infty} x'_{m,k}, \quad x_n = \lim_{k\to\infty} x'_{n,k}, \quad (x'_{m,k}, x'_{n,k} \in A),$$

there exists a sufficiently large index k_0 (dependent on m,n) such that, for $k \geq k_0$,

$$|x_m - x'_{m,k}| < \frac{\delta}{3} \quad \text{and also} \quad |x_n - x'_{n,k}| < \frac{\delta}{3};$$

therefore, if $m,n \geq n_0$, and taking any index $k \geq k_0$, we have

$$|x'_{m,k} - x'_{n,k}| \leq |x'_{m,k} - x_m| + |x_m - x_n| + |x_n - x'_{n,k}| < \frac{\delta}{3} + \frac{\delta}{3} + \frac{\delta}{3} = \delta.$$

Hence, by the uniform continuity of f, $|f(x'_{m,k}) - f(x'_{n,k})| < \varepsilon$ provided that $m,n \geq n_0$ and $k \geq k_0$.

On the other hand, considering any two natural numbers $m,n > n_0$, since

$$g(x_m) = \lim_{k\to\infty} f(x'_{m,k}) \quad \text{and} \quad g(x_n) = \lim_{k\to\infty} f(x'_{n,k}),$$

given the number $\varepsilon > 0$, there is a sufficiently large index k_1, which we may take such that $k_1 \geq k_0$, with the property: if $k \geq k_1$, then $|g(x_m) - f(x'_{m,j})| < \varepsilon/3$ and $|g(x_n) - f(x'_{m,j})| < \varepsilon/3$. Combining these inequalities, if $m,n \geq n_0$ and using $k \geq k_1$, it follows from

$$g(x_m) - g(x_n) = \{g(x_m) - f(x'_{m,k})\} + \{f(x'_{m,k}) - f(x'_{n,k})\} + \{f(x'_{n,k}) - g(x_n)\}$$

that

$$|g(x_m) - g(x_n)| < \frac{\varepsilon}{3} + \frac{\varepsilon}{3} + \frac{\varepsilon}{3} = \varepsilon.$$

This shows that $\{g(x_n)\}_{n=1}^{\infty}$ is a Cauchy sequence. This completes the proof.

Remarks. The result in Problem 62 shows that if f is a uniformly continuous function on the set of all rational points of an interval I, then there exists one and only one continuous extension of f, namely the function g, which is continuous on all of I. As a simple application of the result in Problem 62 we may consider the definition of the exponential function with base $a > 0$, that is,

$$g(x) = a^x, \qquad -\infty < x < \infty.$$

Here we first consider the functions $f_n(r) = a^r$ with rational r and the domain of definition of f_n restricted to the interval $[-n,n]$. Applying the result in Problem 62, we obtain g_n, namely the continuous extension of f_n. Finally, we define g on the whole real line as follows: if x is any real number and n is a positive integer such that $|x| \le n$, we put $g(x) = g_n(x)$; this definition is independent of the integer n.

PROBLEM 63. If, for $a > 0$,

$$f(x) = \cos ax, \qquad \text{resp.} \qquad f(x) = \cosh ax, \tag{E.45}$$

then for any real numbers x and y we have the functional equation

$$f(y + x) + f(y - x) = 2f(x)f(y). \tag{E.46}$$

This follows immediately from the familiar formulas

$$\cos(y \pm x) = \cos x \cos y \mp \sin x \sin y,$$

$$\cosh(y \pm x) = \cosh x \cosh y \pm \sinh x \sinh y.$$

Show that the only nonzero continuous functions on the real line R^1 that satisfy the functional equation (E.46) are the trigonometric and the hyperbolic cosines (E.45).

Solution. Let some function f satisfy the requirements of the claim. We let $x = 0$ and choose y such that $f(y) \neq 0$; then

$$f(0) = 1. \tag{E.47}$$

For $y = 0$ we get

$$f(-x) = f(x), \tag{E.48}$$

that is, f is an even function.

Since the continuous function f is positive for $x = 0$, there exists a positive number c such that $f(x)$ is positive on the entire interval $[0,c]$. The following considerations follow two different lines of reasoning depending on the two cases:

Case 1. $f(c) \leq 1$

Case 2. $f(c) > 1$.

We consider Case 1 first. Since $0 < f(c) \leq 1$, there is a θ with $0 \leq \theta < \pi/2$ such that

$$f(c) = \cos\theta. \tag{E.49}$$

We now write (E.46) in the form

$$f(y + x) = 2f(x)f(y) - f(y - x)$$

and put, successively,

$$x = c, \quad y = c,$$

$$x = c, \quad y = 2c,$$

$$x = c, \quad y = 3c,$$

etc. Using (E.47) and (E.49), we obtain

$$f(2c) = 2\cos^2\theta - 1 = \cos 2\theta,$$

$$f(3c) = 2\cos\theta\cos 2\theta - \cos\theta = \cos 3\theta,$$

$$f(4c) = 2\cos\theta\cos 3\theta - \cos 2\theta = \cos 4\theta,$$

etc. By induction, we get for any positive integer m,

$$f(mc) = \cos m\theta. \tag{E.50}$$

If in (E.46) we put $x = y = c/2$, we get, using (E.47) and (E.49),

$$[f(c/2)]^2 = \frac{f(0) + f(c)}{2} = \frac{1 + \cos\theta}{2} = \{\cos(\theta/2)\}^2;$$

since $f(x)$ is positive between 0 and c and $\cos x$ is positive between 0 and θ, we may take the root and obtain

$$f(c/2) = \cos(/2).$$

In the same fashion we get, putting $x = y = (1/2^2)c$ in (E.46),

$$f(c/2^2) = \cos(\theta/2^2),$$

etc. By induction, for $n = 1,2,3,\ldots$, we get

$$f(c/2^n) = \cos(\theta/2^n). \qquad \text{(E.51)}$$

By repeating the process that got us from (E.49) to (E.50), we obtain from (E.51)

$$f(mc/2^n) = \cos(m\theta/2^n).$$

Thus, if x is of the form $m/2^n$, we have

$$f(cx) = \cos \theta x. \qquad \text{(E.52)}$$

But every positive real number can be represented as a p-adic fraction with $p = 2$. Since f and the trigonometric cosine function are continuous, we see that (E.52) holds for any $x > 0$. That (E.52) also holds for $x = 0$ and $x < 0$, can be seen from (E.47) and (E.48), respectively. If in (E.52) we replace x by x/c and put $\theta/c = a$, we obtain

$$f(x) = \cos ax.$$

In Case 2 we have $f(c) > 1$. Then there exists a θ such that $f(c) = \cosh \theta$. By repeating the reasoning used in Case 1 and applying the corresponding formulas for the hyperbolic cosine we get, for $a > 0$,

$$f(x) = \cosh ax.$$

For $a = 0$, we get, of course, in both cases $f(x) = 1$ for all $x \in R^1$. This completes the proof.

PROBLEM 64. Show that any continuous mapping g of the real line R^1 into itself such that

$$g\left(\frac{x + y}{2}\right) = \frac{g(x) + g(y)}{2}$$

for any $x, y \in R^1$ is of the form

$$g(x) = cx + a,$$

where c and a are constants.

Solution. Indeed, for $y = 0$, we obtain from the functional equation

$$g\left(\frac{x}{2}\right) = \frac{g(x) + g(0)}{2} = \frac{g(x) + a}{2},$$

where $a = g(0)$; thus

$$\frac{g(x) + g(y)}{2} = g\left(\frac{x + y}{2}\right) = \frac{g(x + y) + a}{2},$$

that is, $g(x + y) = g(x) + g(y) - a$, and, with $f(x) = g(x) - a$,

$$f(x + y) = f(x) + f(y) \quad \text{for any } x, y \in R^1.$$

But g is continuous; hence f is continuous and so, by the result in Problem 48, f is of the form $f(x) = cx$ with $c = f(1)$. Thus g is of the form $g(x) = cx + a$, where c and a are constants.

PROBLEM 65. Let f be an everywhere finite, nondecreasing, real-valued function whose domain of definition is the real line R^1. In addition, let f be right continuous at every point; that is,

$$\lim_{x \to y+} f(x) = f(y).$$

Suppose a half-open interval $(a,b]$ of finite length is covered by open intervals (a_i,b_i); that is,

$$(a,b] \subset \sum_{i=1}^{\infty} (a_i,b_i).$$

Show that

$$f(b) - f(a) \le \sum_{i=1}^{\infty} [f(b_i) - f(a_i)].$$

Solution. By right continuity of f, given $\varepsilon > 0$, there exists $\delta > 0$ such that $a + \delta < b$ and

$$f(a + \delta) < f(a) + \varepsilon.$$

By the Heine-Borel theorem, the closed interval $[a + \delta,b]$ is covered by a finite set (a_{i_1},b_{i_1}), ..., (a_{i_m},b_{i_m}) of the given intervals (a_i,b_i). Clearly, for this finite covering

$$a_{i_1} < a + \delta,$$

$$b_{i_m} > b,$$

and the intervals may be so numbered that for each k (k = 1,2,...,m),

$$a_{i_{k+1}} < b_{i_k}.$$

Since f is nondecreasing, we have

$$\sum_{i=1}^{\infty} [f(b_i) - f(a_i)] \geq \sum_{k=1}^{m} [f(b_{i_k}) - f(a_{i_k})]$$

$$= f(b_{i_m}) - f(a_{i_1}) + \sum_{k=1}^{m-1} [f(b_{i_k}) - f(a_{i_{k+1}})]$$

$$\geq f(b) - f(a + \delta) > f(b) - f(a) - \varepsilon$$

for every $\varepsilon > 0$.

Remark. Let A and B be given real numbers and suppose that $A < B + \varepsilon$ for any $\varepsilon > 0$. Does it follow that $A \leq B$? Yes, indeed! Note that $A > B$ implies $A \geq B + \varepsilon$ for some $\varepsilon > 0$.

PROBLEM 66. Let f be a continuous function on the interval [0,1]. Show that

$$\lim_{n\to\infty} \sum_{k=0}^{n} (-1)^k \binom{n}{k} f\left(\frac{k}{n}\right)/2^n = 0.$$

Solution. We note that f is uniformly continuous on [0,1]. Hence, for any given $\varepsilon > 0$, there exists a positive integer $N(\varepsilon)$ such that

$$|f(k/n) - f(\{k + 1\}/n)| < \varepsilon, \quad (n > N;\ k = 0,1,\dots,n-1).$$

For $n > N$, we have

$$S_n = \frac{1}{2^n} \sum_{k=0}^{n} (-1)^k \binom{n}{k} f\left(\frac{k}{n}\right) = \frac{1}{2^n} \sum_{k=0}^{n-1} (-1)^k \binom{n-1}{k} \left\{ f\left(\frac{k}{n}\right) - f\left(\frac{k+1}{n}\right) \right\},$$

$$|S_n| < \frac{\varepsilon}{2^n} \sum_{k=0}^{n-1} \binom{n-1}{k} = \frac{\varepsilon}{2},$$

from which the desired result follows.

Remarks. An alternate way of solving Problem 62 is as follows: Given any positive number ε, we can find a polynomial function P such that $|f(x) - P(x)| \le \varepsilon$ for all x in $[0,1]$. This is the Weierstrass Approximation Theorem (see Problem 40). If n is greater than the degree of P, then by the theory of finite differences (or otherwise)

$$\sum_{k=0}^{n} (-1)^k \binom{n}{k} P\left(\frac{k}{n}\right) = 0.$$

Thus, if n is sufficiently large,

$$\left|2^{-n} \sum_{k=0}^{n} (-1)^k \binom{n}{k} f\left(\frac{k}{n}\right)\right| = \left|2^{-n} \sum_{k=0}^{n} (-1)^k \binom{n}{k} \left\{f\left(\frac{k}{n}\right) - P\left(\frac{k}{n}\right)\right\}\right|$$

$$\le 2^{-n} \sum_{k=0}^{n} \binom{n}{k} \varepsilon = \varepsilon.$$

Since ε is arbitrary, the assertion of the problem is established.

PROBLEM 67. Let the function F(t) have a continuous derivative on $[0,1]$ and put $S_1 = \{t: F'(t) = 0\}$, $S_2 = \{F(t): t \in S_1\}$. Show by an example that the set S_2 may be uncountable.

Solution. Let $g(x) = 0$ on the Cantor set (see Problem 23) and $g(x) = (x - a)(b - x)$ in each interval (a,b) forming the complement of the Cantor set. Let

$$F(t) = \int_0^t g(x)\, dx;$$

Then $F' = g$ is continuous, and S_1 is the uncountable Cantor set. Since $\{x: g(x) > 0\}$ is everywhere dense, F is strictly increasing, and hence one-to-one; therefore S_2 is also uncountable.

PROBLEM 68. The function given by $f(x) = 0$, if x is irrational or $x = 0$, and $1/q$, if $x = p/q$, where $q > 0$, and p and q are integers without common divisors, was considered in Problem 14. We saw in Problem 14 that this function f is not continuous on the set of nonzero rationals of the unit interval, the only possible points of differentiability are the irrationals of the unit interval and zero because f is continuous at these points of the unit interval $[0,1]$.

Show that the function f is not differentiable at its points of continuity on the interval $[0,1]$.

Solution. If $x = 0$ and $h \neq 0$, $[f(x + h) - f(x)]/h = f(h)/h$. Let (h_i) be a sequence of irrationals having zero as limit. Then $\lim_{i \to \infty} f(h_i)/h_i = 0$. Now let $h_i = 1/i$, $i = 1,2,\ldots$ Then $f(h_i)/h_i = (1/i)/(1/i) = 1$. Therefore $\lim_{h \to 0} f(h)/h$ does not exist and f is nondifferentiable at $x = 0$.

If x is an irrational number $0 < x < 1$ and if $h \neq 0$, $[f(x+h) - f(x)]/h = [f(x+h)]/h$. If (h_i) is a sequence of real numbers having zero as limit such that $x + h_i$ is irrational for each i, then $\lim_{i \to \infty} [f(x+h_i)]/h_i = 0$. Let the decimal representation of x be $0.a_1a_2\ldots a_n \ldots$ Choose $h_i = 0.a_1a_2\ldots a_i - x$. Since $x \neq 0$, $a_i \neq 0$ for some i. Let N be the least integer such that $a_N \neq 0$. Then $f(x + h_i) = f(0.a_1a_2\ldots a_i) \geq 10^{-i}$ for all $i \geq N$, and $|h_i| \leq 10^{-i}$. Hence,

$$\left|\frac{f(x + h_i)}{h_i}\right| \geq 1 \quad \text{for all } i \geq N.$$

Therefore $\lim_{h \to 0} [f(x + h)]/h$ does not exist and f is not differentiable.

PROBLEM 69. Find an explicit function f on $[0,1]$ to the real numbers, such that f' is defined on $[0,1]$ and discontinuous on the rationals in $[0,1]$.

Solution. Let $f(x) = x^2 \sin(1/x)$; then (see Problem 10) f is differentiable everywhere, and f' is discontinuous only at $x = 0$. Let $r_1, r_2, \ldots$ be an enumeration of the rationals in $[0,1]$; define

$$g(x) = \sum_{n=1}^{\infty} 2^{-n} f(x - r_n).$$

Then $g'(x) = \Sigma_{n=1}^{\infty} 2^{-n} f'(x - r_n)$, since both series are uniformly convergent. Then g' exists and is discontinuous at each rational in $[0,1]$.

PROBLEM 70. Let $b_{n+1} = \int_0^1 \min(x,a_n)\,dx$ and $a_{n+1} = \int_0^1 \max(x,b_n)\,dx$.

Show that the sequences (a_n) and (b_n) both converge and find their limits.

Solution. For any a_0, b_0, it is easy to see that a_n and b_n both lie between 0 and 1 for all $n \geq 2$. The recurrence formulas then become (for $n \geq 2$)

$$a_{n+1} = (1 + b_n^2)/2, \quad b_{n+1} = a_n - a_n^2/2.$$

If we assume that $\lim_{n\to\infty} a_n = a$, $\lim_{n\to\infty} b_n = b$, they must satisfy

$$a = (1 + b^2)/2, \quad b = a - a^2/2,$$

from which we get

$$a + b - 1 = (a + b - 1)(b - a + 1)/2.$$

Since the factor $(b - a + 1)2 \neq 1$, we have $a + b = 1$, and this yields $a = 2 - \sqrt{2}$, $b = \sqrt{2} - 1$. To show that $a_n \to a$, $b_n \to b$ as $n \to \infty$, we write $a_n = a + t_n$, $b_n = b + s_n$. The recurrence formulas become, after an easy reduction,

$$t_{n+1} = (b + s_n/2)s_n, \quad s_{n+1} = (b - t_n/2)t_n.$$

Now $|s_n| = |b_n - b| \leq \max(b,1-b) = a$, and $|t_n| = |a_n - a| \leq \max(a,1-a) = a$. Hence $|b_n + s_n/2| \leq b + a/2 = 1/\sqrt{2}$, $|b - t_n/2| \leq b + a/2 = 1/\sqrt{2}$. Therefore $|t_{n+1}| \leq |s_n|/\sqrt{2}$ and $|s_{n+1}| \leq |t_n|/\sqrt{2}$. This shows that $t_n \to 0$, $s_n \to 0$ as $n \to \infty$, and the proof is complete.

PROBLEM 71. Let

$$a_{n+1} = \int_0^1 \min(x,b_n,c_n)\, dx, \quad b_{n+1} = \int_0^1 \operatorname{med}(x,c_n,a_n)\, dx,$$

and

$$c_{n+1} = \int_0^1 \max(x,a_n,b_n)\, dx,$$

where $\operatorname{med}(a,b,c) = b$ if $a \leq b \leq c$. Show that the proposed sequences (a_n), (b_n), and (c_n) are convergent, with limits 3/8, 1/2, 5/8, respectively.

Solution. Evidently $\min(x,b_n,c_n) \leq x \leq \max(x,a_n,b_n)$ so that, if $x = \operatorname{med}(x,a_n,c_n)$, we have

$$\min(x,b_n,c_n) \le \operatorname{med}(x,a_n c_n) \le \max(x,a_n,b_n). \tag{E.53}$$

Now, if $\operatorname{med}(x,a_n,c_n) = a_n$, we have either $x \le a_n$ or $c_n \le a_n$, so that $a_n \le \max(x,a_n,b_n)$ implies (E.53) in this case also. A similar argument holds for $\operatorname{med}(x,a_n,c_n) = c_n$, so that (E.53) is true in all cases. By integration there results $a_{n+1} \le b_{n+1} \le c_{n+1}$, $n = 1,2,3,\ldots$

Now we have

$$a_{n+1} = \int_0^1 \min(x,b_n,c_n)\, dx \le \int_0^1 x\, dx = \frac{1}{2},$$

and similarly $c_{n+1} \ge 1/2$. Using this

$$b_{n+2} = \int_0^1 \operatorname{med}(x,a_{n+1},c_{n+1})\, dx$$

$$= \int_0^{1/2} \max(x,a_{n+1})\, dx + \int_{1/2}^1 \min(x,c_{n+1})\, dx \le \int_0^{1/2} \frac{1}{2}\, dx + \int_{1/2}^1 x\, dx = \frac{5}{8}.$$

Dually, $b_{n+2} \ge 3/8$. Since $3/8 \le b_{n+2} \le c_{n+2}$,

$$a_{n+3} = \int_0^1 \min(x,b_{n+2},c_{n+2})\, dx > 0,$$

and similarly $c_{n+3} < 1$.

It is now assumed that n is so large that $0 < a_n \le b_n \le c_n < 1$.

$$a_{n+1} = \int_0^{b_n} x\, dx + \int_{b_n}^1 b_n\, dx = \frac{2b_n - b_n}{2},$$

$$b_{n+1} = \int_0^{a_n} a_n\, dx + \int_{a_n}^{c_n} x\, dx + \int_{c_n}^1 c_n\, dx = \frac{a_n^2 - c_n^2 - 2c_n}{2},$$

$$c_{n+1} = \int_0^{b_n} b_n\, dx + \int_{b_n}^1 x\, dx = \frac{b_n^2 + 1}{2}.$$

Thus

$$b_{n+2} = \frac{1}{2}\left[\left\{\frac{2b_n - b_n^2}{2}\right\}^2 - \left\{\frac{b_n + 1}{2}\right\}^2 + 2\left\{\frac{b_n + 1}{2}\right\}\right]$$

$$= \frac{1}{2} + \frac{(2b_n - 1)(-2b_n^2 + 2b_n + 1)}{8}$$

$$= b_n - \frac{(2b_n - 1)^3 + 5(2b_n - 1)}{16}.$$

Since $0 < -2b_n^2 + 2b_n + 1$ whenever $0 < b_n < 1$, either

$$\frac{1}{2} \le b_{n+2} \le b_n \quad \text{or} \quad \frac{1}{2} > b_{n+2} > b_n.$$

It follows that $\lim_{n\to\infty} b_{2n} = \lim_{n\to\infty} b_{2n+1} = \lim_{n\to\infty} b_n = 1/2$. Then

$$\lim_{n\to\infty} a_n = \lim_{n\to\infty} a_{n+1} = \lim_{n\to\infty} \frac{2b_n - b_n^2}{2} = \frac{3}{8},$$

$$\lim_{n\to\infty} c_n = \lim_{n\to\infty} c_{n+1} = \lim_{n\to\infty} \frac{b_n^2 + 1}{2} = \frac{5}{8}.$$

Remark. Note that

$\text{med}(a,b,c) = \min\{\max(a,b), \max(b,c), \max(c,a)\}$.

PROBLEM 72. Let f be a polynomial of degree n such that

$$\int_0^1 x^k f(x)\, dx = 0 \quad \text{for } k = 1,2,\ldots,n. \tag{E.54}$$

Show that

$$\int_0^1 \{f(x)\}^2\, dx = (n + 1)^2 \left\{\int_0^1 f(x)\, dx\right\}^2. \tag{E.55}$$

Solution. Set $f(x) = a_n x^n + a_{n-1}x^{n-1} + \cdots + a_0$. By (E.54),

$$\int_0^1 \{f(x)\}^2\, dx = a_0 \int_0^1 f(x)\, dx. \tag{E.56}$$

Also

$$\int_0^1 x^k f(x)\, dx = \frac{a_n}{n + k + 1} + \frac{a_{n-1}}{n + k} + \cdots + \frac{a_0}{k + 1};$$

$$\frac{a_n}{n + k + 1} + \frac{a_{n-1}}{n + k} + \cdots + \frac{a_0}{k + 1} = 0, \quad k = 1,2,\ldots,n.$$

Now

$$\frac{a_n}{n + k + 1} + \frac{a_{n-1}}{n + k} + \cdots + \frac{a_0}{k + 1} = \frac{Q(k)}{(k + 1)(k + 2) \cdots (n + k + 1)},$$

where Q is a polynomial of degree at most n. Since $Q(k) = 0$, $k = 1,2,\ldots,n$, $Q(k) = C(k - 1)(k - 2)\cdots(k - n)$ with C a constant. Hence

$$\begin{aligned} &\frac{a_n}{n + k + 1} + \frac{a_{n-1}}{n + k} + \cdots + \frac{a_0}{k + 1} \\ &= \frac{C(k - 1)(k - 2) \cdots (k - n)}{(k + 1)(k + 2) \cdots (n + k + 1)}. \end{aligned} \tag{E.57}$$

For $k = 0$,

$$\int_0^1 f(x)\, dx = \frac{a_n}{n + 1} + \cdots + \frac{a_0}{1} = (-1)^n \frac{C}{n + 1}.$$

Multiply (E.57) by $k + 1$ and set $k = -1$ to obtain

$$a_0 = (-1)^n(n + 1)C \quad \text{or} \quad a_0 = (n + 1)^2 \int_0^1 f(x)\, dx.$$

Thus, by (E.56), we obtain the desired result (E.55).

PROBLEM 73. Let $a < b$ be given numbers and let $f(t)$ be defined, continuous, non-negative and strictly increasing for $a \le t \le b$. By the law of the mean for integrals, for every $p > 0$ there will exist a unique number x_p, $a \le x_p \le b$, such that

$$f^p(x_p) = \frac{1}{b - a} \int_a^b f^p(t)\, dt.$$

Find $\lim_{n \to \infty} x_p$.

Solution. Let ε be a given number such that $0 < \varepsilon < (b - a)/2$. Since f is strictly increasing, $f(b - \varepsilon)/f(b - 2\varepsilon) > 1$. Hence there exists a positive integer P such that for $p > P$,

$$\left\{\frac{f(b - \varepsilon)}{f(b - 2\ )}\right\}^p > \frac{b - a}{\varepsilon}$$

or

$$f^p(b - \varepsilon) > \frac{b - a}{\varepsilon} f^p(b - 2\varepsilon).$$

We have also

$$\int_a^b f^p(t)\, dt > \int_{b-\varepsilon}^b f^p(b - \varepsilon)\, dt,$$

and therefore

$$\frac{1}{b - a}\int_a^b f^p(t)\, dt > \frac{1}{b - a}\int_{b-\varepsilon}^b f^p(b - \varepsilon)\, dt$$

$$= \frac{1}{b - a}\cdot\varepsilon\cdot f^p(b - \varepsilon) > f^p(b - 2\varepsilon).$$

Hence x_p cannot satisfy

$$f^p(x_p) = \frac{1}{b - a}\int_a^b f^p(t)\, dt$$

unless $x_p > b - 2\varepsilon$. It follows that there exists a P such that, whenever $p > P$, $x_p > b - 2\varepsilon$. Hence $\lim_{p\to\infty} x_p = b$.

PROBLEM 74. Consider the function

$$H_n(t,x) = \frac{1}{\pi}\cdot\frac{n}{1 + n^2(t - x)^2}.$$

If n and x are kept fixed, while t vaires between 0 and 1, the function $H_n(t,x)$ is seen to be a continuous function of t. Hence, for any integrable function $f(t)$ $(0 \le t \le 1)$ we can form the integral

$$f_n(x) = \frac{n}{\pi}\int_0^1 \frac{f(t)\, dt}{1 + n^2(t - x)^2}.$$

Show that for any point of continuity x $(0 < x < 1)$ of the function f we have

$$\lim_{n\to\infty} f_n(x) = f(x). \tag{E.58}$$

Solution. First we note that, for $n \to \infty$,

$$\int_0^1 H_n(t,x)\,dt = \frac{n}{\pi}\int_0^1 \frac{dt}{1 + n^2(t - x)^2} = \frac{1}{\pi}\int_{-nx}^{n(1-x)} \frac{dz}{1 + z^2}$$

$$\to \frac{1}{\pi}\int_{-\infty}^{+\infty} \frac{dz}{1 + z^2} = 1.$$

To prove the relation (E.58) it therefore will be sufficient to show that the difference

$$r_n = f_n(x) - f(x)\int_0^1 H_n(t,x)\,dt = \frac{n}{\pi}\int_0^1 \frac{f(t) - f(x)}{1 + n^2(t - x)^2}\,dt$$

tends to zero as $n \to \infty$.

For every $\varepsilon > 0$ there exists a δ such that for $|t - x| < \delta$ the inequality $|f(t) - f(x)| < \varepsilon$ holds. Under the assumption that $0 < x - \delta < x + \delta < 1$ we can express the difference r_n as follows:

$$r_n = \frac{n}{\pi}\int_0^{x-\delta} \frac{f(t) - f(x)}{1 + n^2(t - x)^2}\,dt + \frac{n}{\pi}\int_{x-\delta}^{x+\delta} \frac{f(t) - f(x)}{1 + n^2(t - x)^2}\,dt$$

$$+ \frac{n}{\pi}\int_{x+\delta}^{1} \frac{f(t) - f(x)}{1 + n^2(t - x)^2}\,dt$$

$$= A_n + B_n + C_n.$$

The integral B_n can be estimated as follows:

$$|B_n| \le \int_{x-\delta}^{x+\delta} \frac{|f(t) - f(x)|}{1 + n^2(t - x)^2}\,dt \le \varepsilon\cdot\frac{n}{\pi}\int_{x-\delta}^{x+\delta} \frac{dt}{1 + n^2(t - x)^2}$$

$$< \frac{\varepsilon}{\pi}\int_{-\infty}^{+\infty} \frac{dz}{1 + z^2} = \varepsilon.$$

In the integral A_n we have $|t - x| \ge \delta$; thus

$$|A_n| \le \frac{n}{\pi(1 + n^2\delta^2)}\int_0^{x-\delta} |f(t) - f(x)|\,dt < \frac{A(\delta)}{n},$$

where $A(\delta)$ is independent of n. In a similar fashion we get

$$|C_n| < \frac{C(\delta)}{n};$$

hence

$$|r_n| < \varepsilon + \frac{A(\delta) + C(\delta)}{n}$$

and so, for sufficiently large n,

$$|r_n| < 2\varepsilon.$$

Thus $r_n \to 0$ as $n \to \infty$ and we have the desired result.

PROBLEM 75. A real-valued function f defined on an interval I is said to be *convex* if

$$f[\lambda x + (1 - \lambda)y] \le \lambda f(x) + (1 - \lambda)f(y)$$

whenever x and y belong to I and $0 \le \lambda \le 1$. Geometrically, this means that if P, Q, and R are any three points on the graph of f with Q between P and R, then Q is on or below chord PR.

Prove that if [a,b] is any closed subinterval of the interior of I, then f is bounded on [a,b] and there is a constant K so that for any two points $x, y \in [a,b]$,

$$|f(x) - f(y)| \le K|x - y|.$$

Solution. Observe that $M = \max\{f(a),f(b)\}$ is an upper bound for f on [a,b], since for any point $z = \lambda a + (1 - \lambda)b$ in [a,b]

$$f(z) \le \lambda f(a) + (1 - \lambda)f(b) \le \lambda M + (1 - \lambda)M = M.$$

But f is also bounded below because, writing an arbitrary point in the form $(a + b)/2 + t$, we have

$$f\left(\frac{a + b}{2}\right) \le \frac{1}{2} f\left(\frac{a + b}{2} + t\right) + \frac{1}{2} f\left(\frac{a + b}{2} - t\right)$$

or

$$f\left(\frac{a + b}{2} - t\right) \ge 2f\left(\frac{a + b}{2}\right) + f\left(\frac{a + b}{2} - t\right).$$

Using M as upper bound,

$$-f\left(\frac{a + b}{2} - t\right) \ge -M,$$

so

$$f\left(\frac{a + b}{2} + t\right) \ge 2f\left(\frac{a + b}{2}\right) - M = m.$$

This shows that M and m are upper and lower bounds of f on [a,b], respectively.

We next pick $h > 0$ so that $a - h$ and $b + h$ belong to the interval I, and let m and M be the lower and upper bounds for f on $[a - h, b + h]$. If x and y are distinct points of [a,b], set

$$z = y + \frac{h}{|y - x|}(y - x), \qquad \lambda = \frac{|y - x|}{h + |y - x|}.$$

Then $z \in [a - h, b + h]$, $y = \lambda z + (1 - \lambda)x$, and we have

$$f(y) \le \lambda f(z) + (1 - \lambda)f(x) = \lambda[f(z) - f(x)] + f(x),$$

$$f(y) - f(x) \le \lambda(M - m) < \frac{|y - x|}{h}(M - m) = K|y - x|,$$

where $K = (M - m)/h$. Since this is true for any $x, y \in [a,b]$, we conclude that $|f(y) - f(x)| \le K|y - x|$ as desired.

Remarks. Recalling from Problem 30 the definition of absolute continuity, we see that the choice $\delta = \varepsilon/K$ meets the requirement for asserting that f is absolutely continuous on [a,b]. To see that f must be continuous on the interior of the interval I, we note that [a,b] is an arbitrary closed subinterval in the interior of I. Hence, if f satisfies the conditions in Problem 75, then f is absolutely continuous on any closed interval [a,b] in the interior of I and continuous on the interior of I.

PROBLEM 76. Let f be defined and bounded on [0,1], and $f(ax) = bf(x)$ for $0 \le x \le 1/a$, with a and b numbers larger than 1. Show that $f(0+) = f(0)$.

Solution. Clearly $f(0) = 0$; if M is an upper bound for $|f|$ on [0,1], then, for $0 \le x \le a^{-n}$, $|f(x)| = |b^{-n} f(a^n x)| \le Mb^{-n}$.

PROBLEM 77. A point x is called a *point of condensation of a set* E of real numbers if every neighborhood of x contains uncountably many points of E.

Show that any uncountable set E of real numbers has at least one point of condensation.

Solution. Suppose by way of contradiction that no point of E is a condensation point of E. Then for each x in E, there is an open interval I_x, containing x, such that $I_x \cap E$ is countable. Let J_x be an open interval contained in I_x, and containing x, but having rational endpoints; indeed, $J_x \cap E$ is again countable. Moreover, the collection of all such intervals J_x is countable and we may enumerate it as follows

$$J_1, J_2, \ldots, J_n, \ldots$$

Since each point of E is in some J_k, we may conclude that

$$E = \bigcup_{k=1}^{\infty} (J_k \cap E).$$

The countable union of countable sets being countable, we have the desired contradiction.

PROBLEM 78. Show that any uncountable set of real numbers includes a sequence $(x_n)_{n=1}^{\infty}$ of distinct numbers such that $\Sigma_{n=1}^{\infty} x_n = \pm\infty$.

Solution. Since the set is uncountably infinite, it must have a point of condensation, say a, by Problem 77. For definiteness we assume that $a \geq 0$. If $a > 0$, the neighborhood $(a/2,3a/2)$ contains uncountably many points of the given set and the desired result follows. If $a = 0$, the neighborhood $(-1,1)$ contains uncountably many points of the given set, hence at least one of the intervals $(-1,0)$ or $(0,1)$ must contain uncountably many points of the given set; let $(0,1)$ contain uncountably many points of the given set. Considering all subintervals of the interval $(0,1)$ that are of the form

$$\left(\frac{1}{k+1},\frac{1}{k}\right), \quad k = 1,2,3,\ldots,$$

we see that at least one of them, say the interval

$$\left(\frac{1}{k_0+1},\frac{1}{k_0}\right)$$

must contain uncountably many points of the given set. Picking a sequence $(x_n)_{n=1}^{\infty}$ of the prescribed kind in

$$\left(\frac{1}{k_0+1},\frac{1}{k_0}\right)$$

clearly leads to $\Sigma_{n=1}^{\infty} x_n = \infty$.

PROBLEM 79. Give an example of a continuous curve in the x, y plane that passes through every point of the unit square $[0,1] \times [0,1]$.

Solution. Let g be defined on the interval $[0,2]$ by

$$
\begin{aligned}
g(t) &= 0 && \text{for } 0 \le t \le \tfrac{1}{3} \text{ and for } \tfrac{5}{3} \le t \le 2,\\
&= 3t - 1 && \text{for } \tfrac{1}{3} \le t \le \tfrac{2}{3},\\
&= 1 && \text{for } \tfrac{2}{3} \le t \le \tfrac{4}{3},\\
&= -3t + 5 && \text{for } \tfrac{4}{3} \le t \le \tfrac{5}{3}.
\end{aligned}
$$

We extend the definition of g to all of the real line R^1 by setting

$$g(t + 2) = g(t).$$

Note that g is continuous on R^1 and has period 2.

We now define two functions x and y by the following equations:

$$x(t) = \sum_{n=1}^{\infty} \frac{g(3^{2n-2}t)}{2^n}, \qquad y(t) = \sum_{n=1}^{\infty} \frac{g(3^{2n-1}t)}{2^n}.$$

Both series converge uniformly on R^1; moreover, both functions x and y are continuous on R^1.

Let $z = (x,y)$ and G denote the image of the interval $[0,1]$ under z. We will show that $G = [0,1] \times [0,1]$.

Evidently, $0 \le x(t) \le 1$ and $0 \le y(t) \le 1$ for each t, since the series $\Sigma_{n=1}^{\infty} 2^{-n} = 1$. Hence, G is a subset of $[0,1] \times [0,1]$. It remains to show that $(a,b) \in [0,1] \times [0,1]$. For this purpose we express a and b as p-adic fractions with $p = 2$, that is, we put

$$a = \sum_{n=1}^{\infty} \frac{a_n}{2^n}, \qquad b = \sum_{n=1}^{\infty} \frac{b_n}{2^n},$$

where each a_n and each b_n is either 0 or 1. Let

$$c = 2 \sum_{n=1}^{\infty} \frac{c_n}{3^n},$$

where c_{2n-1} a_n and $c_{2n} = b_n$, for $n = 1,2,\ldots$

Since $2\, \Sigma_{n=1}^{\infty} 3^{-n} = 1$, we see that $0 \le c \le 1$. We will show that $x(c) = a$ and $y(c) = b$.

If we can verify that

$$g(3^k c) = c_{k+1} \tag{E.59}$$

holds for $k = 1,2,\ldots$, then we will have that

$$g(3^{2n-2}c) = c_{2n-1} = a_n \quad \text{and} \quad g(3^{2n-1}c) = c_{2n} = b_n;$$

this, however, will give us $x(c) = a$ and $y(c) = b$. To prove (E.59) we put

$$3^k c = 2\sum_{n=1}^{k} \frac{c_n}{3^{n-k}} + 2\sum_{n=k+1}^{\infty} \frac{c_n}{3^{n-k}} = (\text{an even integer}) + d_k,$$

where

$$d_k = 2\sum_{n=1}^{\infty} \frac{c_{n+k}}{3^n}.$$

Since g has period 2,

$$g(3^k c) = g(d_k).$$

If $c_{k+1} = 0$, we have

$$0 \le d_k \le 2\sum_{n=2}^{\infty} 3^{-n} = \frac{1}{3},$$

and hence $g(d_k) = 0$; thus (E.59) is satisfied in this case. The only other case to consider is $c_{k+1} = 1$. But then we get

$$\frac{2}{3} \le d_k \le 1$$

and so $g(d_k) = 1$. Therefore (E.59) holds in all cases and we have obtained the desired result.

PROBLEM 80. If

$$F(x) = \sum_{n=1}^{\infty} \frac{n^{n-1}x^{n-1}}{n!} e^{-nx}, \tag{E.60}$$

Show that $F(x) = 1$ when x is between 0 and 1, and that $F(x) \neq 1$ when $x > 1$. Find the limit

$$\frac{F(1 + h) - F(1)}{h}$$

as $h \to 0$ through positive values.

Solution. The solution of the problem at hand essentially rests on the fact that the function $F(x)$ in (E.60) is given by

$$\begin{aligned} xF(x) &= x, \qquad 0 < x \leq 1, \\ &= w(x), \qquad 1 \leq x, \end{aligned} \tag{E.61}$$

where $w(x)$ is defined by the equation

$$we^{-w} = xe^{-x} \quad \text{for all } x > 0, \tag{E.62}$$

with

$$\begin{aligned} w(x) &\geq 1 \quad \text{when } 0 < x \leq 1 \\ &\leq 1 \quad \text{when } 1 \leq x. \end{aligned} \tag{E.63}$$

The function $w(x)$ is explicitly given by

$$w(x) = xs(x), \qquad x > 0, \tag{E.64}$$

where s is the inverse function of

$$\frac{\log x}{x - 1},$$

that is,

$$\frac{\log s}{s - 1} = x, \qquad x > 0; \tag{E.65}$$

hence

$$\begin{aligned} F(x) &= 1 \qquad 0 < x \leq 1, \\ &= s(x), \qquad 1 \leq x. \end{aligned} \tag{E.66}$$

Indeed, the expansion of the function $\sigma(\lambda)$ defined by

$$\sigma e^{-\sigma} = \lambda, \quad \text{for } \lambda \leq 1/e,$$

into a Bürmann-Lagrange series being (see, for example, A. Hurwitz & R. Courant: Vorlesungenen über allgemeine Funktionentheorie und elliptische Funtionen, 4th Ed. Berlin-Göttingen-Heidelberg - New York: Springer 1964; pages 138, 141, and 142 are relevant)

$$\sigma(\lambda) = \sum_{n=1}^{\infty} \frac{n^{n-1}}{n!} \lambda^n, \qquad |\lambda| < 1/e,$$

it suffices to set $\lambda = xe^{-x}$ in order to obtain from (E.60) the relation of (E.61), where $w(x)$ is given by (E.62). Now, putting in (E.62)

$$w(x) = xs(x),$$

we obtain the equation (E.65), that is, $s(x)$ is the inverse function of $(\log x)/(x - 1)$, while $w^{-1}(x) = w(x)$; but

$$F(x) = s(x) \quad \text{for } x \geq 1$$

and $s'(1) = -2$. Hence

$$\frac{F(1 + h) - F(1)}{h} = -\frac{1}{2}$$

as $h \to 0$ through positive values.

Remark. The question posed in Problem 80 is due to Ramanujan.

PROBLEM 81. If x is positive, show that

$$R(x) = \frac{1}{x + 1} + \frac{2^0}{(x + 2)^2} + \frac{3^1}{(x + 3)^3} + \frac{4^2}{(x + 4)^4} + \frac{5^3}{(x + 5)^5} + \cdots$$
$$< \frac{1}{x}, \tag{E.67}$$

and find approximately the difference when x is great. Hence show that

$$\frac{1}{1001} + \frac{1}{1002^2} + \frac{3}{1003^3} + \frac{4^2}{1004^4} + \frac{5^3}{1005^5} + \frac{6^4}{1006^6} + \cdots$$

is less than 1/1000 by approximately 10^{-440}.

Solution. We observe that the function $R(x)$ in (E.67) is the Laplace transform of the function $F(x)$ in (E.60) and that $F(x)$ is given by (E.61) and (E.64). It is therefore sufficient to express $R(x)$ by $s(x)$ as follows

$$R(x) = \int_0^\infty e^{-xt}\, F(t)\, dt = \int_0^1 e^{-xt}\, dt + \int_1^\infty e^{-xt}\, s(t)\, dt$$
$$= \frac{1}{x} + \frac{1}{x}\int_1^\infty e^{-xt}\, ds. \tag{E.68}$$

Using the substitution

$$\sigma = s(t), \quad t = s^{-1}(\sigma) = \frac{\log \sigma}{\sigma - 1},$$

we therefore get

$$R(x) = \frac{1}{x} - \frac{e^{-x}}{x} \int_0^1 \{e(1 - \sigma)^{1/\sigma}\}^x \, d\sigma.$$

But

$$e(1 - \sigma)^{1/\sigma} \le (1 - \sigma)^{1/2}, \quad 0 \le \sigma \le 1,$$

and

$$\begin{aligned} e(1 - \sigma)^{1/\sigma} &\ge (1 - 4\sigma/3)^{3/8}, & 0 \le \sigma \le 3/4, \\ &\ge 0, & 3/4 \le \sigma \le 1 \end{aligned}$$

and thus

$$\frac{2}{x + 8/3} \le \int_0^1 \{e(1 - \sigma)^{1/\sigma}\}^x \, d\sigma \le \frac{2}{x + 2}. \tag{E.69}$$

Consequently

$$R(x) = \frac{1}{x} - \frac{2\, e^{-x}}{x(x + 2 + \theta)}, \quad 0 < \theta < 2/3. \tag{E.70}$$

The limits 8/3 and 2 in (E.69) are actually attained for $x = \infty$ and $x = 0$, respectively; that is,

$$\theta(\infty) = \frac{2}{3} \quad \text{and} \quad \theta(0) = 0.$$

where the function $\theta(x)$ is defined by (E.70).

From (E.70) we get, setting $x = 1000$,

$$10^{-439.994608} < \frac{1}{1000} - R(1000) < 10^{-439.994319},$$

that is, the factor multiplying 10^{-440} is situated between

1.0124978 and 1.01249445.

Remark. The question posed in Problem 81 is due to Ramanujan.

PROBLEM 82. Show that the series

$$\frac{x}{1 - x^2} + \frac{x^2}{1 - x^4} + \frac{x^4}{1 - x^8} + \cdots$$

converges to $x/(1 - x)$ if $|x| < 1$ and to $1/(1 - x)$ if $|x| > 1$.

Solution. We observe that

$$\frac{x}{1 - x^2} - \frac{x}{1 - x} = - \frac{x^2}{1 - x^2}.$$

Therefore

$$\frac{x^{2^n}}{1 - x^{2^{n+1}}} - \frac{x^{2^n}}{1 - x^{2^n}} = - \frac{x^{2^{n+1}}}{1 - x^{2^{n+1}}},$$

and by induction

$$\frac{x}{1 - x^2} + \frac{x^2}{1 - x^4} + \frac{x^4}{1 - x^8} + \cdots + \frac{x^{2^n}}{1 - x^{2^{n+1}}} = \frac{x}{1 - x} - \frac{x^{2^{n+1}}}{1 - x^{2^{n+1}}}.$$

The limit of the second term is 0 for $|x| < 1$ and 1 for $|x| > 1$.

PROBLEM 83. For $|x| < 1$, show that

$$P_n(x) = (1 + x)(1 + x^2)(1 + x^4) \cdots (1 + x^{2^{n-1}}) \to \frac{1}{1 - x} \quad \text{as } n \to \infty.$$

Solution. Since

$$(1 - x)P_n(x) = (1 - x^2)(1 + x^2)(1 + x^4) \cdots (1 + x^{2^{n-1}})$$

$$= (1 - x^4)(1 + x^4) \cdots (1 + x^{2^{n-1}})$$

and so forth, we see that

$$(1 - x)P_n(x) = 1 - x^{2^n}$$

or

$$P_n(x) = \frac{1 - x^{2^n}}{1 - x}.$$

Thus, for $|x| < 1$, $\lim_{n\to\infty} P_n(x) = 1/(1 - x)$.

PROBLEM 84. Compute

$$\lim_{t\to\infty} \left\{ \frac{1}{t} + \frac{2t}{t^2 + 1^2} + \frac{2t}{t^2 + 2^2} + \cdots + \frac{2t}{t^2 + n^n} + \cdots \right\}.$$

Solution. Let $f(x) = 2/(1 + x^2)$ and $h = 1/t$. Then

$$\int_h^{(m+1)h} f(x)\,dx \le h[f(h) + f(2h) + \cdots + f(mh)] \le \int_h^{mh} f(x)\,dx$$

and, as $m \to \infty$,

$$\int_h^{\infty} f(x)\,dx \le h \sum_{n=1}^{\infty} f(nh) \le \int_0^{\infty} f(x)\,dx.$$

But

$$\int_h^{\infty} f(x)\,dx = \pi - 2 \arctan h$$

and

$$\int_0^{\infty} f(x)\,dx = \pi$$

and the desired limit equals π.

PROBLEM 85. Show that for all real numbers x we have

$$|e^{\wedge}(12 - 6x + x^2) - (12 + 6x + x^2)| \le \frac{1}{60}|x|^5 e^{|x|}.$$

Solution. Consider the identity

$$\frac{e^x - (1 + x + x^2/2)}{x^3} = \frac{1}{3!} + \frac{x}{4!} + \frac{x^2}{5!} + \cdots.$$

Differentiating twice the above identity, we get

$$\frac{e^x(12 - 6x + x^2) - (12 + 6x + x^2)}{x^5}$$

$$= \frac{2}{5!} + \cdots + \frac{(n+1)(n+2)}{(n+5)!} x^n + \cdots$$

$$= \frac{1}{60} + \cdots + \frac{1}{(n+3)(n+4)(n+5)} \frac{x^n}{n!} + \cdots$$

$$= \frac{1}{60}\left(1 + \cdots + \frac{60}{(n+3)(n+4)(n+5)} \frac{x^n}{n!} + \cdots\right).$$

But

$$\frac{60}{(n+3)(n+4)(n+5)} \frac{|x|^n}{n!} \le \frac{|x|^n}{n!}$$

and so

$$60\left|\frac{e^x(12 - 6x + x^2) - (12 + 6x + x^2)}{x^5}\right| \le 1 + \cdots + \frac{|x|^n}{n!} + \cdots$$

$$= e^{|x|},$$

which yields the desired result.

Remark. The result in Problem 85 can be generalized to:

$$\left|\sum_{k=0}^{n} \{(-1)^k e^x + (-1)^{n+1}\}\binom{n+k}{n}\frac{x^{n-k}}{(n-k)!}\right| \le \frac{1}{(2n+1)!}|x|^{2n+1} e^{|x|}$$

for any real number x.

PROBLEM 86. Show that the product

$$\left(1 + \frac{1}{a-1}\right)\left(1 - \frac{1}{2a-1}\right)\left(1 + \frac{1}{3a-1}\right)\left(1 - \frac{1}{4a-1}\right) \times \cdots$$

$$\times \left(1 + \frac{1}{(2n-1)a-1}\right)\left(1 - \frac{1}{2na-1}\right)$$

tends to the limit $2^{1/a}$ as $n \to \infty$, provided that $a \ne 0, 1, 1/2, 1/3, 1/4, \ldots$

Solution. By Problem 102 of Chapter 1,

$$\left(1 + \frac{1}{a-1}\right)\left(1 - \frac{1}{2a-1}\right)\left(1 + \frac{1}{3a-1}\right)\left(1 - \frac{1}{4a-1}\right) \times \cdots$$

$$\times \left(1 + \frac{1}{(2n-1)a-1}\right)\left(1 - \frac{1}{2na-1}\right)$$

$$= \frac{(n+1)a}{(n+1)a-1} \cdot \frac{(n+2)a}{(n+2)a-1} \cdots \frac{(n+n)a}{(n+n)a-1}$$

$$= \frac{1}{\left\{1 - \frac{1}{n}\frac{1}{\left(1+\frac{1}{n}\right)a}\right\}\left\{1 - \frac{1}{n}\frac{1}{\left(1+\frac{2}{n}\right)a}\right\} \cdots \left\{1 - \frac{1}{n}\frac{1}{\left(1+\frac{n}{n}\right)a}\right\}}.$$

But the foregoing expression tends to the reciprocal of

$$\exp\left\{-\frac{1}{a}\int_0^1 \frac{dx}{1+x}\right\} = \exp\left(-\frac{1}{a}\log 2\right) = 2^{-1/a} \quad \text{as } n \to \infty$$

because

$$\lim_{n\to\infty} \sum_{k=1}^{n} \log\left(1 - \frac{1}{n}\frac{1}{\left(1+\frac{k}{n}\right)a}\right) = \lim_{n\to\infty} \sum_{k=1}^{n} -\frac{1}{n}\frac{1}{\left(1+\frac{k}{n}\right)a}.$$

Indeed,

$$|\log(1+x) - x| \le x^2 \quad \text{for } |x| \le \frac{1}{2}.$$

Hence, letting

$$A_{k,n} = \frac{-1}{\left(1+\frac{k}{n}\right)a} \quad \text{and} \quad B_n = \frac{1}{n},$$

we see that

$$\left|\sum_{k=1}^{n} \log(1 + A_{k,n}B_n) - \sum_{k=1}^{n} A_{k,n}B_n\right| \le B_n \cdot \sum_{k=1}^{n} A_{k,n}^2 B_n$$

whenever $|A_{k,n}|B_n \le 1/2$. By taking n large enough, we get that $|A_{k,n}|B_n \le 1/2$ for $k = 1,2,\ldots,n$ and that

$$B_n \cdot \sum_{k=1}^{n} A_{k,n}^2 B_n$$

differs from 0 by as little as we please.

Remarks. As a particular case of the result in Problem 86 note that for $a = 2$ we have

$$\frac{1}{2}\cdot\frac{3}{2}\cdot\frac{5}{6}\cdot\frac{7}{6}\cdot\frac{9}{10}\cdot\frac{11}{10} \cdots = \frac{1}{\sqrt{2}}.$$

The limit of

$$\frac{(n^2+1)(n^2+2)\cdots(n^2+n)}{(n^2-1)(n^2-2)\cdots(n^2-n)} = \frac{1+\frac{1}{n}\frac{1}{n}}{1-\frac{1}{n}\frac{1}{n}} \cdot \frac{1+\frac{2}{n}\frac{1}{n}}{1-\frac{2}{n}\frac{1}{n}} \cdots \frac{1+\frac{n}{n}\frac{1}{n}}{1-\frac{n}{n}\frac{1}{n}}$$

is

$$\frac{\exp\left\{\int_0^1 x\,dx\right\}}{\exp\left\{-\int_0^1 x\,dx\right\}} = e$$

as $n \to \infty$. The verification is similar to the method used in the Solution of Problem 86.

PROBLEM 87. Find the value of

$$S = \sum_{n=0}^{\infty} (-1)^n \binom{p+n-1}{n} r^n, \qquad |r| < 1$$

as the solution of a differential equation.

Solution. Let u_n denote the $(n + 1)$st term of S; then

$$\frac{u_n}{u_{n-1}} = -r\,\frac{p+n-1}{n}$$

and

$$\sum_{n=0}^{\infty} nu_n = -r \sum_{n=0}^{\infty} (p+n-1)u_{n-1}$$

or

$$\sum_{n=0}^{\infty} nu_n = -r \sum_{n=0}^{\infty} (p+n)u_n. \tag{E.71}$$

But

$$\sum_{n=0}^{\infty} nu_n = r\,\frac{dS}{dr}$$

and so (E.71) becomes

$$\frac{dS}{dr} + \frac{p}{1+r} S = 0,$$

whence

$$\log S + p \log(1 + r) + C = 0 \quad \text{or} \quad S = C(1 + r)^{-p}.$$

If $r = 0$, we have $S = 1$ and $C = 1$; therefore

$$S = \frac{1}{(1 + r)^p}.$$

PROBLEM 88. Express the infinite series

$$\frac{1}{2} + \frac{1\cdot 3}{2\cdot 4}\cdot\frac{1}{2} + \frac{1\cdot 3\cdot 5}{2\cdot 4\cdot 6}\cdot\frac{1}{3} + \frac{1\cdot 3\cdot 5\cdot 7}{2\cdot 4\cdot 6\cdot 8}\cdot\frac{1}{4} + \cdots$$

as a definite integral, and find its value.

Solution. From Problem 87 we get

$$\frac{1}{\sqrt{1 - x}} = 1 + \frac{1}{2}x + \frac{1\cdot 3}{2\cdot 4}x^2 + \frac{1\cdot 3\cdot 5}{2\cdot 4\cdot 6}x^3 + \cdots + \frac{1\cdot 3 \cdots (2n - 1)}{2\cdot 4 \cdots (2n)}x^n + \cdots$$

and so

$$\frac{1}{x}\left\{\frac{1}{\sqrt{1 - x}} - 1\right\}$$

$$= \frac{1}{2} + \frac{1\cdot 3}{2\cdot 4}x + \frac{1\cdot 3\cdot 5}{2\cdot 4\cdot 6}x^2 + \cdots + \frac{1\cdot 3 \cdots (2n - 1)}{2\cdot 4 \cdots (2n)}x^{n-1} + \cdots$$

Thus

$$\frac{1}{2} + \frac{1\cdot 3}{2\cdot 4}\cdot\frac{1}{2} + \frac{1\cdot 3\cdot 5}{2\cdot 4\cdot 6}\cdot\frac{1}{3} + \frac{1\cdot 3\cdot 5\cdot 7}{2\cdot 4\cdot 6\cdot 8}\cdot\frac{1}{4} + \cdots$$

$$= \int_0^1 \frac{1 - \sqrt{1 - x}}{x\sqrt{1 - x}}\,dx = 2\int_0^{\pi/2} \tan\frac{t}{2}\,dt = 2 \log 2.$$

PROBLEM 89. Show that, for $a > 0$,

$$\lim_{n\to\infty} \frac{1^{a-1} - 2^{a-1} + 3^{a-1} - \cdots + (-1)^{n-1} n^{a-1}}{n^a} = 0.$$

Solution. Let $f(x) = x^{a-1}$. Then, for $m = [n/2]$,

$$\frac{1}{n}\sum_{k=1}^{n-1} (-1)^{k-1} f\left(\frac{k}{n}\right) = \frac{2}{n}\sum_{k=1}^{m} f\left(\frac{2k - 1}{n}\right) - \frac{1}{n}\sum_{k=1}^{n-1} f\left(\frac{k}{n}\right)$$

which tends to $\int_0^1 f(x)\,dx - \int_0^1 f(x)\,dx = 0$ as $n \to \infty$

Remark. For a similar question see Problem 43 in Chapter 3.

PROBLEM 90. Show that

$$\lim_{x \to 1-0} \left\{ \frac{x}{1+x} - \frac{x^2}{1+x^2} + \frac{x^3}{1+x^3} - \frac{x^4}{1+x^4} + \cdots \right\} = \frac{1}{4}.$$

Solution. For $|x| < 1$,

$$\frac{x}{1+x} - \frac{x^2}{1+x^2} + \frac{x^3}{1+x^3} - \frac{x^4}{1+x^4} + \cdots = x(1 - 2x + 3x^2 - 4x^3 + \cdots)$$

$$= x\frac{d}{dx}\left(\frac{x}{1+x}\right) = \frac{x}{(1+x)^2} \to \frac{1}{4} \quad \text{as } x \to 1-0.$$

PROBLEM 91. Show that

$$\lim_{x \to 1-0} \sum_{n=1}^{\infty} \frac{(-1)^{n-1}}{n} \frac{x^n}{1+x^n} = \frac{1}{2}\log 2.$$

Solution. Let $0 < x < 1$. The series

$$\sum_{n=1}^{\infty} \frac{(-1)^{n-1}}{n}$$

converges to $\log 2$ and the factor $x^n/(1 + x^n)$ is bounded by 1 and is monotone decreasing as n increases; hence the series

$$\sum_{n=1}^{\infty} \frac{(-1)^{n-1}}{n} \frac{x^n}{1+x^n}$$

converges uniformly on $(0,1)$ by Abel's test for uniform convergence. The desired result is obtained by letting $x \to 1-0$ termwise in the series (this is of course permissible here).

PROBLEM 92. Let $x > 0$. Show that

$$\sum_{n=2}^{\infty} \frac{n}{(1 + 2x)(1 + 3x)\cdots(1 + nx)} = \frac{1}{x}.$$

Solution. Let

$$S = 1 + \frac{r^2}{1 + 2x} + \frac{r^3}{(1 + 2x)(1 + 3x)} + \frac{r^4}{(1 + 2x)(1 + 3x)(1 + 4x)} + \cdots.$$

Then

$$\frac{u_n}{u_{n-1}} = \frac{r}{1 + nx}, \tag{E.72}$$

where u_n is the n-th term of S. Proceeding as in the Solution of Problem 87, we obtain the differential equation

$$\frac{dS}{dr} + \frac{1 - r}{xr} S = \frac{1}{xr}$$

because (E.72) gives

$$\sum_{n=2}^{\infty} (nx + 1)u_n = r \sum_{n=2}^{\infty} u_n + r.$$

But

$$\sum_{n=2}^{\infty} \frac{n}{(1 + 2x)(1 + 3x)\cdots(1 + nx)}$$

is the value of $\frac{dS}{dr}$ when $r = 1$.

PROBLEM 93. Show that

$$S = \sum_{n=2}^{\infty} \frac{n!}{p(p + 1)(p + 2)\cdots(p + n)} = \frac{1}{p - 1}.$$

Solution. We may write

$$S = \frac{1}{p} \sum_{n=0}^{\infty} \frac{1}{\binom{p+n}{n}}.$$

Putting

$$S_1 = \sum_{n=0}^{\infty} \frac{r^n}{\binom{p+n}{n}},$$

then S becomes S_1 when $r = 1$. But from S_1 we have

$$\frac{u_n}{u_{n-1}} = r\frac{n}{p+n}$$

which gives the differential equation

$$\frac{dS_1}{dr} + \frac{n-r}{r(1-r)}S_1 = \frac{p}{r(1-r)}.$$

Therefore

$$S_1 = \frac{p(1-r)^{p-1}}{r^p}\left\{C + \int_0^r \frac{t^{p-1}}{(1-t)^p}\,dt\right\}.$$

But

$$\int_0^r \frac{t^{p-1}}{(1-t)^p}\,dt = \sum_{k=0}^{p-2}(-1)^k\frac{r^{p-k-1}}{(p-k-1)(1-r)^{p-k-1}} + (-1)^p\log(1-r).$$

Thus

$$S_1 = p\left\{\sum_{k=0}^{p-2}(-1)^k\frac{(1-r)^k}{(p-k-1)r^{k+1}} + (-1)^p\frac{(1-r)^{p-1}}{r^p}\log(1-r) + \frac{C(1-r)^{p-1}}{r^p}\right\}.$$

Multiplying both sides of the last equation by r^p, we get $C = 0$ if $r = 0$. Therefore $S = 1/(p-1)$

PROBLEM 94. Show that

$$\frac{1}{2}\left(\frac{2x}{1+x^2}\right) + \frac{1}{2\cdot4}\left(\frac{2x}{1+x^2}\right)^3 + \frac{1\cdot3}{2\cdot4\cdot6}\left(\frac{2x}{1+x^2}\right)^5 + \cdots = x \quad \text{for } |x| < 1,$$

$$= \frac{1}{x} \quad \text{for } |x| > 1.$$

Solution. From the binomial expansion theorem we have

$$\sqrt{1-u} = 1 - \frac{1}{2}u - \frac{1}{2\cdot4}u^2 - \frac{1\cdot3}{2\cdot4\cdot6}u^3 - \cdots - \frac{1\cdot3\cdots(2n-3)}{2\cdot4\cdot6\cdots(2n)}u^n - \cdots$$

for $|u| < 1$. Thus, for $|t| < 1$,

$$\frac{1}{2}t + \frac{1}{2\cdot4}t^3 + \frac{1\cdot3}{2\cdot4\cdot6}t^5 + \cdots = \frac{1-(1-t^2)^{1/2}}{t} \tag{E.73}$$

But $(2x)/(1+x^2) \le 1$ for all real numbers x with equality only if $x = 1$

because $0 \le (1 - x)^2$ for all real x. Setting $t = (2x)/(1 + x^2)$ with $|x| < 1$ in (E.73) we get the first part of our claim because $\{1 - (1 - t^2)^{\frac{1}{2}}\}/t = x$. Finally, for $x \neq 0$ the expression $(2x)/(1 + x^2)$ does not change if we replace x by 1/x and so the second part of our claim follows as well.

PROBLEM 95. Let α be a parameter with $0 < \alpha \le 1$, and define

$$f_\alpha(x) = [\alpha/x] - \alpha[1/x] \quad \text{for } 0 < x < 1,$$

where [t] denotes the integer part of t. Show that

$$\int_0^1 f_\alpha(x)\, dx = \alpha \log \alpha$$

and, for m a positive integer and $\zeta(m + 1) = \Sigma_{k=1}^{\infty} k^{-(m+1)}$,

$$\int_0^1 x^m f_\alpha(x)\, dx = (\alpha^{m+1} - \alpha) \frac{\zeta(m + 1)}{m + 1}.$$

Solution. We write

$$f_\alpha(x) = -(\alpha/x - [\alpha/x]) + \alpha(1/x - [1/x]).$$

Since, for $j = 1,2,3,\ldots,$

$$\int_{\alpha/(j+1)}^{\alpha/j} (\alpha/x - [\alpha/x])\, dx = \alpha\int_{1/(j+1)}^{1/j} (1/x - [1/x])\, dx$$

it follows that

$$\int_0^1 f_\alpha(x)\, dx = -\int_\alpha^1 \frac{\alpha}{x}\, dx = \alpha \log \alpha.$$

We set

$$I(m) = \int_0^1 x^m (\alpha/x - [\alpha/x])\, dx.$$

Since

$$\int_{\alpha/(j+1)}^{\alpha/j} x^m (\alpha/x - [\alpha/x])\, dx = \alpha^{m+1}\int_{1/(j+1)}^{1/j} x^m (1/x - [1/x])\, dx,$$

we obtain

$$\int_0^1 x^m f_\alpha(x)\, dx = (\alpha - \alpha^{m+1}) I(m) - \alpha \int_\alpha^1 x^{m-1}\, dx$$

$$= (\alpha - \alpha^{m+1})(I(m) - 1/m).$$

But

$$(m + 1) \int_0^1 [1/x]x^m\, dx = \sum_{n=1}^{\infty} n \int_{1/(n+1)}^{1/n} (m + 1)x^m\, dx$$

$$= 1 + \frac{1}{2^{m+1}} + \frac{1}{3^{m+1}} + \cdots = \zeta(m + 1)$$

and

$$\frac{1}{m} - I(m) = \int_0^1 [1/x]x^m\, dx.$$

Remarks. Note that if P is a real polynomial, then by Problem 94

$$\int_0^1 P(x)\, f_\alpha(x)\, dx = 0 \quad \text{for all } \alpha \in (0,1]$$

only if $P = 0$.

It is easy to see that

$$\lim_{n\to\infty} \frac{1}{n} \sum_{k=1}^{n} (n/k - [n/k]) = \int_0^1 (1/x - [1/x])\, dx = \lim_{n\to\infty} \int_{1/n}^1 (1/x - [1/x])\, dx$$

$$= 1 - \lim_{n\to\infty} \left(1 + \frac{1}{2} + \frac{1}{3} + \cdots + \frac{1}{n} - \log n\right) = 1 - c,$$

where c is Euler's Constant (see Solution to Problem 14 in Chapter 3 or Problem 113 in Chapter 3).

PROBLEM 96. Let n be a positive integer larger than 1. Show that

$$\int_a^b (x - a)^n (b - x)^n\, dx = 2\cdot\frac{2\cdot4\cdot6}{3\cdot5\cdot7} \cdots \frac{2n}{2n + 1}\cdot\left(\frac{b - a}{2}\right)^{2n+1}$$

and, setting

$$K = 2\cdot\frac{2\cdot4\cdot6}{3\cdot5\cdot7} \cdots \frac{2n}{2n + 1},$$

verify that

$$\sqrt[n+1]{2(n + 1)K} \le 4.$$

Solution. Since, for k an integer larger than 1,

$$\int \sin^k x \, dx = \int \sin^{k-1} x \, d(-\cos x)$$

$$= - \sin^{k-1} x \cos x + \int (k - 1)\sin^{k-2} x \cos^2 x \, dx$$

$$= - \sin^{k-1} x \cos x + (k - 1)\int (\sin^{k-2} x - \sin^k x) \, dx$$

we see that

$$k \int \sin^k x \, dx = - \sin^{k-1} x \cos x + (k - 1)\int \sin^{k-2} x \, dx.$$

Thus

$$\int_0^{\pi/2} \sin^k x \, dx = \frac{k - 1}{k}\int_0^{\pi/2} \sin^{k-2} x \, dx \quad \text{for } k > 1$$

and so

$$\int_0^{\pi/2} \sin^{2n+1} x \, dx = \frac{2n}{2n + 1}\cdot\frac{2n - 2}{2n - 1} \cdots \frac{2}{3}\cdot\int_0^{\pi/2} \sin x \, dx = \frac{K}{2}.$$

Substituting $x = a \cos^2 t + b \sin^2 t$ with $t \in [0,\pi/2]$, we obtain

$$x - a = (b - a)\sin^2 t, \quad b - x = (b - a)\cos^2 t,$$

and $dx = 2(b - a)\sin t \cos t$. Therefore

$$\int_a^b (x - a)^n(b - x)^n \, dx = 2(b - a)^{2n+1}\int_0^{\pi/2} (\sin t \cos t)^{2n+1} \, dt$$

$$= \left(\frac{b - a}{2}\right)^{2n+1}\int_0^{\pi/2} \sin^{2n+1} 2t \, d(2t) = 2\cdot\frac{2\cdot4\cdot6}{3\cdot5\cdot7} \cdots \frac{2n}{2n + 1}\left(\frac{b - a}{2}\right)^{2n+1}.$$

Now, by Problem 15, if f and g are continuous function on [a,b] and g is monotonic, then there exists a point t in [a,b] such that

$$\int_a^b f(x)\ g(x)\ dx = g(a) \int_a^t f(x)\ dx + g(b) \int_t^b f(x)\ dx.$$

Putting $f(x) = (x - a)^n$ and $g(x) = (b - x)^n$, we obtain

$$\int_a^b (x - a)^n (b - x)^n\ dx = (b - a)^n \int_a^t (x - a)^n\ dx = (b - a)^n \frac{(t - a)^{n+1}}{n + 1}.$$

But

$$\int_a^b (x - a)^n (b - x)^n\ dx = K\left(\frac{b - a}{2}\right)^{2n+1},$$

where

$$K = 2\cdot\frac{2\cdot 4\cdot 6}{3\cdot 5\cdot 7} \cdots \frac{2n}{2n + 1}.$$

Hence

$$t = a + \frac{b - a}{4} \sqrt[n+1]{2(n + 1)K}.$$

Putting $f(x) = (b - x)^n$, $g(x) = (x - a)^n$, we obtain

$$\int_a^b (x - a)^n (b - x)^n\ dx = (b - a)^n \int_t^b (b - x)^n\ dx$$

and the intermediate value

$$t' = b - \frac{b - a}{4} \sqrt[n+1]{2(n + 1)K}.$$

The intermediate values to and t' are connected by the relation

$$t + t' = a + b.$$

Since $a \le t \le b$ and $a \le t' \le b$, we see immediately that

$$\sqrt[n+1]{2(n + 1)K} \le 4.$$

There is equality when $n = 0$ and $\sqrt[n+1]{2(n + 1)K} \to 1$ as $n \to \infty$.

PROBLEM 97. Show that

$$\lim_{n\to\infty} \left\{ \left(\frac{1}{n}\right)^n + \left(\frac{2}{n}\right)^n + \left(\frac{3}{n}\right)^n + \cdots + \left(\frac{n - 1}{n}\right)^n \right\} = \frac{1}{e - 1}.$$

Solution. Let

$$a_{n,k} = 1^n + 2^n + 3^n + \cdots + (k-1)^n \quad \text{and} \quad b_{n,k} = k^n.$$

Then

$$\frac{a_{n,k+1} - a_{n,k}}{b_{n,k+1} - b_{n,k}} = \frac{k^n}{(k+1)^n - k^n} = \frac{1}{\left(1 + \frac{1}{k}\right)^n - 1}.$$

Since

$$\frac{1}{k+1} < \log\left(1 + \frac{1}{k}\right) < \frac{1}{k} \quad \text{or} \quad \frac{n}{k+1} < \log\left(1 + \frac{1}{k}\right)^n < \frac{n}{k},$$

and get that

$$\lim_{k\to\infty} \log\left(1 + \frac{1}{k}\right)^n = \lim_{k\to\infty} \frac{n}{k} \quad \text{or} \quad \lim_{k\to\infty} \left(1 + \frac{1}{k}\right)^n = \exp\left(\lim_{k\to\infty} \frac{n}{k}\right).$$

Letting $n \to \infty$ in such a way that n/k tends to 1 as $k \to \infty$, we get the desired limit with the help of the result in Problem 39 of Chapter 3.

PROBLEM 98. Let

$$T_n = \left\{1 + \frac{1}{n^2}\right\}\left\{1 + \frac{2}{n^2}\right\}\left\{1 + \frac{3}{n^2}\right\} \cdots \left\{1 + \frac{n}{n^2}\right\}.$$

Show that $T_n \to \sqrt{e}$ as $n \to \infty$.

Solution. Let

$$S_n = \log T_n = \sum_{p=1}^{n} \log\left(1 + \frac{p}{n^2}\right).$$

Since

$$x - \frac{x^2}{2} < \log(1 + x) < x \quad \text{for } x > 0$$

we get, taking $x = p/n^2$,

$$\frac{p}{n^2} - \frac{p^2}{2n^4} < \log\left(1 + \frac{p}{n^2}\right) < \frac{p}{n^2},$$

$$\sum_{p=1}^{n} \frac{p}{n^2} - \frac{1}{2}\sum_{p=1}^{n} \frac{p^2}{n^4} < \sum_{p=1}^{n} \log\left(1 + \frac{p}{n^2}\right) < \sum_{p=1}^{n} \frac{p}{n^2}.$$

But

$$\sum_{p=1}^{n} p = \frac{n(n+1)}{2} \quad \text{implying} \quad \sum_{p=1}^{n} \frac{p}{n^2} = \frac{n+1}{2n} \to \frac{1}{2} \text{ as } n \to \infty$$

and

$$\sum_{p=1}^{n} p^2 < n \cdot n^2 \quad \text{implying} \quad \sum_{p=1}^{n} \frac{p^2}{n^4} < \frac{n^3}{n^4} = \frac{1}{n} \to 0 \text{ as } n \to \infty.$$

Thus, the sequence $(S_n)_{n=1}^{\infty}$, being bounded below and above by sequences that converge to 1/2, must itself converge to 1/2. Therefore $T_n = \exp(S_n)$ converges to $e^{\frac{1}{2}}$.

PROBLEM 99. For $n = 1,2,3,\ldots$, let g_n be continuous functions on a closed, bounded interval [a,b] such that

$$g_n(x) \geq 0 \quad \text{for } a \leq x \leq b \text{ and } \int_a^b g_n(x)\, dx = 1.$$

If f is a continuous function on [a,b], show that all terms of the numerical sequence ${}_a^b\, g_n(x)\, f(x)\, dx$, $n = 1,2,3,\ldots$ are situated between the smallest and the largest values of f on [a,b].

Solution. Clearly,

$$m \int_a^b g_n(x)\, dx \leq \int_a^b g_n(x)\, f(x)\, dx \leq M \int_a^b g_n(x)\, dx$$

whenever $m \leq f(x) \leq M$ for all $x \in [a,b]$. Alternately, since f is continuous, there is a point t in [a,b] such that (see Problem 17)

$$\int_a^b g_n(x)\, f(x)\, dx = f(t) \int_a^b g_n(x)\, dx.$$

PROBLEM 100. Let $a_1, a_2, \ldots, a_p$ be positive. Show that

$$\lim_{n \to \infty} \left(\frac{\sqrt[n]{a_1} + \sqrt[n]{a_2} + \cdots + \sqrt[n]{a_p}}{p} \right)^n = \sqrt[p]{a_1 a_2 \cdots a_p}.$$

Solution. In Problem 77 of Chapter 3 we took up the case p = 2. The proof of the present claim is completely analogous to the proof given in the special case p = 2.

PROBLEM 101. Show that

$$\left\{\frac{(4n + 3)(2n + 1)}{4n + 4}\,\frac{\pi}{2}\right\}^{\frac{1}{2}} > \frac{2\cdot 4 \cdots (2n)}{1\cdot 3 \cdots (2n - 1)} > \left\{\frac{2n(2n + 1)}{4n + 1}\cdot\pi\right\}^{\frac{1}{2}}.$$

Solution. In the Solution of Problem 96 we noted that, for k an integer larger than 1,

$$\int_0^{\pi/2} \sin^k x\, dx = \frac{k - 1}{k}\int_0^{\pi/2} \sin^{k-2} x\, dx;$$

hence

$$\int_0^{\pi/2} \sin^k x\, dx = \frac{(k - 1)(k - 3) \cdots 4\cdot 2}{k(k - 2) \cdots 5\cdot 3\cdot 1} \qquad \text{if k is odd,}$$

$$= \frac{(k - 1)(k - 3) \cdots 3\cdot 1}{k(k - 2) \cdots 4\cdot 2}\cdot\frac{\pi}{2} \qquad \text{if k is even.}$$

Since $(\sin x - 1)^2 \geq 0$, we have the inequality $1 \geq 2 \sin x - \sin^2 x$. Multiplying this inequality by $\sin^{2n-1} x$ and by $\sin^{2n} x$, and integrating between 0 and $\pi/2$ yields the desired result.

PROBLEM 102. Show that

$$\frac{1}{2n + 2} + \frac{1}{2}\cdot\frac{1}{2n + 4} + \frac{1\cdot 3}{2\cdot 4}\cdot\frac{1}{2n + 6} + \frac{1\cdot 3\cdot 5}{2\cdot 4\cdot 6}\cdot\frac{1}{2n + 8} + \cdots$$

$$= \frac{2\cdot 4\cdot 6 \cdots (2n)}{3\cdot 5\cdot 7 \cdots (2n + 1)}.$$

Solution. We first note that, for $|x| < 1$,

$$\frac{1}{\sqrt{1 - x^2}} = 1 + \frac{1}{2}x^2 + \frac{1\cdot 3}{2\cdot 4}x^4 + \frac{1\cdot 3\cdot 5}{2\cdot 4\cdot 6}x^6 + \cdots$$

and so

$$\frac{x^{2n+1}}{\sqrt{1 - x^2}} = x^{2n+1} + \frac{1}{2}x^{2n+3} + \frac{1\cdot 3}{2\cdot 4}x^{2n+5} + \frac{1\cdot 3\cdot 5}{2\cdot 4\cdot 6}x^{2n+7} + \cdots.$$

Hence

$$\int_0^1 \frac{x^{2n+1}}{\sqrt{1 - x^2}}\,dx = \frac{1}{2n + 2} + \frac{1}{2}\cdot\frac{1}{2n + 4} + \frac{1\cdot 3}{2\cdot 4}\cdot\frac{1}{2n + 6} + \frac{1\cdot 3\cdot 5}{2\cdot 4\cdot 6}\cdot\frac{1}{2n + 8} + \cdots.$$

On the other hand, setting $x = \sin t$,

$$\int_0^1 \frac{x^{2n+1}}{\sqrt{1 - x^2}}\,dx = \int_0^{\pi/2} \sin^{2n} t\,dt = \frac{2\cdot 4\cdot 6 \cdots (2n)}{3\cdot 5\cdot 7 \cdots (2n + 1)}.$$

PROBLEM 103. Show that

$$\frac{1}{2\cdot 5} + \frac{1}{8\cdot 11} + \frac{1}{14\cdot 17} + \cdots = \frac{1}{9}\left(\frac{\pi}{3} - \log 2\right).$$

Solution. We have, for $|x| < 1$,

$$\frac{x}{1 + x^3} = x(1 - x^3 + x^6 - x^9 + x^{12} - x^{15} + \cdots)$$

$$= x - x^4 + x^7 - x^{10} + x^{13} - x^{16} + \cdots.$$

Thus

$$\int_0^1 \frac{x}{1 + x^3}\,dx = \frac{1}{2} - \frac{1}{5} + \frac{1}{8} - \frac{1}{11} + \frac{1}{14} - \frac{1}{17} + \cdots$$

$$= 3\left(\frac{1}{2\cdot 5} + \frac{1}{8\cdot 11} + \frac{1}{14\cdot 17} + \cdots\right).$$

But

$$\int_0^1 \frac{x\,dx}{1 + x^3} = \frac{1}{6}\left\{\log \frac{x^2 - x + 1}{(x + 1)^2} + 2\sqrt{3}\,\arctan \frac{2x - 1}{\sqrt{3}}\right\}\Bigg|_0^1$$

$$= \frac{1}{3}\left(\frac{\pi}{3} - \log 2\right).$$

PROBLEM 104. Solve the system of ten equations

$$x + y + z + u + v = 2$$

$$px + qy + rz + su + tv = 3$$

$$p^2x + q^2y + r^2z + s^2u + t^2v = 16$$

$$p^3x + q^3y + r^3z + s^3u + t^3v = 31$$

$$p^4x + q^4y + r^4z + s^4u + t^4v = 103$$

$$p^5x + q^5y + r^5z + s^5u + t^5v = 235$$

$$p^6x + q^6y + r^6z + s^6u + t^6v = 674$$

$$p^7x + q^7y + r^7z + s^7u + t^7v = 1669$$

$$p^8x + q^8y + r^8z + s^8u + t^8v = 4526$$

$$p^9x + q^9y + r^9z + s^9u + t^9v = 11595.$$

Solution. First we consider the general system

$$x_1 + x_2 + x_3 + \cdots + x_{n-1} + x_n = a_1$$

$$x_1y_1 + x_2y_2 + \cdots + x_ny_n = a_2$$

$$x_1y_1^2 + x_2y_2^2 + \cdots + x_ny_n^2 = a_3$$

...

$$x_1y_1^{2n-1} + x_2y_2^{2n-1} + \cdots + x_ny_n^{2n-1} = a_{2n}.$$

We let

$$F(\theta) = \frac{x_1}{1 - \theta y_1} + \frac{x_2}{1 - \theta y_2} + \cdots + \frac{x_n}{1 - \theta y_n}.$$

But

$$\frac{x_1}{1 - \theta y_1} = x_1(1 + \theta y_1 + \theta^2 y_1^2 + \theta^3 y_1^3 + \cdots),$$

$$\frac{x_2}{1 - \theta y_2} = x_2(1 + \theta y_2 + \theta^2 y_2^2 + \theta^3 y_2^3 + \cdots),$$

...

$$\frac{x_n}{1 - \theta y_n} = x_n(1 + \theta y_n + \theta^2 y_n^2 + \theta^3 y_n^3 + \cdots).$$

Thus

$$F(\theta) = (x_1 + x_2 + \cdots + x_n) + (x_1y_1 + x_2y_2 + \cdots + x_ny_n)\theta$$

$$+ (x_1y_1^2 + \cdots + x_ny_n^2)\theta^2 + \cdots$$

$$+ (x_1y_1^{2n-1} + x_2y_2^{2n-1} + \cdots + x_ny_n^{2n-1})\theta^{2n-1}$$

$$+ (x_1y_1^{2n} + \cdots + x_ny_n^{2n})\theta^{2n} + \cdots$$

and so

$$F(\theta) = a_1 + a_2\theta + a_3\theta^2 + \cdots + a_{2n}\theta^{2n-1} + \cdots$$

Reducing the fractions to a common denominator, we find

$$F(\theta) = \frac{A_1 + A_2\theta + A_3\theta^2 + \cdots + A_n\theta^{n-1}}{1 + B_1\theta + B_2\theta^2 + \cdots + B_n\theta^n}.$$

Hence

$$(a_1 + a_2\theta + a_3\theta^2 + \cdots + a_{2n}\theta^{2n-1} + \cdots)(1 + B_1\theta + B_2\theta^2 + \cdots + B_n\theta^n)$$

$$= A_1 + A_2\theta + \cdots + A_n\theta^{n-1}$$

Therefore

$$A_1 = a_1,$$

$$A_2 = a_2 + a_1B_1,$$

$$A_3 = a_3 + a_2B_1 + a_1B_2,$$

$$\cdots$$

$$A_n = a_n + a_{n-1}B_1 + a_{n-2}B_2 + \cdots + a_1B_{n-1},$$

$$0 = a_{n+1} + a_nB_1 + \cdots + a_1B_n,$$

$$0 = a_{n+2} + a_{n+1}B_1 + \cdots + a_2B_n,$$

$$\cdots$$

$$0 = a_{2n} + a_{2n-1}B_1 + \cdots + a_nB_n.$$

Since the quantities $a_1, a_2, \ldots, a_n, a_{n+1}, \ldots, a_{2n}$ are known, the last equations enable us to first find $B_1, B_2, \ldots, B_n$ and then $A_1, A_2, \ldots, A_n$. Knowing the A_i and B_i, we can construct a rational function $F(\theta)$ and then expand it into partial fractions. Doing so, we get

$$F(\theta) = \frac{p_1}{1 - q_1\theta} + \frac{p_2}{1 - q_2\theta} + \frac{p_3}{1 - q_3\theta} + \cdots + \frac{p_n}{1 - q_n\theta}.$$

It is clear that

$$x_1 = p_1, \quad y_1 = q_1;$$

$$x_2 = p_2, \quad y_2 = q_2;$$

$$\cdots$$

$$x_n = p_n, \quad y_n = q_n.$$

The general system is solved.

For the given case we have

$$F(\theta) = \frac{2 + \theta + 3\theta^2 + 2\theta^3 + \theta^4}{1 - \theta - 5\theta^2 + \theta^3 + 3\theta^4 - \theta^5}.$$

Expanding into partial fractions, we obtain the following values for the unknowns:

$$x = -\frac{3}{5}, \qquad p = -1,$$

$$y = \frac{18 + \sqrt{5}}{10}, \qquad q = \frac{3 + \sqrt{5}}{2},$$

$$z = \frac{18 - \sqrt{5}}{10} \qquad r = \frac{3 - \sqrt{5}}{2},$$

$$u = -\frac{8 + \sqrt{5}}{2\sqrt{5}}, \qquad s = \frac{\sqrt{5} - 1}{2},$$

$$v = \frac{8 - \sqrt{5}}{2\sqrt{5}}, \qquad t = -\frac{\sqrt{5} + 1}{2}.$$

Remark. Problem 104 and its solution is due to Ramanujan.

PROBLEM 105. Let k be a nonnegative integer and $0 < a < b$. Evaluate $\int_a^b x^k \, dx$ as limit of a sum.

Solution. Put $q = \sqrt[n]{b/a}$ and consider the partition

$$a < aq < aq^2 < \cdots < aq^{n-1} < aq^n = b$$

of the interval [a,b]. We have to sum

$$a^k(aq - a) + (aq)^k(aq^2 - aq) + \cdots + (aq^{n-1})^k(aq^n - aq^{n-1})$$

$$= a^{k+1}(q - 1)\{1 + q^{k+1} + q^{2(k+1)} + \cdots + q^{(n-1)(k+1)}\}$$

$$= a^{k+1}(q - 1)\frac{q^{n(k+1)} - 1}{q^{k+1} - 1} = (b^{k+1} - a^{k+1})\frac{q - 1}{q^{k+1} - 1}$$

$$= (b^{k+1} - a^{k+1})\frac{1}{q^k + q^{k-1} + \cdots + 1}.$$

But $\lim_{n\to\infty} \sqrt[n]{b/a} = 1$ (see Solution of Problem 18 of Chapter 3) and so

$$\lim_{n\to\infty} (q^k + q^{k-1} + \cdots + 1) = k + 1.$$

Remark. To show that $\int_1^b 1/x\, dx = \log b$ for $b > 1$, we can proceed as follows: we choose partition $x_i = (b)^{i/n}$ with $i = 0,1,2,\ldots,n$ and consider the sum

$$\sum_{i=1}^{n} \frac{(b)^{1/n} - (b)^{(i-1)/n}}{(b)^{(i-1)/n}} = \sum_{i=1}^{n} [(b)^{1/n} - 1] = n[(b)^{1/n} - 1].$$

But

$$\lim_{n\to\infty} n[(b)^{1/n} - 1] = \log b.$$

PROBLEM 106. Give an example of a series

$$\sum_{k=1}^{\infty} f_k(x)$$

of functions continuous on a closed bounded interval [a,b] that converges absolutely and uniformly for which the Weierstrass M-test fails.

Solution. On the interval [0,1] let the function f_k be defined by

$f_k(x) = 0 \quad$ for $0 \le x \le \dfrac{1}{2k+1}$ and $\dfrac{1}{2k-1}$;

$f_k(x) = \dfrac{1}{k} \quad$ for $x = \dfrac{1}{2k}$;

$f_k(x)$ is defined linearly in the intervals $\left[\dfrac{1}{2k+1}, \dfrac{1}{2k}\right]$

and $\left[\dfrac{1}{2k}, \dfrac{1}{2k-1}\right]$.

It is easy to see that the series $\Sigma_{k=1}^{\infty} f_k(x)$ satisfies the following conditions:

(a) the series is uniformly convergent,

(b) any rearrangement of the series is uniformly convergent (i.e., the series is uniformly and absolutely convergent),

(c) $\Sigma_{k=1}^{\infty} M_k$ diverges, where M_k is the upper bound of $|f_k(x)|$ as x ranges over the interval $[0,1]$.

PROBLEM 107. Let $b_1 \ge b_2 \ge \cdots \ge b_n \ge 0$. Show that a necessary and sufficient condition that the series $\Sigma_{n=1}^{\infty} b_n \sin nx$ should be uniformly convergent throughout any interval is that $nb_n \to 0$ as $n \to \infty$.

Solution. To see that the condition is necessary, observe that, if $x = \pi/(2p)$, and $n = [\frac{1}{2}p + 1]$, where $[x]$ means the integral part of x,

$$b_n \sin nx + b_{n+1} \sin(n+1)x + \cdots + b_p \sin px$$

$$> b_p(\sin nx + \cdots + \sin px) > b_p(\tfrac{1}{2}p - 1)\sin \tfrac{1}{2}\pi,$$

since there are at least $\frac{1}{2}p - 1$ terms in the bracket, in each of which $mx > \frac{1}{4}\pi$. Since the given series is uniformly convergent in an interval including the origin, the left-hand side of the above inequality tends to zero as $p \to \infty$. Hence $pb_p \to 0$.

Next we show the sufficiency of the condition. It will be enough to consider the interval $[0,\pi]$, since each term of the series is an odd function and has the period 2π. Consider the sum

$$s_{n,p} = b_n \sin nx + \cdots + b_p \sin px,$$

where now n and p are unconnected. Let $T_n = \sup\{mb_m : m \ge n\}$, so that $T_n \to 0$.

If $x \geq \pi/n$, we apply Abel's Inequality (see Problem 58 of Chapter 2). We have

$$|\sin nx + \cdots + \sin rx| = \left|\frac{\cos(n - \frac{1}{2})x - \cos(r + \frac{1}{2})x}{2 \sin \frac{1}{2}x}\right| \leq \frac{1}{\sin \frac{1}{2}x}$$

for all values of n and r, and, since $(\sin \theta)/\theta$ is steadily decreasing for $0 < \theta < \frac{1}{2}\pi$, $1/(\sin \frac{1}{2}x) \leq \pi/x$, and we deduce that

$$|s_{n,p}| \leq \frac{b_n \pi}{x} \leq nb_n \leq T_n.$$

If $x \leq \pi/p$, we have, since $\sin \theta \leq \theta$,

$$|s_{n,p}| \leq b_n nx + \cdots + b_p px \leq pT_n x \leq \pi T_n.$$

If $\pi/p < x < \pi/n$, we combine the two arguments. We have

$$|s_{n,p}| \leq |s_{n,k}| + |s_{k+1,p}|,$$

and, applying Abel's Inequality (see Problem 58 of Chapter 2) to the second part, and the other method to the first part, obtain

$$|s_{n,p}| \leq kT_n x + b_{k+1}\pi/x \leq T_n[kx + \pi/\{(k + 1)x\}].$$

Taking $k = [\pi/x]$, we have $|s_{n,p}| \leq T_n(\pi + 1)$. Hence in any case $|s_{n,p}| \leq AT_n$, and, since $T_n \to 0$ as $n \to \infty$, the solution is complete.

PROBLEM 108. Let

$$f(x) = e^{-1/x^2} \quad \text{for } x \neq 0, \text{ and } f(0) = 0.$$

Show that all the derivatives of f have the value 0 at $x = 0$.

Solution. Clearly, f is continuous at $x = 0$. By definition,

$$f'(0) = \lim_{x \to 0} \frac{f(x) - f(0)}{x - 0} = \lim_{x \to 0} \frac{e^{-1/x^2}}{x}.$$

But

$$\lim_{x \to 0} \frac{e^{-1/x^2}}{x} = \lim_{x \to 0} \frac{\frac{1}{x}}{e^{1/x^2}} = \lim_{x \to 0} \frac{\frac{1}{x^2}}{-\frac{2}{x^3}e^{1/x^2}} = \lim_{x \to 0} \tfrac{1}{2} x e^{-1/x^2} = 0.$$

This shows that $f'(0) = 0$.

To deal with higher derivatives we note that, if $x \neq 0$,

$$f'(x) = 2x^{-3} e^{-1/x^2}, \qquad f''(x) = 4x^{-6} e^{-1/x^2} - 6x^{-4} e^{-1/x^2}.$$

It is easy to see, by induction, that $f^{(n)}(x)$ is a linear combination of terms of the form $(e^{-1/x^2})/x^m$ with $0 < m \leq 3n$. Consequently, to see that $f^{(n)}(0) = 0$, as well as to see that $f^{(n)}$ is continuous at $x = 0$, it will be sufficient to show that

$$\lim_{x \to 0} \frac{e^{-1/x^2}}{x^m} = 0$$

for all positive integers m. But

$$\lim_{x \to 0} \frac{x^{-m}}{e^{1/x^2}} = \lim_{x \to 0} \frac{-m x^{-m-1}}{-2x^{-3} e^{1/x^2}} = \frac{m}{2} \lim_{x \to 0} \frac{x^{-m+2}}{e^{1/x^2}};$$

after a finite number of steps the exponent in the numerator will be positive, and then the limit is seen to be 0.

PROBLEM 109. Let $(a_j)_{j=0}^{\infty}$ be a given sequence of real numbers. Show that there is an infinitely often differentiable real-valued function f on $[0,\infty)$ whose support is contained in the interval $[0,1]$ such that $f^{(j)}(0) = a_j$ for $j = 0,1,2,\ldots$ - By the support of f we mean the closure of the set $\{x: f(x) \neq 0\}$.

Solution. We first consider two lemmas.

LEMMA 1: There is an infinitely often differentiable function $g: [0,\infty) \to R^1$ with $g(0) = 1$, $0 \leq g \leq 1$, and whose support is contained in the interval $[0,a]$ with $a > 0$.

Proof. Let $h: R^1 \to R^1$ be defined by $h(x) = \exp(-1/x^2)$ when $x > 0$ and $h(x) = 0$ when $x \leq 0$. Then define

$$g(x) = \frac{h(a - x)}{h(a - x) + h(x - a/2)}.$$

Clearly, g is infinitely often differentiable, $g(x) = 0$ for $x \geq a$, $g(x) = 1$ for $x \leq a/2$, and $0 \leq g(x) \leq 1$. This proves the lemma.

LEMMA 2: For any integer $n \geq 1$ and any $\varepsilon > 0$, there is an infinitely often differentiable function $f\colon [0,\infty) - R^1$ whose support is contained in the interval $[0,1]$ such that

(i) $f^{(n)}(0) = 1, \quad f^{(k)}(0) = 0$ for $k < n$;

(ii) $|f^{(k)}| < \varepsilon$ for $k < n$.

Proof. Take g as in *LEMMA* 1 with support contained in $[0,a]$ and also a function v as in *LEMMA* 1 with support contained in $[0,1]$. Define

$$f(s) = v(s)\int_0^s \int_0^{t_n} \cdots \int_0^{t_3}\int_0^{t_2} g(t_1)dt_1 \cdots dt_n$$

and choose the number a small enough such that (ii) holds. Obviously (i) holds. This proves *LEMMA* 2.

Returning now to the problem, we define

$$f = \sum_{j=0}^{\infty} c_j f_j,$$

where f_j and c_j are defined recursively: $f_j\colon [0,\infty) \to R^1$, f_j infinitely often differentiable and with support contained in $[0,1]$, f_0 as in *LEMMA* 2, c_0 arbitrary. Then by *LEMMA* 2 we can choose f_j in the following way:

(1) $f_n^{(n)}(0) = 1, \quad f_n^{(k)}(0) = 0$ for $k < n$;

(2) $c_n = a_n - [c_{n-1}f_{n-1}^{(n)}(0) + \cdots + c_0 f_0^{(n)}(0)]$;

(3) $|c_n f_n^{(k)}| \leq 1/2^n$ for $k < n$.

By (3), the series $\Sigma_{j=0}^{\infty} c_j f_j^{(k)}$ converges uniformly for every k. On the other hand $\Sigma_{j=0}^{\infty} c_j f_j(0) = 0$. Therefore, f is infinitely often differentiable and its support is contained in $[0,1]$; moreover, $f^{(k)}(0) = \Sigma_{j=0}^{\infty} c_j f_j^{(k)}(0)$. By (1),

$$f^{(k)}(0) = \sum_{j=0}^{k} c_j f_j^{(k)}(0)$$

$$= [c_0 f_0^{(k)}(0) + \cdots + c_{k-1}f_{k-1}^{(k)}(0)] + c_k f_k^{(k)}(0) = (a_k - c_k) + c_k = a_k.$$

Remarks. The result in Problem 109 can readily be extended to the following claim: Given a sequence of real numbers $(a_j)_{j=0}^{\infty}$, there exists an infinitely often differentiable function $f\colon R^1 \to R^1$ with support contained in the interval $[-1,1]$ such that $f^{(j)}(0) = a_j$ for $j = 0,1,2,\ldots$

By taking a sequence $(a_j)_{j=0}^{\infty}$ such that

$$\limsup_{n\to\infty} (|a_n|)^{1/n} = +\infty$$

and constructing f as in the foregoing claim, it follows that there exist infinitely often differentiable functions $f\colon R^1 \to R^1$ that cannot be represented by power series.

PROBLEM 110. Show that

$$\int_0^1 x^{-x}\,dx = \sum_{m=1}^{\infty} \frac{1}{m^m}.$$

Solution. Since

$$x^{-x} = e^{-x \log x},$$

we see that

$$x^{-x} = 1 + \sum_{n=1}^{\infty} (-1)^n \frac{x^n \log^n x}{n!};$$

the series is uniformly convergent for $0 < x \le 1$ as it can be majorized by the series

$$\sum_{n=0}^{\infty} \frac{1}{n!}\left(\frac{1}{e}\right)^n.$$

Observing that

$$\int_0^1 x^n \log^n x\,dx = (-1)^n \frac{n!}{(n+1)^{n+1}},$$

we obtain what we have set out to do.

Remarks. In an entirely similar fashion one can show that

$$\int_0^1 x^x\,dx = \sum_{m=1}^{\infty} (-1)^{m+1} \frac{1}{m^m}.$$

Using Simpson's rule, the integrals

$$\int_0^1 x^{-x}\, dx \quad \text{and} \quad \int_0^1 x^x\, dx$$

can be calculated with a high degree of accuracy.

PROBLEM 111. Let f be a continuous function on a closed bounded interval [a,b] and let f_h denote its Steklow function (see Problem 77 of Chapter 2). Show that

$$\lim_{h \to 0} \int_a^b |f_h(x) - f(x)|\, dx = 0. \tag{E.74}$$

Solution. Let h be sufficiently small so that the interval [x - h,x + h] is contained in the interval [a,b]. Then, by the result in Problem 15,

$$f_h(x) = \frac{1}{2h} \int_{x-h}^{x+h} f(t)\, dt = f(s).$$

where $s \in [x - h, x + h]$. Thus, for $x \in (a,b)$, $\lim_{h \to 0} f_h(x) = f(x)$. Since f is continuous on [a,b], it is bounded. But $|f(x)| \le M$ for all $x \in [a,b]$ implies $|f_h(x)| \le M$ for all $x \in [a,b]$. Hence $|f_h(x) - f(x)| \le 2M$ for all x $\in$ [a,b], independent of the parameter h. Using the result in Problem 35 we see that the order of taking the limit and integration can be interchanged in (E.74), completing the proof.

PROBLEM 112. Define a sequence $v_n = v_n(x)$ recursively by

$$v_1 = x, \quad v_{n+1} = \left(2 + \frac{1}{n}\right)v_n - 1, \; n \ge 1.$$

It is not hard to show that the sequence $v_1, v_2, v_3, \ldots$ converges for at most one real value of x. Find x such that $v_1, v_2, v_3, \ldots$ converges.

Solution. By induction we have

$$v_n = \frac{1 \cdot 3 \cdots (2n - 1)(x - s_n)}{(n - 1)!}, \qquad n = 1,2,3,\ldots,$$

where

$$s_1 = 0 \quad \text{and} \quad s_{n+1} = s_n + \frac{n!}{1\cdot 3 \cdots (2n+1)}.$$

If $v_1, v_2, v_3, \ldots$ converges, then $x - s_n \to 0$ as $n \to \infty$, whence

$$x = \lim_{n\to\infty} s_n = \sum_{n=1}^{\infty} \frac{n!}{1\cdot 3 \cdots (2n+1)} = \frac{\pi}{2} - 1.$$

(See the Remark at the end of the Solution.) For this x we find that

$$v_n = \frac{n}{2n+1} + \frac{n(n+1)}{(2n+1)(2n+3)} + \cdots,$$

and since the k-th term here is increasing and tends to 2^{-k}, it follows that v_n tends to $\frac{1}{2} + \frac{1}{4} + \cdots = 1$.

Remark. To see that

$$\sum_{n=1}^{\infty} \frac{n!}{1\cdot 3 \cdots (2n+1)} = \frac{\pi}{2} - 1$$

we note that, for $m = 1,2,3,\ldots,$

$$\int_0^{\pi/2} \sin^{2m+1} x \, dx = \frac{2\cdot 4\cdot 6 \cdots 2m}{1\cdot 3\cdot 5 \cdots (2m+1)} = \frac{2^m m!}{1\cdot 3\cdot 5 \cdots (2m+1)}$$

(see the Solution of Problem 96) and so

$$\frac{1}{1\cdot 3} + \frac{1\cdot 2}{1\cdot 3\cdot 5} + \frac{1\cdot 2\cdot 3}{1\cdot 3\cdot 5\cdot 7} + \cdots$$

$$= \int_0^{\pi/2} \left(\frac{\sin^3 x}{2} + \frac{\sin^5 x}{2^2} + \frac{\sin^7 x}{2^3} + \cdots \right) dx$$

$$= \int_0^{\pi/2} \sin x \left(\frac{1}{1 - \frac{\sin^2 x}{2}} - 1 \right) dx = 2\int_0^{\pi/2} \frac{\sin x}{1 + \cos^2 x} dx - 1.$$

Setting $\tan \frac{x}{2} = t$, the last integral becomes $4 \int_0^1 (1 + t^4)^{-1} t \, dt = \pi/2$ and we have obtained the desired result.

PROBLEM 113. Show that if $g(2t) = g(t) \cos t$, then

$$g(t) = g(0)\cdot\frac{\sin t}{t}.$$

Solution. If in $g(2t) = g(t) \cos t$ we replace successively t by t/2, t/4, ..., $t/2^n$, we obtain the following equalities:

$$g(t) = g(t/2)\cos(t/2),$$

$$g(t/2) = g(t/4)\cos(t/4),$$

$$\ldots$$

$$g(t/2^{n-1}) = g(t/2^n)\cos(t/2^n).$$

Hence

$$g(t/2^n)\cos(t/2)\cos(t/4) \cdots \cos(t/2^n) = g(t).$$

But

$$\cos(t/2)\cos(t/4) \cdots \cos(t/2^n) = \frac{\sin t}{t} \frac{t/2^n}{\sin(t/2^n)}$$

(see Solution of Problem 99 in Chapter 1). Thus

$$g(t/2^n) \frac{\sin t}{t} \frac{t/2^n}{\sin(t/2^n)} = g(t).$$

As $n \to \infty$, $t/2^n$ tends to zero; $g(t/2^n)$ tends to $g(0)$ and

$$\frac{t/2^n}{\sin(t/2^n)}$$

tends to 1. Thus $g(0)\{\sin t\}/t = g(t)$.

PROBLEM 114. If $a > 0$, $x_n = a^{x_{n-1}}$, $x_1 = 0$ show that

(i) $x_n \to \infty$ when $a > e^{\frac{1}{e}}$

(ii) x_n tends to a finite limit when $e^{-e} \le a \le e^{\frac{1}{e}}$;

(iii) x_n does not tend to a limit when $a < e^{-e}$.

Moreover, in (iii), $x_{2n} \to A$, $x_{2n+1} \to B$, where A, B satisfy

$$a^A = B, \quad a^B = A.$$

Solution. We have

$$\frac{x_{n+1}}{x_n} = \frac{a^{x_n}}{a^{x_{n-1}}} = a^{x_n - x_{n-1}}.$$

Suppose $a > 1$. Then $x_{n+1} > x_n$ if $x_n > x_{n-1}$. Now $x_2 = 1 > 0 = x_1$, so that x_n increases with n and must tend either to a limit L or to infinity. Suppose k exists such that $a^k = k$. Clearly $k > 0$ and

$$\frac{k}{x_{n+1}} = a^{k - x_n}.$$

Thus $k > x_{n+1}$ if $k > x_n$. But $k > 0 = x_1$. Hence if there exists any root k of $a^x = x$ it follows that $x_n \to L \le k$. But

$$L = \lim_{n\to\infty} x_{n+1} = \lim_{n\to\infty} a^{x_n} = a^L.$$

Thus if there is any real root of the equation $a^x = x$, x_n tends to the least root; if there is no root, $x_n \to \infty$.

The equation $a^x = x$ $(x > 0, a > 0)$.

Putting

$$y = \frac{a^x}{x}$$

we have

$$\frac{1}{y}\frac{dy}{dx} = \log a - \frac{1}{x}.$$

Thus if $a > 1$, $\frac{dy}{dx} < = > 0$ according as $x < = > \frac{1}{\log a}$ and $x = \frac{1}{\log a}$ gives a minimum value e log a for y. If $a > 1$, then for large x, and for small positive x, $y > 1$. Hence $a^x = x$ has 2, 1, or 0 roots according as $e \log a < = > 1$; i.e., as $a < = > e^{1/e}$. If $a < 1$, $\frac{dy}{dx}$ is always negative, and the equation $a^x = x$ has one and only one root.

The case $a = 1$ gives $x_n = 1$ and therefore $x_n \to 1$.

We now turn to the case $a < 1$. Our original equation shows that if $x_n > x_{n-1}$ then $x_{n+1} < x_n$ and so $x_{n+2} > x_{n+1}$. We are led to consider separately the odd and even successions $x_1, x_3, x_5, \ldots$ and $x_2, x_4, x_6, \ldots$ The equation

$$\frac{x_{n+2}}{x_n} = a^{x_{n+1}-x_{n-1}} = a^{a^{x_n} - a^{x_{n-2}}}$$

shows (by induction) that these are respectively increasing and decreasing. Since x_n is essentially positive it follows that $x_{2n-1} \to B$ and hence that $x_{2n} \to A \geq B$. We have

$$A = \lim_{n\to\infty} x_{2n+2} = \lim_{n\to\infty} a^{a^{x_{2n}}} = a^{a^A} \quad \text{and similarly } B = a^{a^B}.$$

If k is any root of $a^{a^k} = k$ we have clearly $0 < k < 1$. Also

$$\frac{k}{x_n} = a^{a^k - a^{x_{n-2}}},$$

so that $x_n < = > k$ according as $x_{n-2} < = > k$. Thus A, B are respectively the greatest and least roots of $a^{a^x} = x$.

The equation $a^{a^x} = x$ $(a < 1,\ x > 0)$.

There is always one root of this equation - the root k of $a^x = x$ which we have shown to exist. For convenience we put $a = e^{-b}$. Taking logarithms twice we have

$$-bx + \log b = \log\log\frac{1}{x}.$$

Put

$$y = bx + \log\log\frac{1}{x} - \log b,$$

so that

$$\frac{dy}{dx} = b - \frac{1}{x\log\frac{1}{x}}.$$

Now

$$\frac{d}{dx}\left(x\log\frac{1}{x}\right) = \log\frac{1}{x} - 1,$$

so that $x\log\frac{1}{x}$ has a maximum value $\frac{1}{e}$ when $\log\frac{1}{x} = 1$, $x = \frac{1}{e}$. Hence $\frac{dy}{dx}$ has a maximum value $b - e$. If $b \leq e$, $\frac{dy}{dx}$ is always negative, and there is one root only of the equation $y = 0$. Thus $A = B$ if $a \geq e^{-e}$. If $b > e$ consider $\frac{dy}{dx}$ at the point $x = k$, where k is the root of $a^x = x$ (and therefore of $a^{a^x} = x$) already referred to. Then

$$k \log a = \log k, \quad \text{that is,} \quad -kb = \log k,$$

and so

$$\frac{dy}{dx} = b - \frac{1}{x \log \frac{1}{x}} = b - \frac{1}{k^2 b}.$$

To decide the sign of this expression we note that the increasing function $bx + \log x$ is negative when $x = 1/b$ and $b > e$. Thus $1/b < k$, the only value of x for which $bx + \log x = 0$. Thus, when $x = k$, $\frac{dy}{dx} > 0$. It follows that $y > 0$ for $x = k + \delta$ and $y < 0$ for $x = k - \delta$, δ being small. But $y > 0$ for small values of x and $y < 0$ when x approaches 1. Thus there is a real root of the equation $y = 0$ in each of the intervals $(0,k)$, $(k,1)$, and it follows that $A > B$. Thus x_n does not tend to a limit. The properties (i), (ii), and (iii) are therefore proved.

From these results and the fact that

$$x_{n+1} = a^{x_n},$$

it follows at once that $A = a^B$ and $B = a^A$.

Remark. Compare Problem 115 with Problem 20 in Chapter 3.

PROBLEM 115. Let f be a polynomial without constant term, defined on the interval $[a,x]$. We insert, between a and x, the $n - 1$ geometric means

$$p_1 = a \sqrt[n]{x/a}, \quad p_2 = a \sqrt[n]{(x/a)^2}, \quad \ldots, \quad p_{n-1} = a \sqrt[n]{(x/a)^{n-1}},$$

and denote by A the arithmetic mean of the values

$$f(a), \quad f(p_1), \quad \ldots, \quad f(p_{n-1}), \quad f(x).$$

On the other hand, we insert, between a and x, the $n - 1$ arithmetic means

$$q_1 = a + \frac{x - a}{n}, \quad q_2 = a + 2\cdot\frac{x - a}{n}, \quad \ldots, \quad q_{n-1} = a + (n - 1)\cdot\frac{x - a}{n},$$

and denote by B the arithmetic mean of the values

$$\frac{f(a)}{a}, \quad \frac{f(q_1)}{q_1}, \quad \ldots, \quad \frac{f(q_{n-1})}{q_{n-1}}, \quad \frac{f(x)}{x}.$$

Show that, as $n \to \infty$, the fraction B/A tends to a limit which is entirely independent of the polynomial f; this limit is

$$\frac{\log \frac{x}{a}}{x - a}.$$

Solution. Suppose first that

$$f(x) = x^{m+1},$$

where m is a positive integer. Then

$$B = \frac{1}{n + 1} \quad a^m + \left(a + \frac{x - a}{n}\right)^m + \cdots + \left(a + n\,\frac{x - a}{n}\right)^m ,$$

or, expanding and setting

$$S_p = 1^p + 2^p + 3^p + \cdots + n^p,$$

we have

$$B = a^m + \frac{m}{1}\,\frac{S_1}{(n + 1)n}\,a^{m-1}\,(x - a)$$

$$+ \frac{m(m - 1)}{1\cdot 2}\,\frac{S_2}{(n + 1)n^2}\,a^{m-2}\,(x - a)^2 + \cdots$$

$$+ \frac{m(m - 1)\,\cdots\,2\cdot 1}{1\cdot 2\,\cdots\,m}\,\frac{S_m}{(n + 1)n^m}\,(x - a)^m.$$

By the result in Problem 15 of Chapter 1, we have

$$(p + 1)S_p = (n + 1)^{p+1} - \frac{(p + 1)p}{1\cdot 2}\,S_{p-1}$$

$$- \frac{(p + 1)p(p - 1)}{1\cdot 2\cdot 3}\,S_{p-2} - \cdots$$

$$- (p + 1)S_1 - (n + 1),$$

hence

$$\frac{(p + 1)S_p}{(n + 1)n^p} = \left(\frac{n + 1}{n}\right)^p - \frac{\frac{(p + 1)p}{1\cdot 2}\,S_{p-1} + \cdots + (p + 1)S_1 + (n + 1)}{(n + 1)n^p}.$$

The numerator of the last fraction is of degree p in n; thus the limit, as $n \to \infty$, of this fraction vanishes, and we have

$$\lim_{n\to\infty} \frac{(p+1)S_p}{(n+1)n^p} = \lim_{n\to\infty} \left(\frac{n+1}{n}\right)^p = 1,$$

hence

$$\lim_{n\to\infty} \frac{S_p}{(n+1)n^p} = \frac{1}{p+1},$$

and

$$\lim_{n\to\infty} B = a^m + \frac{m}{1\cdot 2} a^{m-1}(x-a) + \frac{m(m-1)}{1\cdot 2\cdot 3} a^{m-2}(x-a)^2 + \cdots$$

$$+ \frac{1}{m+1}(x-a)^m.$$

Multiplying both sides of this equality by $(m+1)(x-a)$, and adding a^{m+1} on both sides, we see that

$$(m+1)(x-a)\left(\lim_{n\to\infty} B\right) + a^{m+1}$$

$$= a^{m+1} + \frac{m+1}{1} a^m (x-a) + \frac{(m+1)m}{1\cdot 2} a^{m-1}(x-a)^2 + \cdots + (x-a)^{m+1}$$

$$= [a + (x-a)]^{m+1} = x^{m+1},$$

hence

$$\lim_{n\to\infty} B = \frac{x^{m+1} - a^{m+1}}{(m+1)(x-a)}.$$

On the other hand,

$$A = \frac{1}{n+1}\left[a^{m+1} + a^{m+1}\left(\frac{x}{a}\right)^{\frac{m+1}{n}} + a^{m+1}\left(\frac{x}{a}\right)^{\frac{2(m+1)}{n}} + \cdots + x^{m+1}\right]$$

$$= \frac{x^{m+1}\left(\frac{x}{a}\right)^{\frac{m+1}{n}} - a^{m+1}}{(n+1)\left[\left(\frac{x}{a}\right)^{\frac{m+1}{n}} - 1\right]}.$$

As $n \to \infty$, the numerator reduces to $x^{m+1} - a^{m+1}$; the denominator presents itself in the indeterminate form $\infty \times 0$. But we can write the denominator in the form

$$\frac{\left(\frac{x}{a}\right)^{\frac{m+1}{n}} - 1}{\frac{1}{n+1}}$$

and, using l'Hospital's Rule twice, we get

$$\lim_{n\to\infty} \frac{\left(\frac{x}{a}\right)^{\frac{m+1}{n}} - 1}{\frac{1}{n+1}} = \lim_{n\to\infty} \frac{-\frac{m+1}{n^2}\left(\frac{x}{a}\right)^{\frac{m+1}{n}} \log \frac{x}{a}}{-\frac{1}{(n+1)^2}}$$

$$= \lim_{n\to\infty} (m+1)\left(\frac{n+1}{n}\right)^2 \log \frac{x}{a} = (m+1) \log \frac{x}{a}.$$

Thus

$$\lim_{n\to\infty} A = \frac{x^{m+1} - a^{m+1}}{(m+1) \log \frac{x}{a}},$$

and, consequently,

$$\lim_{n\to\infty} \frac{B}{A} = \frac{\log \frac{x}{a}}{x - a}.$$

This limit is independent of m.

If we take $f(x) = K \cdot x^{m+1}$, where K is a constant, the fraction B/A is clearly not affected; the factor K introduced in the numerator and denominator in B/A cancels out. In the case $f(x) = x$, the fraction B/A also converges to the limit $(\log \frac{x}{a})/(x - a)$ as $n \to \infty$, as a simple calculation shows.

Finally, to see that the claim is valid when f is a polynomial without constant term, in other words, when f(x) is a linear combination of x, x^2, x^3, ..., x^k, we may invoke the result in Problem 28 of Chapter 2.